Sustainable Pavement Materials: Design, Application and Performance Evaluation

Sustainable Pavement Materials: Design, Application and Performance Evaluation

Guest Editor

Jiaqi Chen

MDPI

Basel • Beijing • Wuhan • Barcelona • Belgrade • Novi Sad • Cluj • Manchester

Guest Editor
Jiaqi Chen
Department of Civil
Engineering
Central South University
Changsha
China

Editorial Office
MDPI AG
Grosspeteranlage 5
4052 Basel, Switzerland

This is a reprint of the Special Issue, published open access by the journal *Materials* (ISSN 1996-1944), freely accessible at: https://www.mdpi.com/journal/materials/special_issues/5K6D2W971G.

For citation purposes, cite each article independently as indicated on the article page online and as indicated below:

Lastname, A.A.; Lastname, B.B. Article Title. *Journal Name* **Year**, *Volume Number*, Page Range.

ISBN 978-3-7258-5085-3 (Hbk)
ISBN 978-3-7258-5086-0 (PDF)
https://doi.org/10.3390/books978-3-7258-5086-0

Contents

About the Editor

Jiaqi Chen

Jiaqi Chen, Ph.D., has served as an associate professor at the Department of Civil Engineering, Central South University since 2015. His general research is in the field of pavement engineering, including multi-scale virtual simulation of granular materials for transportation infrastructure, dynamic response and life-cycle evaluation of pavement structure, meso-mechanical behavior of reclaimed asphalt pavement, and pavement–environment interaction. He has authored or co-authored more than 40 journal and conference publications on pavement engineering, and has more than 2000 citations.

materials

MDPI

Editorial

Sustainable Pavement Materials: Design, Application, and Performance Evaluation

Jiaqi Chen

Department of Civil Engineering, Central South University, Changsha 410075, China; chenjiaqi@csu.edu.cn

The study of pavement sustainability integrates environmental, economic, and social considerations across the pavement life cycle, with material selection profoundly influencing durability, resource efficiency, safety and maintenance strategies. As global demand for sustainable infrastructure grows, recent research has prioritized innovative road materials and design methodologies to enhance pavement sustainability. Despite significant progress, challenges such as optimizing material performance, reducing environmental footprint, and ensuring cost-effectiveness persist, necessitating further exploration.

This Special Issue—Sustainable Pavement Materials: Design, Application, and Performance Evaluation—in *Materials* (2024–2025) presents eleven research articles that advance sustainable pavement technologies, addressing durability, environmental impact, and resource efficiency in road engineering. In their work, Ma et al. [1] employed multi-scale finite element modeling to reveal that coupled salt–thermal–mechanical effects reduced the cohesive strength of asphalt pavements by 23.5~63.8% and the adhesive strength by 25~71.6%, emphasizing the need for durable designs in harsh environments. Li et al. [2] found that aging and salt–alkali coupling effects decreased the fatigue life and self-healing properties of graphene oxide-modified asphalt, critical for saline–alkaline and high-altitude regions. Meanwhile, Zhou et al. [3] demonstrated that industrial animal oil regenerated aged SBS-modified asphalt, reducing sulfoxide index by up to 38.88% and aromaticity index by up to 63.77%, enhancing recyclability.

In cementitious materials, Li et al. [4] developed sulfur-modified alite calcium sulfoaluminate cement, which achieved 30.45% higher compressive strength at 1 day compared to Portland cement, providing an environmentally friendly alternative. Tang et al. [5] showed that calcined coal-series kaolinite and limestone, replacing 45% of Portland cement, produced a compressive strength of 47.2 MPa at 28 days, improving pore structure and lowering CO_2 emissions. In addition, Zhang et al. [6] enhanced calcined coal gangue-blended cement with Stöber nano-SiO_2 particles, increasing compressive strength and refining microstructure. Li et al. [7] improved the compressive strength of cement-stabilized iron tailing soil by using ionic soil stabilizers to increase compactness, yielding an increase of 35.3% after 7 days of testing.

Rudziewicz et al. [8] optimized foam polypropylene fiber-reinforced concrete for 3D printing, achieving a compressive strength of 28.76 MPa with cement of reduced CO_2 emissions. Gierasimiuk et al. [9] conducted a comparative study on texture measurement techniques for exposed aggregate concrete pavements, providing a reliable assessment of skid resistance. Luo et al. [10] optimized cement content detection in controlled low-strength soils, achieving errors below 2.4% by analyzing water content and hydration time effects. Hou et al. [11] developed a cement-emulsified asphalt composite for negative-temperature spraying, which exhibited superior frost resistance, making it suitable for plateau environments.

Collectively, these studies advance sustainable pavement solutions through their exploration of innovative materials, recycling options, and robust performance evaluation, laying the foundation for durable, eco-friendly road infrastructure. To further their adoption, future research should focus on long-term field performance, cost-effectiveness, and standardized testing.

Funding: This research received no external funding.

Acknowledgments: Thanks to all the authors and peer reviewers for their valuable contributions to this Special Issue 'Sustainable Pavement Materials: Design, Application and Performance Evaluation'. I would also like to express my gratitude to all the staff and people involved in this Special Issue.

Conflicts of Interest: The author declares no conflicts of interest.

References

1. Ma, J.; Chen, J.; Tang, M.; Liu, Y. Multi-Scale Modeling and Damage Mechanisms of Asphalt Pavements Under Coupled Salt–Thermal–Mechanical Effects. *Materials* **2025**, *18*, 2337. [CrossRef] [PubMed]
2. Li, R.; Zhu, W.; Chen, H.; Li, X.; Sun, L.; Yue, J. Influence of Aging and Salt–Alkali Coupling on the Fatigue and Self-Healing Behavior of Graphene Oxide-Modified Asphalt. *Materials* **2025**, *18*, 1808. [CrossRef] [PubMed]
3. Zhou, S.; Wang, D.; Wu, L.; Maimaitisidike, A.; Wang, Z.; Zhao, H.; Ren, J. Thermal-Ultraviolet-Humidness Coupling Ageing and Regeneration Properties and Mechanisms of SBS-Modified Asphalt Under Hot–Wet Environment Conditions. *Materials* **2025**, *18*, 1731. [CrossRef] [PubMed]
4. Li, X.; Kang, G.; Dou, S.; Ma, B.; Tang, J.; Zhou, H.; Zhang, H.; Wang, J.; Shen, X. Preparation and Properties of Sulfur-Modified Alite Calcium Sulfoaluminate Cement. *Materials* **2024**, *17*, 6258. [CrossRef] [PubMed]
5. Tang, J.; Yu, Y.; Bu, Y.; Ma, B.; Zhou, H.; Zhou, R.; Wang, J.; Zhang, H. The Synergistic Effect of Calcined Coal-Series Kaolinite and Limestone on the Hydration of Portland Cement. *Materials* **2024**, *17*, 4512. [CrossRef] [PubMed]
6. Zhang, N.; Zhou, H.; Hu, Y.; Wang, J.; Hou, G.; Ma, J.; Jiang, R. Effect of Stöber Nano-SiO_2 Particles on the Hydration Properties of Calcined Coal Gangue-Blended Cement. *Materials* **2024**, *17*, 4218. [CrossRef] [PubMed]
7. Li, H.; Jia, J.; Lu, X.; Cheng, X.; Zhu, J.; Zhang, L.; Guo, P.; Zhai, G. The Effect of Ionic Soil Stabilizer on Cement and Cement-Stabilized Iron Tailings Soil: Hydration Difference and Mechanical Properties. *Materials* **2025**, *18*, 1444. [CrossRef] [PubMed]
8. Rudziewicz, M.; Maroszek, M.; Setlak, K.; Góra, M.; Hebda, M. Optimization of Foams—Polypropylene Fiber-Reinforced Concrete Mixtures Dedicated for 3D Printing. *Materials* **2024**, *17*, 4106. [CrossRef] [PubMed]
9. Gierasimiuk, P.; Wasilewska, M.; Gardziejczyk, W. A Comparative Study on the Texture of Exposed Aggregate Concrete (EAC) Pavements Using Different Measurement Techniques. *Materials* **2024**, *17*, 4147. [CrossRef] [PubMed]
10. Luo, Y.; Jiang, L.; Qin, L.; Luo, Q.; Connolly, D.P.; Wang, T. Optimizing Cement Content in Controlled Low-Strength Soils: Effects of Water Content and Hydration Time. *Materials* **2024**, *17*, 5915. [CrossRef] [PubMed]
11. Hou, Y.; Niu, K.; Tian, B.; Chen, J.; Li, X. Development of Negative-Temperature Cement Emulsified Asphalt Spraying Materials Based on Spraying Performance and Rheological Parameters. *Materials* **2024**, *17*, 3137. [CrossRef] [PubMed]

materials

MDPI

Article

Influence Mechanism of Waterborne Polyurethane on the Properties of Emulsified Asphalt

Jian Tan [1], Shuguang Hou [2,*], Rui Jin [2], Xiao Zhong [3] and Xiaoxi Zou [3]

1 Zhaotong Yizhao Expressway Investment and Development Co., Ltd., Zhaotong 657000, China
2 College of Transportation Engineering, Nanjing Tech University, Nanjing 211816, China; jrzs@njtech.edu.cn
3 Zhaotong Transportation Bureau, Zhaotong 657099, China
* Correspondence: houshuguang@njtech.edu.cn

Abstract

To elucidate the modification mechanism of waterborne polyurethane (WPU) on emulsified asphalt, anionic and cationic WPUs are utilized as modifiers. As well, their effects on physical properties, microstructure, and compatibility are characterized using basic performance tests, Fourier transform infrared spectroscopy (FTIR), and atomic force microscopy (AFM). The results show that WPU-modified emulsified asphalt exhibited a higher softening point, reduced penetration, and decreased ductility, suggesting enhanced high-temperature stability but diminished low-temperature flexibility. Among all samples, the combination of cationic WPU with cationic emulsified asphalt shows the highest softening point (54.1 °C), whereas cationic emulsified asphalt alone exhibits the lowest one (52.9 °C). Anionic emulsified asphalt demonstrates the highest penetration (79 mm), while non-ionic WPU combined with cationic emulsified asphalt shows the lowest one (59.3 mm). The ductility decreases from 90.3 cm to 28.7 cm. The storage stability varies with WPU ion type. Cationic WPU-modified samples showed the poorest storage stability (0.7% residue), while anionic-modified samples exhibit the best one (0.4% residue). FTIR analysis confirms the presence of characteristic WPU absorption peaks, indicating that physical blending occurs, and chemical interaction is limited. AFM observations reveal that anionic WPUs provide superior compatibility, forming fine, uniformly distributed particles with the lowest surface roughness (5.655 nm). In contrast, cationic WPUs form chain-like structures that cure effectively but exhibit poor dispersion. This study provides a basis for the development of high-performance WPU-modified emulsified asphalt.

Keywords: modified emulsified asphalt; waterborne polyurethane; ionic emulsifiers; influence mechanism; chemical composition; microphase structure

1. Introduction

The cold recycling of asphalt pavement using emulsified asphalt offers several advantages, including high reclaimed asphalt pavement (RAP) utilization, construction at ambient temperature, extended construction seasons, and minimal environmental pollution. Recognized as a green technology with low emissions and minimal environmental impact, cold recycling aligns with China's strategic goals of carbon peaking and carbon neutrality, and is receiving increasing attention in the highway engineering field. Its application in engineering practice is expanding. As the primary binder in cold recycling, emulsified asphalt is widely used in asphalt pavement maintenance and repair [1]. Through the action of surfactants, emulsified asphalt forms a stable emulsion system in water, improving both its workability and environmental properties.

However, both practical experience and research demonstrated that traditional emulsified asphalt cold-recycled pavements were associated with several critical limitations, including inadequate mechanical performance, low toughness, accelerated deterioration, and shorter service life [2]. These pavements are prone to cracking and loosening, which undermine overall mechanical properties and road performance [3], ultimately reducing their service life. Therefore, the development of high-performance modified emulsified asphalt has become a critical focus in pavement engineering.

Extensive studies were conducted to enhance performance, aging resistance, and storage stability of emulsified asphalt. Among various modifiers, waterborne polyurethane (WPU) has gained some attention due to its favorable mechanical strength, weather resistance, and environmental friendliness. WPU has been shown to improve the rheological properties of asphalt, particularly its high-temperature deformation resistance [4–6]. Zhang et al. [7] reported that 15% WPU content offers optimal performance and significantly improves storage stability. Hesami et al. [8] concluded that WPU enhanced the ductility but reduced the adhesion of emulsified asphalt, and the modification process was predominantly physical. The WPU content of less than 6% was therefore recommended. Similarly, Zhao et al. [9] observed a positive correlation between WPU dosage and performance when the dosage remains under 9 wt%. In contrast, Mu et al. [10] argued that limited compatibility between WPU and asphalt led to inconsistent modification effects. Therefore, further research was required to better understand the compatibility, structural integrity, and microstructural evolution of WPU-modified emulsified asphalt [11].

WPU has been classified into cationic, anionic, and non-ionic types [12]. The stability of emulsified asphalt is influenced differently by emulsifiers of various ionic types [13]. Anionic polyurethane has been reported to enhance the rheological behavior of emulsified asphalt, shifting its characteristics from viscous to elastic performance [14]. FTIR analysis revealed that the modification of cationic emulsified asphalt by WPU primarily involved physical blending [15]. Mu et al. [16] examined the performance of cationic WPU-modified emulsified asphalt across different temperatures and identified 110 °C as the optimal evaporation residue temperature, where FTIR absorption peak was the most intense and key performance indicators—penetration, softening point, and ductility—were significantly improved. Although various modification effects were observed among different WPU ionic types [17], the underlying mechanisms remain insufficiently studied [18,19], particularly concerning the compatibility behavior between emulsified asphalt and WPU with different ionic natures.

The incorporation of WPU into emulsified asphalt effectively improves its performance. However, the poor compatibility between WPU and asphalt limited further enhancement of modification effects. To address this issue, researchers commonly regulated WPU structure by introducing ionic groups. Most existing studies focused on modification mechanisms of cationic WPU, while limited attention was given to anionic and non-ionic types. The specific roles and comparative effects of different WPU types on emulsified asphalt remained unclear, and a systematic understanding of their interaction mechanisms, interfacial behavior, and impact on asphalt microstructure is still lacking. Therefore, this study aimed to develop WPU-modified emulsified asphalt using different ionic types of WPU and to clarify their modification effects, revealing the mechanisms by which WPU influences the performance and structure of emulsified asphalt.

In this study, anionic, cationic, and non-ionic WPUs were incorporated into anionic and cationic emulsified asphalt systems to prepare modified samples. These samples were first evaluated for basic metrics, including penetration, softening point, ductility and storage stability. FTIR analysis was then conducted to investigate the effects of WPU type on chemical composition and key functional groups, providing insights into the physical and chemical interactions between WPU and asphalt. Finally, AFM tests were used to examine the microstructural characteristics of modified samples, revealing the compatibility mechanism and microstructural evolution of WPU-modified emulsified asphalt. A comprehensive framework linking performance, structure, and mechanism was established, clarifying the regulatory role of WPU ionic structure in interfacial compatibility and modification effectiveness. This work provides theoretical support and technical guidance for the design and optimization of high-performance WPU-modified emulsified asphalt materials.

2. Materials and Methods

2.1. Materials

2.1.1. Base Asphalt

Emulsified asphalt was prepared using 70# base asphalt with a penetration grade of 60–80, and its basic properties were tested according to JTG E20-2011 [20]. The results as shown in Table 1 indicated that all material properties met the requirements of JTG F40-2004 [21]. The test methods used in Table 1 were all conducted in accordance with JTG E20-2011 [20]. The asphalt used in this study was supplied by Zhonghai Asphalt Co., Ltd., Binzhou, Shandong, China.

Table 1. Basic technical indicators of 70# base asphalt.

Technical Indexes	Unit	Test Results	Standard Requirements	Experimental Methods
Penetration (25 °C, 5 s, 100 g)	0.1 mm	70	60–80	T0604
Ductility (10 °C)	cm	18	≥15	T0605
Softening point	°C	50	≥47	T0606

2.1.2. Anionic/Cationic Emulsifiers

Slow-cracking fast-setting cationic emulsifier and slow-cracking fast-setting anionic emulsifier produced by Rongcheng Road and Bridge Co., Ltd., Nantong, China, were selected, and the performance indicators are listed in Table 2.

Table 2. Parameters of slow-cracking and fast-setting anionic and cationic emulsifiers.

Parameter		Specification Requirements	Test Results
Appearance	Cationic	Brown viscous liquid	
	Anionic	Light yellow liquid	
Active substance content	Cationic	70 ± 2	69
	Anionic		70
pH value	Cationic	6–8	7.6
	Anionic		
Free amine content (%)	Cationic	≤2	1.25
	Anionic		1.23
Density (g/cm^3)	Cationic	1.0	1.0
	Anionic	1.05	1.05
Effective ingredient (%)	Cationic	50	50
	Anionic	40	40

2.1.3. Experimental Water

The water used in this experiment was laboratory tap water without any impurity. The main water quality parameters were as follows. The pH value ranged from 7.5 to 8.5, electrical conductivity ranged from 5 to 8 μS/cm, hardness ranged from 0.1 to 0.3 mmol/L, and chloride ion concentration ranged from 10 to 30 mg/L. This water quality ensured that there was no additional interference in the emulsification process.

2.1.4. Hydrochloric Acid Modifier

Cationic emulsifiers require an acidic environment when used as modifiers, while anionic emulsifiers require an alkaline environment. Therefore, when using cationic emulsifiers, hydrochloric acid was used to adjust the pH value to 2~3, and when using anionic emulsifiers, alkali was used to adjust the pH value to 12~13. The basic indicators of hydrochloric acid and sodium hydroxide are shown in Table 3. Both materials were purchased from Sinopharm Chemical Reagent Co., Ltd., Nanjing, China.

Table 3. Technical index of hydrochloric acid and sodium hydroxide.

Test Specimens	Acid	Alkali
Chemical formula	HCl	NaOH
Molecular weight	36.5	40.0
Concentration/%	36–38	5%
Color	Colorless	White
Density/(g/cm^3)	1.18	1.05

2.1.5. Emulsifier

Two types of anionic and cationic emulsified asphalts were used, and their basic performance indicators are shown in Table 4. They were supplied by Yutuo Chemical Co., Ltd., Panjin, Liaoning, China.

Table 4. Performance index of cationic/anionic emulsified asphalt.

Test Items	Quality Indicators		Measured
Residual on sieve (1.18 mm)/%	Cationic	<0.1	0.005
	Anionic		0.003
Particle Polarity	Cationic	Cationic	Cationic
	Anionic	Anionic	Anionic
Particle size/μm	Cationic	≤7	4.41
	Anionic		4.30
Standard viscosity/(Pa.s)	Cationic	8–25	12
	Anionic		15
Storage stability (1d, 25 °C)/%	Cationic	<1	0.4
	Anionic		0.3
Content/%	Cationic	≥60	63.7
	Anionic		62.8
Penetration/(0.1 mm)	Cationic	40–120	75
	Anionic	50–300	79
Softening point/°C	Cationic	≥42	52.9
	Anionic		53.1
Ductility/cm	Cationic	≥40	52.5
	Anionic		90.3

2.1.6. Different Ionic WPUs

The performance parameters of different ionic WPUs are presented in Table 5. The waterborne polyurethane was purchased from Shanghai Sisheng Polymer Materials Co., Ltd., Shanghai, China. The content of each type of polyurethane in this study was set at 4 wt%.

Table 5. Performance indexes of different ionic WPUs.

Type	Designation	Solids Content (%)	pH Value	Viscosity Value (mPa.s)	Color	Specific Gravity (g/cm^3)
A	Cationic WPU	38 ± 1	3–7	<200	Milky white	1.06 ± 0.02
B	Anionic WPU	30 ± 1	6–9	≥100 20 °C		1.05
C	Non-ionic WPU		6–8	>100		

2.2. *Sample Preparation*

2.2.1. Preparation of Matrix Emulsified Asphalt

The emulsification equipment selected for this experiment was the JM-1 small emulsifier from Wuxi Petroleum Instrument Equipment Co., Ltd., Wuxi, China. Sample preparation was performed as follows. To ensure the reliability and reproducibility of test results, parallel experiments were carried out, and no fewer than three effective samples were tested for each group. The allowable error was limited to 15%. If this threshold was exceeded, the samples were re-prepared and retested. The procedure was as follows:

(1) Heating of asphalt

To ensure proper flowability, base asphalt, which was solid at room temperature, was preheated in an oven at 140 °C. This step allowed asphalt to flow smoothly through the colloid mill during its emulsification.

(2) Preparation of soap solution

The soap solution was prepared by dissolving the anionic/cationic emulsifier in water in a predetermined ratio. The solution temperature of soap solution was controlled at approximately 60 °C.

(3) Preparation of emulsified asphalt

Prior to emulsification, hot water (60 °C) was circulated through colloid mill to minimize heat loss during its emulsification. Once preheating was complete, soap solution was cycled into colloid mill. After no significant bubbles were generated, the heated asphalt was slowly added. The mixture was sheared for 1 min, and then the three-way valve was opened to release emulsified asphalt.

(4) Storage of emulsified asphalt

Immediately after preparation, emulsified asphalt was allowed to cool to room temperature while being gently stirred with a glass rod. After cooling, it was filtered using a 1.18 mm filter screen. Once filtration was completed, it was sealed for storage.

2.2.2. Preparation of WPU Modified Emulsified Asphalt

The preparation equipment for WPU-modified emulsified asphalt was selected to be an FM 300 intermittent high-speed shear disperser from Shanghai Fluke Company, Shanghai, China.

The preparation methods for modified emulsified asphalt are generally categorized into two approaches, including modification prior to emulsification and emulsification followed by modification. In this study, the former approach was initially used. The soap solution was prepared and emulsified by a colloid mill, after which WPU was added, after which WPU was added, followed by the gradual incorporation of heated base asphalt. However, emulsified asphalt showed poor flowability and solidified after being left for a period, which led to a poor emulsification effect. Therefore, the method of emulsification followed by modification was adopted. Base emulsified asphalt was first prepared. After that, WPU and emulsified asphalt were blended using a high-speed shear mixer. During this process, a glass rod was used for slowly stirring until the sample cooled to room temperature, after which it was sealed and stored.

2.3. *Test Methods*

2.3.1. Basic Performance Test

Basic properties of prepared WPU-modified emulsified asphalt were evaluated according to JTG E20-2011 [20], including penetration, ductility, and softening point.

2.3.2. FTIR Test

To analyze the effect of WPU on primary chemical components and functional groups of emulsified asphalt, samples were prepared by depositing a thin film onto specialized silicon wafers (10 mm × 10 mm) using a pipette, and the film thickness was controlled to less than 10 mm. After film formation, infrared spectroscopy tests were conducted. FTIR spectroscopy test was performed using a Thermo Scientific Nicolet iS50 FTIR spectrometer (Thermo Fisher Scientific, Madison, WI, USA). The front surface was selected for testing, and the test mode was set to ATR with absorbance. Each sample and background were scanned 32 times. The resolution was set at 4 cm^{-1}, with a gain of 1.000, mirror speed of 0.4747, and an aperture of 100. The spectral range was set from 4000 cm^{-1} to 400 cm^{-1}.

2.3.3. AFM Test

To investigate the effect of WPU on micromorphology and microstructure of emulsified asphalt, samples were prepared by pipetting the material onto specialized silicon wafers (10 mm × 10 mm) using a pipette, and each sample thickness was controlled to less than 10 mm. After curing, AFM analysis was performed using a Dimension FastScan Atomic Force Microscope (Bruker Corporation, Billerica, MA, USA). The micro-cantilever was characterized by a tip radius of 650 nm, a length of 115 μm, and a width of 25 μm. The scanning area was set to 10 μm × 10 μm, with a resolution of 256 × 256. The tip operated at a frequency of 70 kHz and possessed an elasticity coefficient of 0.4 N/m.

3. Results and Discussion

3.1. *WPU-Modified Emulsified Asphalt Basic Properties*

To evaluate the effect of different types of WPU on basic properties of emulsified asphalt, a series of fundamental property tests were performed on the prepared samples. The corresponding results are presented in Table 6. For clarity, the abbreviations used are based on the first letters of sample names.

Table 6. Performance indexes of WPU-modified emulsified asphalt samples.

Test Items	Cationic Emulsified Asphalt	Cationic WPU + Cationic Emulsified Asphalt	Non-ionic WPU + Cationic Emulsified Asphalt	Anionic Emulsified Asphalt	Anionic WPU + Anionic Emulsified Asphalt	Non-ionic WPU + Cationic Emulsified Asphalt
Residual on sieve (1.18 mm)/%	0.005	0.006	0.003	0.003	0.015	0.002
Particle polarity	C	C	C	A	A	A
Standard viscosity/ (Pa.s)	12	16	15	15	17	16

Table 6. *Cont.*

Test Items	Cationic Emulsified Asphalt	Cationic WPU + Cationic Emulsified Asphalt	Non-ionic WPU + Cationic Emulsified Asphalt	Anionic Emulsified Asphalt	Anionic WPU + Anionic Emulsified Asphalt	Non-ionic WPU + Cationic Emulsified Asphalt
Storage stability (1d, 25 °C)/%	0.4	0.7	0.5	0.3	0.4	0.5
Solid content (%)	65	63.7	62.8	65	62.8	62.8
Penetration/ 0.1 mm	75	60.3	59.3	79	70	63.3
Softening point/°C	52.9	54.1	53.8	53.1	54	53.9
Ductility (15 °C)/cm	52.5	43.2	36.4	90.3	44.7	28.7

In Table 6, the residual amount on the sieve for all systems remains very low (0.002–0.015%), indicating that emulsified asphalt particles are fine and the sieve loss is minimal, reflecting better emulsion quality and system stability. The viscosity ranges from 12 Pa.s to 17 Pa.s. After modification by WPU, the standard viscosity of emulsified asphalt increases to varying extents. This suggests that the incorporation of WPU enhances the viscosity of emulsified asphalt, thereby improving its adhesion and high-temperature stability.

As shown in Table 6, the penetration values of four WPU-modified emulsified asphalt samples decrease when compared to that of base emulsified asphalt, indicating an improvement in high-temperature stability. This trend corresponds to the observed increase in standard viscosity. The decreased penetration correlates with increased viscosity, consequently elevating the softening point while reducing ductility at 15 °C. Compared to the base cationic or anionic emulsified asphalt, WPU-modified samples demonstrate lower storage stability, primarily affected by the pH and particle size of WPU. WPU is slightly alkaline, and when mixed with acidic emulsified asphalt, heat is generated, which accelerates moisture evaporation and decreases stability [22]. Larger particle sizes promote faster settling, further compromising stability. The use of anionic emulsifiers helps reduce particle size and improve stability. However, the dosage must be optimized, as both insufficient and excessive amounts of WPU negatively affect performance of modified emulsified asphalt.

Compared to the control sample (unmodified cationic emulsified asphalt), the cationic WPU-modified emulsified asphalt is characterized by a 19.6% reduction in penetration, a 2.27% increase in softening point, and a 17.7% reduction in ductility. In the non-ionic WPU-modified system, the penetration is reduced by 20.9%, the softening point is reduced by 1.7%, and the ductility decreases by 30.7%. Overall, WPU modification improves the hardness and high-temperature stability of emulsified asphalt but compromises its low-temperature toughness. Among the two, cationic WPU modification provides an improved rutting resistance while maintaining a certain level of low-temperature flexibility, making it suitable for regions with wide temperature fluctuations. In contrast, non-ionic WPU modification produces a harder emulsified asphalt with improved high-temperature stability but increases the susceptibility to low-temperature cracking. This is more appropriate for heavy-load environment at high temperature, but may increase cracking risk under cold conditions.

Compared to anionic emulsified asphalt, the penetration of anionic WPU-modified emulsified asphalt decreases by 11.4%, the softening point increases by 1.7%, and the ductility decreases by 50.5%. For non-ionic WPU-modified anionic emulsified asphalt, the penetration is reduced by 19.9%, the softening point is increased by 1.5%, and the ductility is decreased by 68.2%. Overall, WPU modification enhances the hardness and high-temperature stability of emulsified asphalt, while significantly compromising its low-temperature flexibility—particularly in the non-ionic system, where the reduction in ductility is more pronounced than in the anionic system. The anionic WPU-modified asphalt demonstrates an increased hardness and softening point, yet the ductility decreases by 50.5%, indicating the reduced low-temperature crack resistance. The non-ionic WPU-modified asphalt shows even greater hardness (19.9% reduction in penetration) and a 68.2% reduction in ductility, resulting in increased brittleness at low temperatures and making it more suitable for high-temperature applications. Although the anionic WPU-modified system demonstrates slightly better low-temperature crack resistance, further improvements remain necessary.

In summary, WPU modification enhances the hardness and high-temperature stability of emulsified asphalt but reduces its low-temperature ductility and storage stability. After modification, penetration values are reduced, the softening point increases, and the ductility decreases, indicating rutting resistance is improved but low-temperature ductility is diminished. Among the modified systems, cationic WPU-modified emulsified asphalt achieves a balance between high-temperature stability and moderate low-temperature flexibility, rendering it suitable for regions with large temperature fluctuations. The non-ionic WPU-modified system exhibits an increased hardness but suffers from pronounced low-temperature brittleness, rendering it more suitable for heavy-load applications in high-temperature environments. The storage stability generally decreases in all WPU-modified emulsified asphalt samples. Additionally, storage stability is generally reduced across all WPU-modified emulsified asphalts. Specifically, the cationic WPU-modified emulsified asphalt shows the lowest storage stability (0.7% sieve residue), while the anionic WPU-modified sample exhibits the highest stability (0.4% sieve residue). Overall, the anionic WPU system performs slightly better under low-temperature conditions but still requires further optimization to enhance crack resistance and storage stability.

3.2. *Effect of WPU on Chemical Composition and Functional Groups of Emulsified Asphalt*

WPU resin consists of two main components of polyisocyanates and polyether/polyester polyols [23]. It readily reacts with phenols, anhydrides, carboxylic acids, and even water present in emulsified asphalt, leading to the formation of polyurethane carbamate acid (Equation (1)). Isocyanates also react with phenols to generate carboxylic acids, which subsequently undergo further reactions to form anhydrides (Equation (2)). These anhydrides react with water to produce amides and carbon dioxide. When aromatic carboxylic acids are present in asphalt or aromatic groups exist within the isocyanates, such stable products as anhydrides and ureas are formed (Equation (3)). In addition, isocyanates react with anhydrides to yield acylamides (Equation (4)).

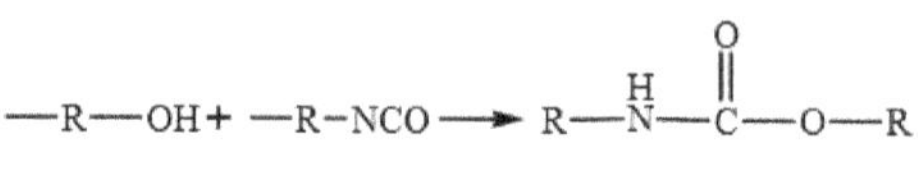
(1)

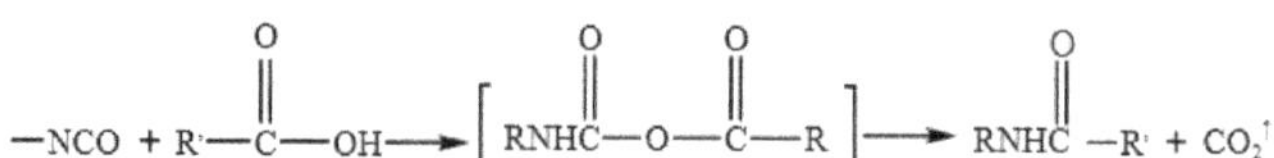
(2)

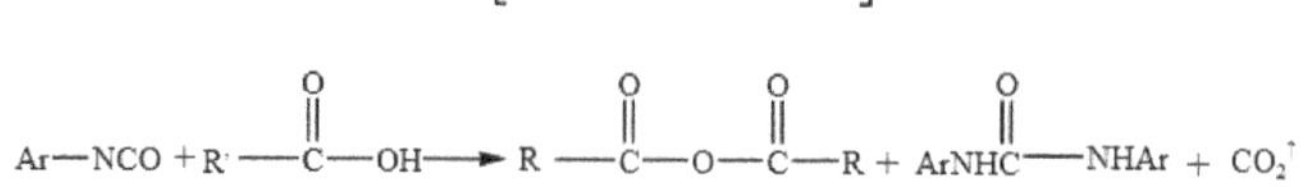
(3)

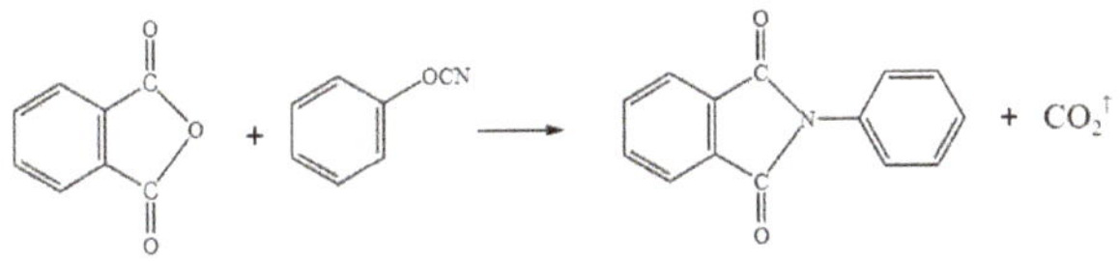

(4)

To investigate the curing behavior of WPU in emulsified asphalt and its interaction with asphalt matrix, FTIR is employed to analyze cationic and anionic emulsified asphalt, non-ionic WPU-modified emulsified asphalt, as well as cationic and anionic WPU-modified cationic and anionic emulsified asphalt. The functional groups and their variations in these blended systems are identified by examining characteristic peak positions and intensity changes intensity in the infrared spectra of each sample. Furthermore, FTIR is also used to characterize chemical structure and ionic properties of cationic, anionic, and non-ionic WPUs as shown in Figure 1.

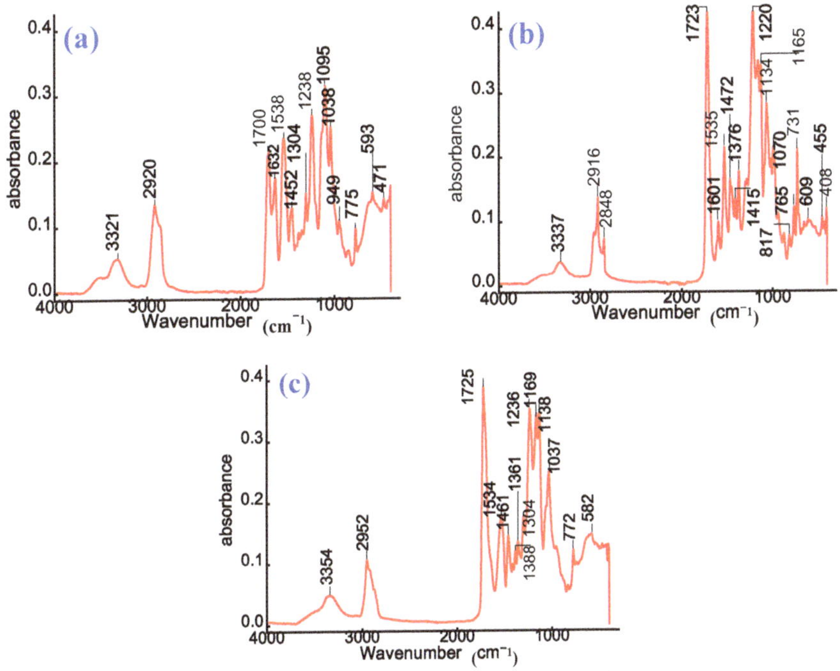

Figure 1. Infrared spectra of (**a**) non-ionic WPU, (**b**) cationic WPU, and (**c**) anionic WPU.

As shown in Figure 1, the infrared spectra of cationic, anionic, and non-ionic WPUs are characterized by absorption peaks corresponding to carbamate (-NHCOO-) and ester (C=O) groups. The N-H and O-H stretching vibrations observed at 3321–3337 cm^{-1} appear more intense in non-ionic WPU, whereas the peak of anionic WPU broadens due to the presence of carboxyl and sulfonic acid groups. These variations in hydrogen bond interactions may influence the stability of WPU within the emulsified asphalt system. The enhanced hydrogen bonding in non-ionic WPU is considered to improve emulsion cohesion, whereas polar functional groups in anionic WPU are thought to affect the dispersion through electrostatic interactions.

The C-H stretching vibrations at 2920–2952 cm^{-1} correspond to the predominant soft segment structure in WPU. From Figure 1a, non-ionic WPU exhibits a stronger peak because of the presence of polyether or polyester segments, while cationic and anionic WPUs display a marginal reduction in peak intensity, which is attributed to the influence of polar groups. These differences in soft segment composition are considered to affect the compatibility and stability between WPUs and asphalt. Non-ionic WPU is considered to enhance wettability and dispersibility, while cationic and anionic WPUs may modify emulsification performance via polar group interactions.

In Figure 1, the C=O stretching vibration at 1700–1725 cm^{-1} is attributed to an ester functional group, with a slight red shift observed in cationic WPU. This shift is considered indicative of stronger interactions with acidic emulsifiers in emulsified asphalt, affecting the stability of emulsion, while anionic WPU interacts with alkaline components in asphalt, influencing the emulsification effect [24]. The benzene ring skeletal vibrations at 1534–1601 cm^{-1} are detected in all three samples but may be more pronounced in anionic WPU. This enhancement is thought to increase the rigidity of emulsified asphalt film, improving mechanical strength but potentially decreasing flexibility.

To investigate chemical structural evolution and modification effects of cationic emulsified asphalt, non-ionic WPU-modified cationic emulsified asphalt, and cationic WPU-modified cationic emulsified asphalt, FTIR characterization is conducted on prepared samples. The corresponding results are presented in Figure 2.

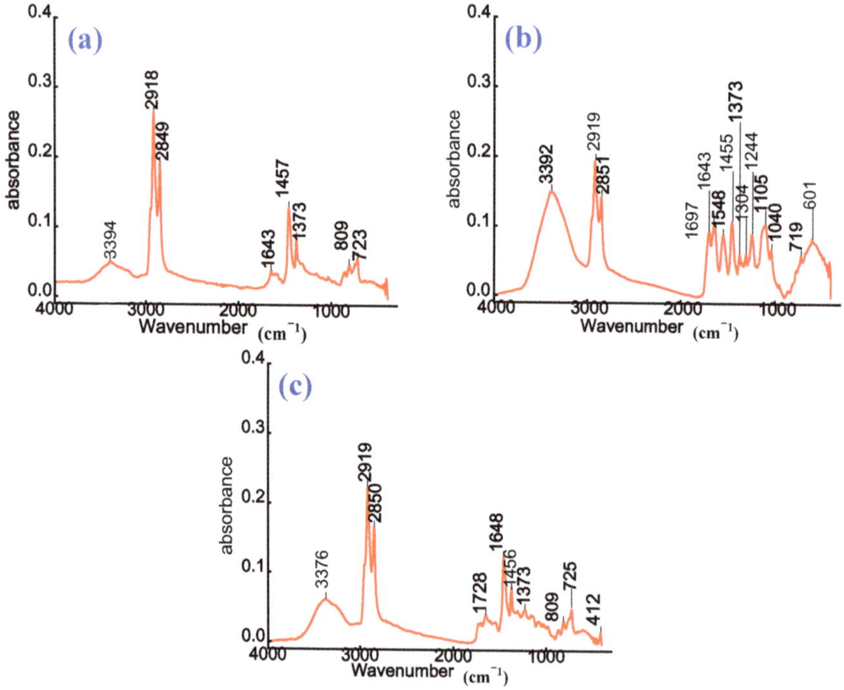

Figure 2. Infrared spectra of (**a**) cationic, (**b**) non-ionic WPU-modified cationic, and (**c**) cationic WPU-modified cationic emulsified asphalt.

As shown in Figure 2, the absorption peaks at 3394 cm^{-1}, 3392 cm^{-1}, and 3376 cm^{-1} are attributed to N–H stretching vibrations of quaternary ammonium salts and O–H stretching vibrations of emulsified asphalt. This enhancement suggests that the introduction of WPU increases hydrogen bond interactions, thereby strengthening the association between emulsified particles and ions, improving the stability of emulsified system [25]. The ester

C=O absorption peaks at 1697 cm^{-1} in Figure 2b and 1728 cm^{-1} in Figure 2c, along with the amide C=O peak at 1648 cm^{-1} in Figure 2c indicate that WPU chemically reacts with asphalt, leading to the formation of new ester and amide bonds [26]. These reactions improve the stability of polyurethane within asphalt matrix, with a more pronounced effect observed in the cationic WPU-modified system.

In Figure 2b, the C–O stretching vibration at 1244 cm^{-1}, along with ether C–O–C and alcohol C–O stretching vibrations at 1105 cm^{-1} and 1040 cm^{-1}, respectively, suggest that the polyether or polyester segments in WPU's soft phase affect the flexibility of emulsified asphalt. Among these, non-ionic WPU exhibits a more significant contribution, enhancing both the ductility and aging resistance of asphalt. The aromatic ring C-H out-of-plane bending vibrations at 900–700 cm^{-1} and long-chain methylene in-plane rocking vibration at 723 cm^{-1} in Figure 2a further confirm the dispersion of WPU within the asphalt matrix. Additionally, the C-N and O-H out-of-plane bending vibration at 601 cm^{-1} in Figure 2b indicates the presence of amide bonds and carboxyl groups, which influence the interfacial interactions of emulsified asphalt.

Overall, the formations of amide and carboxylic ester groups are more pronounced in cationic WPU-modified asphalt, indicating stronger chemical interactions that improve emulsion stability. In contrast, the incorporation of non-ionic WPU enhances the flexibility and wettability of asphalt, primarily due to the contribution of its polyether structure. The characteristic peaks in unmodified cationic emulsified asphalt appear generally weaker, indicating that WPU modification significantly improves the chemical structure and overall performance of emulsified asphalt.

To investigate chemical structure evolution and modification effects of anionic emulsified asphalt, non-ionic WPU-modified anionic emulsified asphalt, and anionic WPU-modified anionic emulsified asphalt, FTIR is employed to characterize three prepared samples. The corresponding results are presented in Figure 3.

As shown in Figure 3a, the absorption peak at 3394 cm^{-1} is assigned to the O–H stretching vibration, indicating the presence of polar functional groups in both emulsified asphalt and the WPU system. The absorption peaks at 2919 cm^{-1} and 2850 cm^{-1} are attributed to the antisymmetric and symmetric stretching vibrations of methylene groups, confirming the presence of alkane structures. A comparison of the spectra for WPU-modified anionic emulsified asphalt in Figure 3b,c with that of anionic emulsified asphalt in Figure 3a shows a decrease in the intensity of absorption peaks at 2919 cm^{-1}, 2850 cm^{-1}, 1457 cm^{-1}, and 1373 cm^{-1}. This reduction suggests a relative decrease in alkyl group content after modification. This is due to the dilution of saturated aliphatic components in asphalt by the introduced WPU or partial chemical reactions [14].

Furthermore, the absorption peak at 1642 cm^{-1} in Figure 3a is attributed to the bending vibration of water molecules (H-O-H) and the stretching vibration of C=C bonds, indicating the coexistence of moisture and aromatic ring structures. In Figure 3c, a weak absorption peak appears at 1734 cm^{-1}, and at 1720 cm^{-1} in Figure 3b, and both correspond to the stretching vibration of carbonyl (C=O) groups. These peaks suggest that the incorporation of WPU introduces additional oxygen-containing functional groups [27], such as ester or amide groups. This effect is more prominent in Figure 3b,c, further indicating that chemical interactions between WPU and asphalt lead to the formation of ester or amide derivatives. These findings further confirm the structural stability of the asphalt matrix.

In summary, WPU incorporation not only reduces the alkyl content of emulsified asphalt but also introduces new polar functional groups, thereby enhancing the overall polarity. Non-ionic WPU improves the wettability and flexibility of asphalt, whereas anionic WPU influences emulsion dispersion and stability through the electrostatic effects of its polar groups.

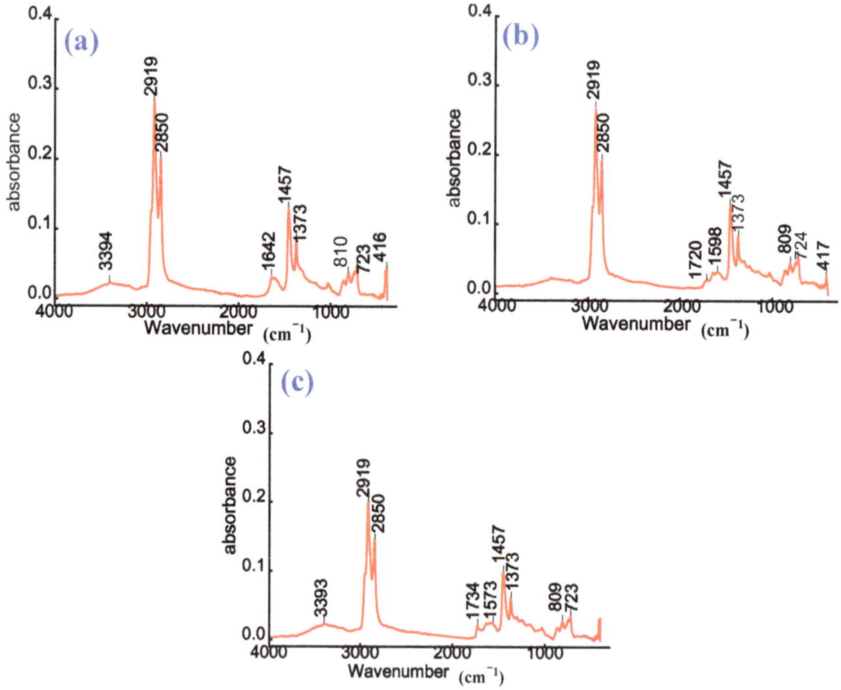

Figure 3. Infrared spectra of (**a**) anionic, (**b**) non-ionic WPU-modified anionic, and (**c**) anionic WPU-modified anionic emulsified asphalt.

3.3. *Influence of WPU on Micromorphology and Microstructure of Emulsified Asphalt*

AFM is utilized to characterize the surface morphology, microstructure, and surface roughness parameters of anionic and cationic emulsified asphalt, non-ionic WPU-modified emulsified asphalt, and anionic/cationic WPU-modified anionic/cationic emulsified asphalt. The effects of various modification formulations are systematically investigated.

3.3.1. Influence of WPU on Surface Morphology of Emulsified Asphalt

The impact of cationic emulsifiers on microstructural characteristics of base asphalt is examined through AFM. The corresponding results are presented in Figure 4.

In AFM images, lighter regions indicate higher surface elevations, whereas darker regions correspond to lower elevations. This topographical pattern is also observed in the 3D morphology. As shown in Figure 4a, the sample surface exhibits a honeycomb-like structure with a relatively uniform distribution of continuous phase and a vertical height variation of approximately 52.3 nm. Figure 4b further reveals that the surface of cationic emulsified asphalt is characterized by noticeable roughness and an undulating morphology.

The phase image in Figure 4c shows a phase angle variation ranging from $-3.9°$ to $4.6°$, with a relatively uniform color distribution, indicating minimal phase separation and homogeneous surface properties. Figure 4d quantifies the surface height variation within the range of 0.5–1.5 μm, where a downward trend suggests the presence of local depressions or valleys. Beyond 1.5 μm, the surface height increases sharply, implying the existence of protrusions or peak-like structures, with a total height difference nearing 25 nm.

To assess the influence of non-ionic WPU on microscopic morphology of cationic emulsified asphalt, AFM characterizations are conducted on non-ionic WPU–modified samples, and the corresponding results are shown in Figure 5.

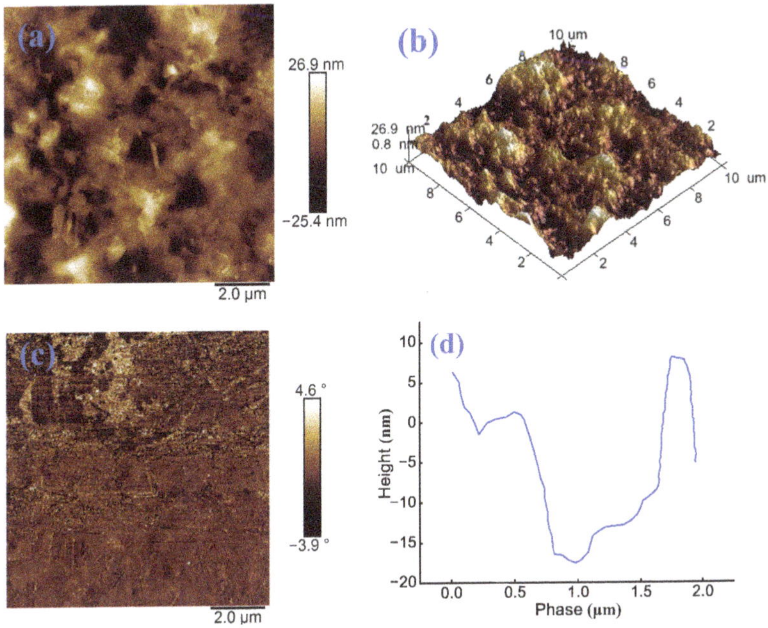

Figure 4. (**a**) Height image, (**b**) 3D height image, (**c**) Phase image, and (**d**) Treatment image of cationic emulsified asphalt.

As shown in Figure 5a, the sample surface exhibits a maximum height of 57.0 nm and a minimum of −62.2 nm, resulting in a total height difference of 119.2 nm, which is increased by 66.9 nm when compared to that in Figure 4a. Uneven color distribution is observed, and brighter regions likely correspond to WPU phase, indicating that the introduction of WPU induces microphase separation among cationic groups [28]. This separation leads to the aggregation of WPU and the formation of larger phase domains, thereby increasing surface roughness. Figure 5b further reveals more distinct surface protrusions and depressions, along with reduced uniformity, supporting the occurrence of microphase separation.

In Figure 5c, the phase angle ranges from −7.3° to 8.7°, exceeding the range in Figure 4c, and is accompanied by an enhanced color contrast. This indicates greater surface hardness and more pronounced phase separation. These observations indicate greater surface stiffness and more evident phase separation, likely due to limited interaction between WPU and the cationic emulsified asphalt matrix, which reduces overall compatibility [29]. Figure 5d shows a surface height difference exceeding 60 nm, further confirming that WPU incorporation enhances surface undulations and roughness. These morphological changes influence key material properties, such as adhesion performance and mechanical stability.

To further examine the influence of cationic WPU on the microstructure of cationic emulsified asphalt, a systematic analysis is conducted on the evolution of ionic structure within asphalt microstructure. AFM characterization results of modified samples are presented in Figure 6.

Figure 6a shows that the sample exhibits a bright and uniform color distribution, with a phase structure more homogeneous than those shown in Figures 4a and 5a. The maximum height difference is 34.1 nm, representing the smallest surface fluctuation among the three sample groups and indicating relatively low surface roughness. Figure 6b further reveals a more uniform surface morphology, confirming that the incorporation of cationic WPU improves the microstructure of emulsified asphalt.

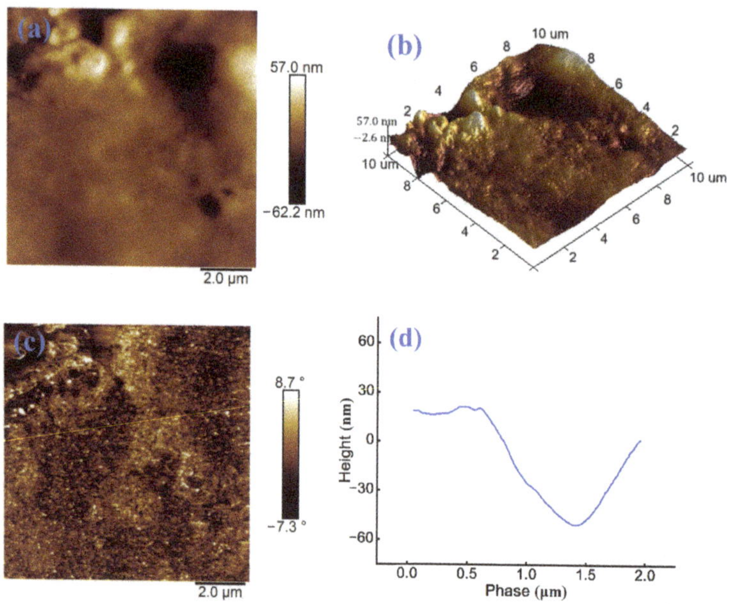

Figure 5. (**a**) Height image, (**b**) 3D height image, (**c**) phase image, and (**d**) treatment image of non-ionic WPU-modified cationic emulsified asphalt.

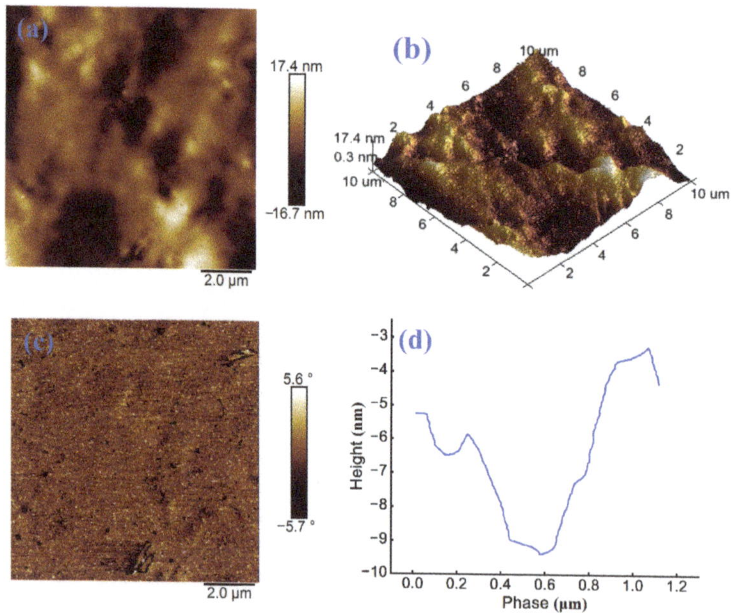

Figure 6. (**a**) Height image, (**b**) 3D height image, (**c**) phase image, and (**d**) treatment image of cationic WPU-modified cationic emulsified asphalt.

The modification effect of cationic WPU primarily results from the introduction of cationic groups, which regulates the molecular structure of WPU. Hydrogen bond donors (-NH, -OH) and acceptors (C=O) in the cationic WPU interact with polar groups (carboxyl and ester) in the cationic emulsified asphalt through hydrogen bonding [30], which is consistent with FTIR analysis results. These interactions enhance intermolecular forces and improve interfacial compatibility among different components. As a result, the asphalt matrix shows an improved uniformity, reduced microscopic phase separation, and decreased surface roughness.

Figure 6c,d further validate these findings. The phase angle variation remains minimal, with a balanced color distribution and smooth transitions, indicating a highly uniform surface. A phase height difference of approximately 5 nm is observed, providing further evidence of an enhanced surface uniformity and a refined microstructure. These results clearly demonstrate that cationic WPU modification significantly improves interfacial compatibility, uniformity, and surface characteristics of cationic emulsified asphalt, offering experimental support for further optimization of modification system.

To investigate the influence of microstructure on anionic emulsified asphalt, AFM characterization is performed on modified samples as shown in Figure 7.

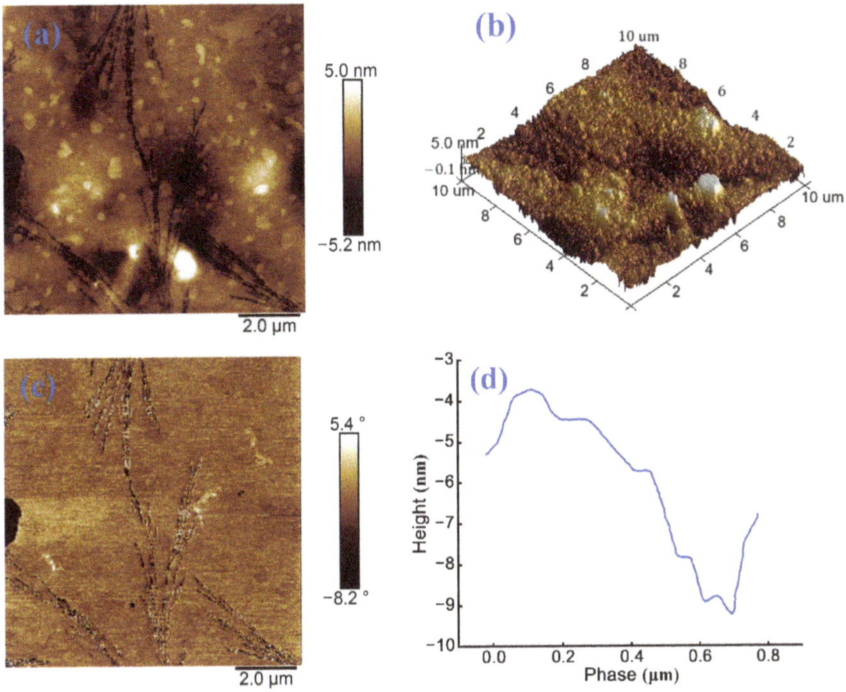

Figure 7. (**a**) Height image, (**b**) 3D height image, (**c**) phase image, and (**d**) treatment image of anionic emulsified asphalt.

Figure 7a shows that the sample displays a relatively uniform honeycomb-like morphology, characterized by a balanced color distribution and minimal bright spots. The surface height variation ranges from −5.2 nm to 5.0 nm, which is significantly lower than that observed in cationic emulsified asphalt. Figure 7b also reveals a smoother surface compared to Figure 4a. This difference is primarily attributed to the charge properties and particle sizes of anionic and cationic emulsifiers [31]. In anionic emulsified asphalt, negatively charged particles repel each other, promoting a more uniform dispersion and resulting in a smoother surface. In contrast, positively charged particles in cationic emulsified asphalt interact electrostatically with negatively charged mineral aggregates, causing particle aggregation and greater surface undulations. Moreover, as shown in Figure 5, the particles in anionic emulsified asphalt are generally smaller and more uniformly dispersed within the matrix, thereby reducing the overall surface roughness.

Figure 7c presents a phase angle variation ranging from −8.2° to 5.4°, along with a relatively a smooth color transition, indicating minimal variations in surface physical properties. A low degree of phase separation is detected. The height variation in Figure 7d approaches 6 nm, reflecting reduced surface roughness, which is consistent with the previous analytical results.

To investigate modification effect of non-ionic WPU on the microstructure of anionic emulsified asphalt, AFM characterization is conducted on the modified samples. The corresponding results are shown in Figure 8.

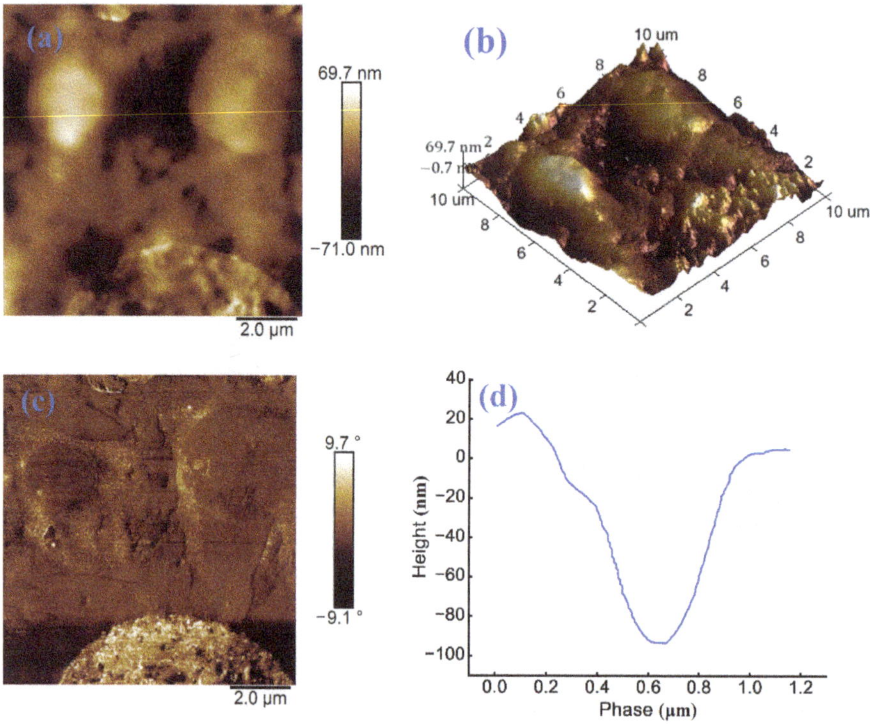

Figure 8. (**a**) Height image, (**b**) 3D height image, (**c**) phase image, and (**d**) treatment image of non-ionic WPU-modified anionic emulsified asphalt.

Figure 8a shows a reduced honeycomb structure, accompanied by an increase in both the number and area of bright regions. The maximum surface height point reaches 69.7 nm, while the minimum is −71.0 nm, respectively, yielding a total height variation of 140.7 nm. Compared to Figure 5a, the surface roughness increases by 21.5 nm, indicating that the roughness of WPU-modified anionic emulsified asphalt exceeds that of cationic emulsified asphalt. This increase is likely caused by weak interactions between WPU and anionic groups, leading to WPU aggregation, non-uniform dispersion, and elevated surface roughness [32]. This observation is further supported by Figure 8b.

Figure 8c displays a broader range of phase angle variations when compared to Figure 7c, indicating an enhanced phase separation after the addition of WPU into anionic emulsified asphalt. The height difference in Figure 8d is approximately 110 nm, and the surface appears uneven, further confirming poor compatibility of WPU within anionic emulsified asphalt system.

To examine the effect of anionic WPU modification on the microstructure of anionic emulsified asphalt and to facilitate the comparison with previously modified samples, AFM tests are employed for surface characterization. The detailed results are presented in Figure 9.

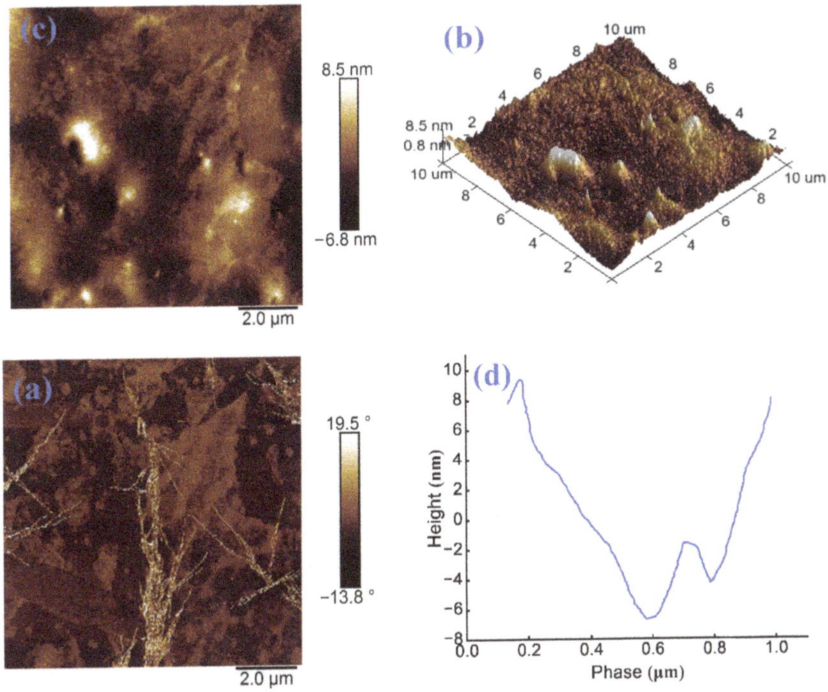

Figure 9. (**a**) Height image, (**b**) 3D height image, (**c**) phase image, and (**d**) treatment image of anionic WPU-modified anionic emulsified asphalt.

As shown in Figure 9a, the continuous and dispersed phases of the sample are uniformly distributed, and only a few bright regions are observed. The maximum surface height variation is measured at 15.3 nm. Compared to Figure 8a, an increased number of honeycomb-like structures increases and exhibits a more balanced distribution. Further analysis of Figure 9b reveals a more uniform surface morphology when compared to 3D images of other anionic emulsified asphalt samples, indicating that the incorporation of anionic WPU enhances microstructural homogeneity.

Figure 9c presents a phase angle range from $-13.8°$ to $19.5°$, with minimal overall color variation. However, a small "dendritic" bright region is observed, indicating a localized area of significant variation in physical properties within the scanned region. This phenomenon may be attributed to sample selection or the preparation process [33].

As shown in Figure 9d, the phase height difference is approximately 16 nm, which is considerably lower than that in Figure 8d and comparable to the value in Figure 7d. These findings suggest that anionic WPU achieves a more effective modification of anionic emulsified asphalt than non-ionic WPU. Overall, these findings strongly support the effectiveness of anionic WPU in enhancing interfacial compatibility, surface uniformity, and microstructural performance of anionic emulsified asphalt.

3.3.2. Influence of WPU on Surface Roughness of Emulsified Asphalt

Although the honeycomb structure partially captures surface undulations, its characterization range is limited, making it difficult to comprehensively describe the overall morphology and surface microfeatures. In contrast, root mean square (RMS) roughness, a key parameter for quantifying surface irregularities [34,35], is calculated using Equations (5) and (6). The RMS roughness of both base emulsified asphalt and WPU-modified emulsified asphalt is quantified, providing a more precise representation of microstructural characteristics and local heterogeneity. Therefore, introducing RMS roughness as a primary characterization metric enhances the understanding of surface morphology and its influence on the functional performance of asphalt materials.

$$R_q = \left[\frac{\iint [h(x,y) - h_0]^2 dA}{\iint dA} \right]^{1/2} \tag{5}$$

$$h_0 = \frac{\iint h(x,y) dA}{\iint dA} \tag{6}$$

where R_q—RMS roughness, nm; $h(x,y)$—height function, nm; A—scanned area; h_0—reference height, nm.

The RMS roughness of sample is calculated using NanoScope Analysis 3.0 software, and the results are presented in Table 7.

Table 7. Rq values of different emulsified asphalt samples.

Sample Name	Cationic Emulsified Asphalt	Cationic WPU + Cationic Emulsified Asphalt	Non-ionic WPU + Cationic Emulsified Asphalt	Anionic Emulsified Asphalt	Anionic WPU + Anionic Emulsified Asphalt	Non-ionic WPU + Cationic Emulsified Asphalt
Surface Rq/nm	7.32	4.73	13.6	1.75	1.63	19.7

As shown in Table 7, the lowest surface roughness of 1.75 nm is observed in anionic emulsified asphalt, whereas the highest value (19.7 nm) is recorded for non-ionic WPU-modified anionic emulsified asphalt. The incorporation of WPU generally increases the surface roughness of samples. However, RMS roughness of cationic emulsified asphalt modified with cationic WPU decreases to 4.73 nm. This behavior is attributed to the intrinsic leveling and filling properties of polyurethane, which allow it to fill microscale surface irregularities and produce smoother surfaces. Additionally, WPU covers or fills uneven substrate areas, thereby reducing surface granularity and overall roughness.

Upon addition to emulsified asphalt, WPU cures and interacts with phenolic and nitrogen-, oxygen-, and hydrogen-containing functional groups in asphalt through hydrogen bonding and other intermolecular interactions [36,37]. These interactions enhance the compatibility between WPU and asphalt, promoting uniform dispersion of WPU within the asphalt matrix. Consequently, the modifying effects of WPU are enhanced, improving the flexibility, viscosity, and adhesiveness of the asphalt mixture. As these properties improve, surface roughness decreases due to the formation of a more uniform and continuous asphalt coating, which reduces surface irregularities.

AFM test results show that anionic WPU-modified anionic emulsified asphalt exhibits the lowest surface roughness, indirectly indicating superior compatibility between anionic WPU and base asphalt. However, the impact of WPU addition on surface roughness remains uncertain. To optimize the modification effect and achieve controlled surface roughness, further comprehensive studies are required, including selection of suitable WPU types, dosage optimization, and refinement of emulsified asphalt formulation and preparation conditions.

4. Conclusions

Based on the above experimental results and discussion, the main conclusions are summarized as follows:

(1) This study compares three ionic types of WPU based on fundamental performance indicators, clarifying the influence of ionic structure differences on the properties of emulsified asphalt. Compared to base emulsified asphalt, all WPU-modified samples exhibit a reduced penetration, an increased softening point, and a decreased ductility. Among them, the non-ionic WPU-modified system enhances material hardness and high-temperature stability but results in the most pronounced low-temperature brittleness, while the cationic WPU maintains a balance between high-temperature stability and low-temperature flexibility.

(2) The intrinsic effect of WPU's ionic structure on the storage stability of emulsified asphalt is identified. WPU modification reduces storage stability, which is primarily affected by the pH value and particle size of WPU. The cationic WPU-modified emulsified asphalt shows the poorest stability with a residue of 0.7% on the sieve, whereas the anionic WPU-modified sample exhibits the best stability with a residue of 0.4% on the sieve.

(3) The modification mechanisms by different ionic WPU structures are systematically investigated through FTIR analysis. The results indicate that ionic WPUs alter the functional group composition and polarity distribution of emulsified asphalt via a combination of physical blending and partial chemical reactions. Cationic WPU promotes the formation of ester groups and amide bonds to improve structural stability, non-ionic WPU enhances the flexibility and wettability, and anionic WPU improves the dispersion.

(4) The microstructural and interfacial morphological effects of ionic WPU on emulsified asphalt are examined using AFM. The results confirm that WPU significantly influences the microstructure and surface roughness of emulsified asphalt. Anionic WPU demonstrates the best dispersion, forming fine and uniformly distributed particles with the lowest surface roughness (5.655 nm, Ra = 1.63 nm). Cationic WPU induces chain-like structures with good curing performance, while non-ionic WPU increases the toughness but leads to greater surface undulation and reduces the uniformity. Distinct regulatory effects on interfacial morphology are observed across different WPU ionic types.

5. Future Directions

This study investigates the properties and compatibility of different ionic WPU-modified emulsified asphalt from both macro- and microscale perspectives, elucidating the modification mechanisms and providing a theoretical foundation for WPU applications in emulsified asphalt. However, current research still exhibits limitations, including that key macroscopic performance indicators (e.g., rheological and aging properties) need to be sufficiently explored, and a comprehensive analysis of WPU's influence on asphalt mixture performance is lacking. Future research should focus on the following directions:

(1) Expanding the research scope. Shifting from material properties to performance evaluation of asphalt mixtures under actual engineering conditions to enhance its applicability.

(2) Enhancing durability assessment. Incorporating critical durability tests (e.g., moisture damage, aging, and freeze–thaw resistance) to evaluate long-term performance.

(3) Deepening mechanistic understanding. Integrating multiscale characterizations and computational models to clarify the interfacial interaction mechanisms between WPU and emulsified asphalt, guiding material design optimization.

These advancements are expected to improve the overall performance of WPU-modified emulsified asphalt and facilitate its broader engineering applications.

Author Contributions: Conceptualization, J.T. and S.H.; methodology, S.H.; software, R.J.; validation, S.H. and R.J.; formal analysis, S.H. and R.J.; investigation, X.Z. (Xiao Zhong); resources, S.H.; data curation, X.Z. (Xiaoxi Zou); writing—original draft preparation, R.J.; writing—review and editing, S.H.; visualization, R.J.; supervision, J.T.; project administration, J.T. and S.H.; funding acquisition, J.T. All authors have read and agreed to the published version of the manuscript.

Funding: This research was funded by the Science and Technology Innovation Demonstration Project of the Department of Transport of Yunnan Province, China, grant number 202159.

Institutional Review Board Statement: Not applicable.

Informed Consent Statement: Not applicable.

Data Availability Statement: The original contributions presented in the study are included in the article, further inquiries can be directed to the corresponding authors.

Conflicts of Interest: Author Jian Tan was employed by the company Zhaotong Yizhao Expressway Investment and Development Co., Ltd. The remaining authors declare that the research was conducted in the absence of any commercial or financial relationships that could be construed as a potential conflict of interest.

References

1. Meng, Y.; Chen, J.; Kong, W.; Hu, Y. Review of emulsified asphalt modification mechanisms and performance influencing factors. *J. Road Eng.* **2023**, *3*, 141–155. [CrossRef]
2. Latifi, A.; Hayati, P.; Goli, A. Advancing Pavement Preservation: Comprehensive Analysis and Optimization of Microsurfacing Mixtures with Stabilizers, Bitumen Types, Emulsifiers, and Fiber Impacts. *Case Stud. Constr. Mater.* **2025**, *22*, e04622. [CrossRef]
3. Lu, P.; He, C.; Wang, J.; Wu, Y.; Ding, J. Polyurethane-modified asphalt mechanism. *Case Stud. Constr. Mater.* **2025**, *22*, e04247. [CrossRef]
4. Carrera, V.; Cuadri, A.A.; García-Morales, M.; Partal, P. The development of polyurethane modified bitumen emulsions for cold mix applications. *Mater. Struct.* **2015**, *48*, 3407–3414. [CrossRef]
5. Huang, G.; Yang, T.; He, Z.; Yu, L.; Xiao, H. Polyurethane as a modifier for road asphalt: A literature review. *Constr. Build. Mater.* **2022**, *356*, 129058. [CrossRef]
6. Keymanesh, M.R.; Ziari, H.; Zalnezhad, H.; Zalnezhad, M. Effects of lead time and manufacturing methods applied for polymer-modified bitumen emulsion (PMBE) on microsurfacing performance. *Road Mater. Pavement Des.* **2022**, *23*, 2271–2292. [CrossRef]
7. Zhang, H.; Zhang, D.; Liu, F.; Wei, T. Investigation of a novel blocked waterborne polyurethane emulsified asphalt mixture for pothole repair: Material development, road performance and construction parameters. *Road Mater. Pavement Des.* **2025**, 1–35. [CrossRef]
8. Hesami, S.; Sadeghi, V.; Azizi, A. Investigation of modified bitumen's rheological properties with synthesized polyurethane by MDI-PPG reactive prepolymers. *J. Thermoplast. Compos. Mater.* **2021**, *34*, 614–632. [CrossRef]
9. Zhao, Y.; Gong, X.; Liu, Q. Research on rheological properties and modification mechanism of waterborne polyurethane modified bitumen. *Constr. Build. Mater.* **2023**, *371*, 130775. [CrossRef]

10. Mu, M.; Liu, X.; Bi, H.; Wang, Z.; Qian, C.; Liu, X.; Liu, Y.; Li, X. Study on Preparation and Properties of Polyurethane-Epoxy Composite Emulsion Modified Emulsified Asphalt. In Proceedings of the 2024 7th International Symposium on Traffic Transportation and Civil Architecture (ISTTCA 2024), Tianjin, China, 21–23 June 2024.
11. Xu, P.; Wang, Y.; Cheng, P.; Cong, P.; Li, D.; Zhang, Z.; Hui, J.; Ye, M. Toughness modification of waterborne epoxy emulsified asphalt by waterborne polyurethane elastomer. *Constr. Build. Mater.* **2023**, *386*, 131547. [CrossRef]
12. Mucci, V.L.; Hormaiztegui, M.E.V.; Amalvy, J.I.; Aranguren, M.I. Formulation, structure and properties of waterborne polyurethane coatings: A brief review. *J. Adhes. Sci. Technol.* **2024**, *38*, 489–516. [CrossRef]
13. Chen, R.; Xu, W.; Chen, Y. Preparation and Properties of Waterborne Polyurethane and SBS Composite-Modified Emulsified Asphalt. *Appl. Sci.* **2024**, *14*, 7288. [CrossRef]
14. Hadavand, B.S.; Jola, B.G.; Didehban, K.; Mirshokraie, A. Modified bitumen emulsion by anionic polyurethane dispersion nanocomposites. *Road Mater. Pavement Des.* **2020**, *21*, 1763–1774. [CrossRef]
15. Zhou, X.; Wang, Z.; Guo, H.; Wang, X.; Chen, W.; Liu, J.; Zhang, H.; Wan, C. Property improvement of epoxy emulsified asphalt modified by waterborne polyurethane in consideration of environmental benefits. *Case Stud. Constr. Mater.* **2024**, *21*, e03559. [CrossRef]
16. Mu, M.; Liu, X.; Bi, H.; Wang, Z.; Qian, C.; Liu, Y.; Liu, X.; Li, X. Effect of evaporation temperature on evaporative residue properties of waterborne polyurethane modified emulsified asphalt. *J. Phys. Conf. Ser.* **2024**, *2873*, 012022. [CrossRef]
17. Zheng, H.; Pang, B.; Jin, Z.; Zhang, Y.; Hou, D.; Bi, J.; Zhang, W.; Yuan, L. Durability enhancement of cement-based repair mortars through waterborne polyurethane modification: Experimental characterization and molecular dynamics simulations. *Constr. Build. Mater.* **2024**, *438*, 137204. [CrossRef]
18. Wang, S.; Chen, X.; Zhang, X.; Liu, F.; Gao, Y.; Li, F. Effect of ionic emulsifiers on the properties of emulsified asphalts: An experimental and simulation study. *Constr. Build. Mater.* **2022**, *347*, 128503. [CrossRef]
19. Kazemi, M.; Karimi, A.; Goli, A.; Hajikarimi, P.; Mohammadi, A.; Doctorsafaei, A.; Fini, E. Biobased polyurethane: A sustainable asphalt modifier with improved moisture resistance. *J. Mater. Civ. Eng.* **2024**, *36*, 04023505. [CrossRef]
20. *JTG E20-2011*; Standard Test Methods of Asphalt and Asphalt Mixtures for Highway Engineering. Ministry of Transport of the People's Republic of China: Beijing, China, 2011.
21. *JTG F40-2004*; Technical Specifications for Construction of Highway Asphalt Pavements. Ministry of Transport of the People's Republic of China: Beijing, China, 2004.
22. Honarkar, H. Waterborne polyurethanes: A review. *J. Dispers. Sci. Technol.* **2018**, *39*, 507–516. [CrossRef]
23. Zhang, J.; Yeoh, G.H.; Kabir, I.I. Polyurethane Materials for Fire Retardancy: Synthesis, Structure, Properties, and Applications. *Fire* **2025**, *8*, 64. [CrossRef]
24. Fang, L.; Zhou, J.; Yang, Z.; Yuan, Q.; Que, Y. Interaction between cement and asphalt emulsion and its influences on asphalt emulsion demulsification, cement hydration and rheology. *Constr. Build. Mater.* **2022**, *329*, 127220. [CrossRef]
25. Patti, A.; Acierno, D. Structure-property relationships of waterborne polyurethane (WPU) in aqueous formulations. *J. Vinyl Addit. Technol.* **2023**, *29*, 589–606. [CrossRef]
26. Cao, Z.; Yuan, L.; Yi, J. Preparation and characterization of polyurethane-modified asphalt containing dynamic covalent bonds. *Constr. Build. Mater.* **2023**, *404*, 133303. [CrossRef]
27. Zhi, W.; Xiang, S.; Bian, R.; Lin, R.; Wu, K.; Wang, T.; Cai, D. Study of MXene-filled polyurethane nanocomposites prepared via an emulsion method. *Compos. Sci. Technol.* **2018**, *168*, 404–411. [CrossRef]
28. Farzaneh, A.; Rostami, A.; Nazockdast, H. Thermoplastic polyurethane/multiwalled carbon nanotubes nanocomposites: Effect of nanoparticle content, shear, and thermal processing. *Polym. Compos.* **2021**, *42*, 4804–4813. [CrossRef]
29. Zhang, Q.; Huang, T.; Wu, J.; Zhang, L.; Lan, Y. Cohesion Performance of Tack Coat Materials between Polyurethane Mixture and Asphalt Mixture. *J. Mater. Civ. Eng.* **2024**, *36*, 04023534. [CrossRef]
30. Wu, C.; Chen, J.; Long, Q.; Sun, D.-X.; Qi, X.-D.; Yang, J.-H.; Wang, Y. Healable, Recyclable, and Ultra-Tough Waterborne Polyurethane Elastomer Achieved through High-Density Hydrogen Bonding Cross-Linking Strategy. *ACS Appl. Mater. Interfaces* **2024**, *16*, 64333–64344. [CrossRef]
31. Yao, Y.; Xiao, M.; Liu, W. Turning things around: From cationic/anionic complexation-induced nanoemulsion instability to toughened water-resistant waterborne polyurethanes. *J. Mater. Chem. A* **2022**, *10*, 18408–18421. [CrossRef]
32. Vaidya, S.M.; Jadhav, S.M.; Patil, M.J.; Mestry, S.U.; Mahajan, U.R.; Mhaske, S.T. Recent developments in waterborne polyurethane dispersions (WPUDs): A mini-review on thermal and mechanical properties improvement. *Polym. Bull.* **2022**, *79*, 5709–5745. [CrossRef]
33. Chen, W.; Wu, Z.; Xie, Y.; He, X.; Su, Y.; Qin, Y.; Tang, D.; Oh, S.-K. Fabrication of silane and nano-silica composite modified Bio-based WPU and its interfacial bonding mechanism with cementitious materials. *Constr. Build. Mater.* **2023**, *371*, 130819. [CrossRef]

34. Wang, T.; Sun, Z.; Liang, F.; Fan, H.; Xiang, J.; Chen, Y. Poly(methylmethacrylate) microspheres with matting characteristic prepared by dispersion polymerization. *Int. J. Polym. Anal. Charact.* **2019**, *24*, 731–740. [CrossRef]
35. Meng, Y.; Ouyang, J.; Ou, J. Investigation on the wetting behavior of asphalt emulsion on aggregate for asphalt emulsion mixture. *Constr. Build. Mater.* **2023**, *400*, 132844. [CrossRef]
36. Shirzad, S.; Idris, I.I.; Hassan, M.; Mohammad, L.N. Self-healing capability and mechanical properties of asphalt mixtures prepared with light-activated polyurethane prepolymer modified asphalt binder. *Transp. Res. Rec.* **2025**, *2679*, 710–728. [CrossRef]
37. Yu, X.; Chen, B.; Dong, F.; Zu, Y.; Wang, S.; Si, J.; Huang, J. Repurposing waste polyurethane into cleaner asphalt pavement materials. *J. Mater. Civ. Eng.* **2023**, *35*, 04023296. [CrossRef]

materials

MDPI

Article

Multi-Scale Modeling and Damage Mechanisms of Asphalt Pavements Under Coupled Salt–Thermal–Mechanical Effects

Jin Ma [1,2,3], Jiaqi Chen [4,*], Mingfeng Tang [1,2,3] and Yu Liu [1,2,3]

1 Guangxi Communication Investment Technology Co., Ltd., Nanning 530001, China; mailmajin@163.com (J.M.); mailmingfeng@163.com (M.T.); hitrain@163.com (Y.L.)
2 Guangxi Expressway Maintenance Engineering Technology Research Center, Nanning 530001, China
3 Key Laboratory of Intelligent Road Maintenance in Guangxi Transportation Industry, Nanning 530001, China
4 Department of Civil Engineering, Central South University, Changsha 410075, China
* Correspondence: chenjiaqi@csu.edu.cn

Abstract: Salts can have detrimental effects on asphalt pavements, leading to permanent damage that compromises their durability and sustainability. This study investigates the damage mechanisms of asphalt pavements under coupled salt–thermal–mechanical effects using multi-scale modeling. Pull-off and semicircular bending (SCB) tests were conducted to determine material parameters and validate numerical models. Experimental data demonstrated that after 48 h of salt treatment at −10 °C, specimens exhibited reductions in cohesive strength ranging from 23.5% to 26% and adhesive strength decreasing by 25% to 44% compared to untreated controls. More severe degradation was observed at 0 °C, with cohesive strength diminishing by up to 63.8% and adhesive strength declining by up to 71.6%. A multi-scale finite element (FE) pavement model incorporating cohesive zone modeling (CZM) was developed to simulate damage behavior within asphalt concrete. Salt diffusion analysis revealed limited penetration depth within short exposure periods, and results showed that salt penetration reached only about 10 mm into the pavement layers after 48 h. Results indicated significant reductions in adhesive and cohesive strengths due to salt exposure, with damage susceptibility increasing under combined thermal fluctuations and mechanical loading. Additionally, the effects of moving load magnitude and speed on pavement damage were examined, showing higher damage accumulation at lower speeds and heavier loads. This research provides insights into pavement deterioration mechanisms, contributing to improved durability and maintenance strategies for asphalt pavements in salt environments.

Keywords: salt erosion; asphalt pavement; cohesive zone model; thermal properties; finite element analysis

1. Introduction

The use of de-icing salts is a common and economical strategy for ensuring safe road conditions in winter across many countries [1–4]. However, these chemicals can adversely affect asphalt pavements, resulting in permanent damage that undermines their durability and sustainability [5,6]. Previous studies have shown that chemical reactions initiated by de-icing salts not only accelerate the deterioration of pavement materials but also pose challenges to the long-term resilience of road infrastructure. Therefore, understanding how salt impacts asphalt performance is essential for developing strategies to enhance the longevity and sustainability of pavements.

The effects of salt on asphalt pavement can be analyzed at three main scales: micro-scale, meso-scale, and macro-scale. At the micro-scale, techniques such as infrared spectroscopy (FTIR) analysis and scanning electron microscopy (SEM) are used to observe changes in both the chemical composition and the morphological characteristics of asphalt before and after salt exposure [3]. These techniques can accurately analyze changes in the chemical composition of asphalt and the effects of salt-induced chemical reactions on its polymer chains. Furthermore, transmission electron microscopy (TEM) and atomic force microscopy (AFM) are employed to investigate the mechanisms of salt-induced damage to the microstructure of asphalt. These techniques enable the identification of changes in interfacial interactions between asphalt particles as well as between asphalt and fillers during salt erosion. Additionally, molecular dynamics (MD) simulations are utilized to investigate the erosion process induced by salts, revealing the molecular interactions and structural changes that occur during this process [7–9].

The meso-scale studies primarily focus on the fine asphalt matrix (FAM) and asphalt mixtures. Traditionally, experimental methods have been the principal approach for investigating the mechanical behavior of asphalt mixtures under the influence of salts. Semicircular bending (SCB) tests [10–12], indirect tensile (IDT) strength tests [13], and fatigue tests, have been used to measure the mechanical properties, and failure mechanisms of asphalt concrete under different conditions. Additionally, various laboratory tests that consider the effects of salt have been conducted to assess the damage to asphalt concrete [14,15]. These tests provide intuitive parameters and results to evaluate the performance of asphalt concrete and inform pavement construction practices. Furthermore, macro-scale experiments that simultaneously consider the effects of temperature and salt have been performed to investigate the deterioration of mechanical properties in asphalt concrete [14,16]. However, while experimental methods are important for understanding fundamental behaviors at a single scale, deriving the mechanisms by which environmental factors affect macro-scale pavement deterioration from these results is challenging. Therefore, integrating research across different scales is essential for a better understanding of the combined effects of salt and temperature on the long-term performance of asphalt materials.

Due to the limitations of experimental methods, numerical simulations have increasingly been employed in pavement research. Methods such as the discrete element method (DEM), finite element method (FEM), and finite difference method (FDM) have been successfully applied to simulate temperature fields and stress distributions within pavement structures [17–23]. These approaches facilitate detailed analyses at both micro and macro scales, contributing to a better understanding of the damage mechanisms of asphalt pavements. Additionally, numerical simulations offer the advantage of being able to model complex interactions and predict the long-term performance of materials. However, currently numerical studies mainly focus on the thermal–mechanical coupling effect on asphalt concrete [24–26] or focus on isolated factors by changing some specific parameters of the model, which may not fully capture the coupled interactions among salt diffusion, thermal effects, and mechanical loading.

Consequently, the main objective of this study is to investigate damage mechanisms of asphalt pavements under coupled salt–thermal–mechanical effects based on multi-scale modeling analysis. By systematically integrating experimental measurements with advanced numerical simulations, the research aims to provide fundamental insights into the complex environmental factors that influence asphalt material performance and structural durability. A multi-scale pavement model involving a local heterogeneous asphalt concrete course was developed. The damage and fracture of the asphalt concrete were simulated by embedding adhesive and cohesive elements. Various laboratory experiments were conducted to determine the material parameters for the asphalt concrete and multi-scale pavement FE model. The accuracy of the models was validated by three-point bending tests of asphalt concrete. Additionally, with the validated model, the damage mechanisms of pavement under coupled salt–thermal–mechanical effects were investigated in this study. The flowchart of this study is shown in Figure 1.

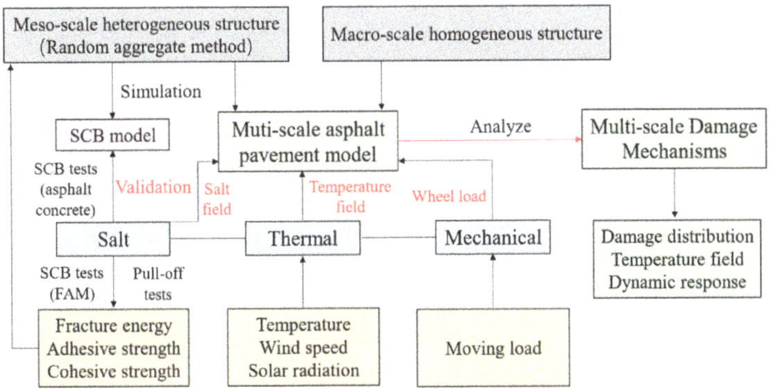

Figure 1. Flowchart of multi-scale modeling and damage mechanisms analysis.

2. Theories

For a deeper understanding of the mechanisms affecting pavement performance, the penetration of salt ions, the thermal behavior of the pavement, and the material degradation process on pavement response were considered, as shown in Figure 2. Three theories were employed to investigate the damage mechanisms in this study.

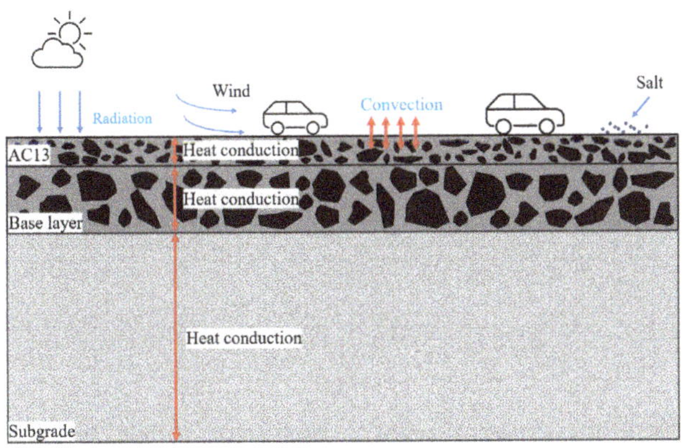

Figure 2. Schematic diagram of environmental and mechanical effects of pavement.

2.1. *Diffusion Theory*

During winter road de-icing operations, salt ions can penetrate the interior of asphalt pavement, causing deterioration of pavement materials. In this process, the non-stationary diffusion of salt ions can be characterized by Fick's second law, as shown in Figure 3. Assuming that the diffusion coefficient is independent of concentration and only vertical diffusion is considered, the following equation can be used to characterize it:

$$\frac{\partial c}{\partial t} = D\frac{\partial^2 c}{\partial x^2} \quad (1)$$

where c is the concentration; t is time; x is the position coordinate, presenting the direction of diffusion; D is the diffusion coefficient. By setting the initial concentration and boundary conditions and using numerical methods for calculation, the concentration distribution of salt ions at different times can be obtained.

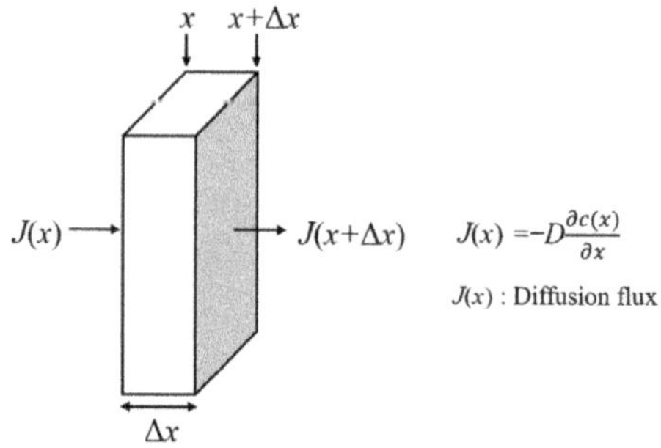

Figure 3. Schematic diagram of Fick's second law.

2.2. Damage Theory

Due to the fracture strength of aggregates being much larger than that of the fine aggregate matrix (FAM) and the interface, cracks were commonly assumed that only appeared in the FAM and the FAM-aggregate interface in previous models [14,27]. In this study, cohesive elements were incorporated into FE models to simulate crack initiation and evolution in materials. To simulate the process of fracture in asphalt concrete, the Bilinear Traction-Separation Law was used to describe the behavior within the cohesive elements, as illustrated in Figure 4. This law defines the relationship between the traction force and the relative displacement of the crack surfaces.

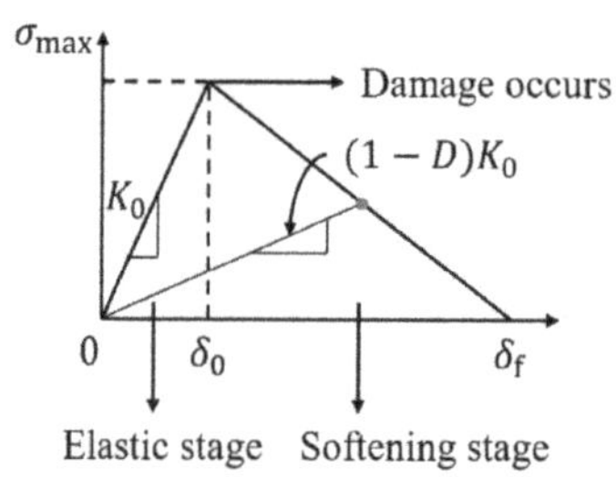

Figure 4. Schematic diagram of bilinear traction-separation law.

Initially, when the relative displacement is less than the initial damage displacement δ_0, a linear relationship exists between traction force and displacement, indicating the elastic stage of the cohesive element. As the crack displacement exceeds δ_0, the material undergoes irreversible softening, resulting in a progressive reduction in stiffness. Ultimately, when the crack displacement surpasses the final displacement (δ_f), the stiffness drops to zero, signifying complete material failure.

2.3. Heat Transfer Theory

The pavement temperature field is influenced by three primary modes of heat transfer: thermal radiation, thermal convection, and thermal conduction. In this study, based on existing research, the combined effects of these heat transfer modes were considered, and the following pavement heat equations, as the control equation for the pavement temperature field are shown below.

(1) Thermal Radiation

When solar radiation reaches the surface, it is not fully absorbed. The solar radiation that can enter the surface is shown in the Equation.

$$Q_{in} = (1 - \widetilde{\alpha})Q_s \tag{2}$$

where Q_{in} is the solar radiation entering the surface, W/m^2; $\widetilde{\alpha}$ is the surface reflectivity; Q_s is the solar radiation reaching the surface, W/m^2.

The surface absorbs radiant heat under sunlight while also radiating heat to the surrounding environment. The radiation effect is typically influenced by the reflectivity and absorptivity of the surface material. According to the Stefan–Boltzmann law, the radiative heat flux density can be expressed as follows:

$$Q_{out} = \varepsilon\sigma(T^4 - T_{ambient}^4) \tag{3}$$

where Q_{out} is the relative heat flux, W/m^2; ε is the emissivity; σ is the Stefan-Boltzmann constant, 5.67×10^{-8} W/m^2·K^4 [20]; T is the pavement temperature, K; $T_{ambient}$ is the ambient temperature, K.

(2) Thermal Convection

There is also convective heat transfer between the surface and the atmospheric environment, where the air carries away heat from the road surface through its flow. Thermal convection can be expressed using Newton's law of cooling:

$$q_c = h(T - T_{ambient}) \tag{4}$$

where q_c is the convective heat flux, W/m^2; h is the convective heat transfer coefficient, W/m^2·K, $h = 5.6 + 4 \times v_{wind}$ [20]; T is the pavement temperature, K; $T_{ambient}$ is the ambient temperature, K.

(3) Thermal Conduction

Within the road structure, heat flow is transferred through conduction due to temperature gradients, which can be calculated using Fourier's law:

$$q = -k\nabla T \tag{5}$$

where q is the heat flux, W/m^2; k is the thermal conductivity, W/mK; ∇T is the temperature gradient, K/m.

Therefore, the balance equation in this study is as follows:

$$Q = \varepsilon\sigma(T^4 - T_{ambient}^4) + h(T - T_{ambient}) - k\nabla T \tag{6}$$

3. Materials and Experiments

3.1. Experimental Research Program

A series of laboratory experiments was conducted to obtain the material parameters required for the FE model and to validate the model. The adhesive strength and cohesive strength of the asphalt concrete were determined through pull-off tests, while the fracture energy parameters of the FAM and the force-displacement curves of the asphalt concrete were obtained through SCB tests. The test details are shown in Table 1. Detailed methodological procedures are elaborated in Sections 3.3 and 3.4.

Table 1. Experimental research program information.

Type of Tests		Salt Concentration	Temperature	Immersion Duration	Sample Number
Pull-off test (FAM)		0, 3, 6, 9%	−10, 0 °C	0, 12, 24, 36, 48 h	120
SCB test	FAM	0, 3, 6, 9%	−10, 0 °C	0, 12, 24, 36, 48 h	120
	Asphalt concrete	0, 3, 6, 9%	−10, 0 °C	0, 48 h	48

3.2. Materials

In this study, asphalt concrete samples were prepared with basalt aggregates, mineral filler, and SBS-modified asphalt. The particle size of coarse aggregate in this study was 2.36–16 mm, and the FAM is a mixture of fine aggregate (particle size $\leq$ 2.36 mm), mineral filler, and SBS-modified asphalt. The gradation of aggregates used in the experiment is shown in Table 2. Moreover, the optimum binder content of asphalt concrete specimens was determined as 4.6% by total weight. The target air void content of asphalt concrete and FAM was 4%.

Table 2. Aggregate gradations of asphalt concrete and FAM.

Sieve Size (mm)		16	13.2	9.5	4.75	2.36	1.18	0.6	0.3	0.15	0.075
Passing percentage (%)	Asphalt concrete	100	94.7	75.3	57.7	37.6	25.2	15.7	9.3	6.7	5.0
	FAM	100	100	100	100	100	67.0	41.9	24.8	17.7	13.3

3.3. Pull-Off Test

The pull-off tests were conducted to measure the adhesive strength of the FAM-coarse aggregate interface, as well as the cohesive strength within the FAM. To investigate the adhesive and cohesive strengths of the FAM and coarse aggregate under different salt solution conditions, pull-off specimens were prepared and subjected to different experimental conditions. Given that salt intrusion is a long-term and complex process that is difficult to fully simulate real-world conditions, this study simplifies the model calculations and analysis by referencing previous studies [5,28]. Therefore, specimens were immersed in salt solutions with concentrations of 0, 3, 6, and 9%, at a temperature of 0 °C for durations of 0, 12, 24, 36, and 48 h. The fracture of asphalt concrete may occur in the following locations: (1) within the coarse aggregate; (2) at the interface between the coarse aggregate and the FAM; and (3) within the FAM itself. Generally, the fracture strength of the aggregates is significantly greater than that of the aggregate–FAM interface and the FAM [29–31]. Therefore, this study focuses exclusively on cases (2) and (3).

To quantify the adhesive strength at the aggregate–FAM interface ($\sigma_{\text{FAM-aggregate}}$) and the cohesive strength within the FAM (σ_{FAM}), an automatic adhesion tester (DeFelsko, Ogdensburg, NY, USA) was used to apply a controlled pull-off force to the specimens until failure occurred, as shown in Figure 5.

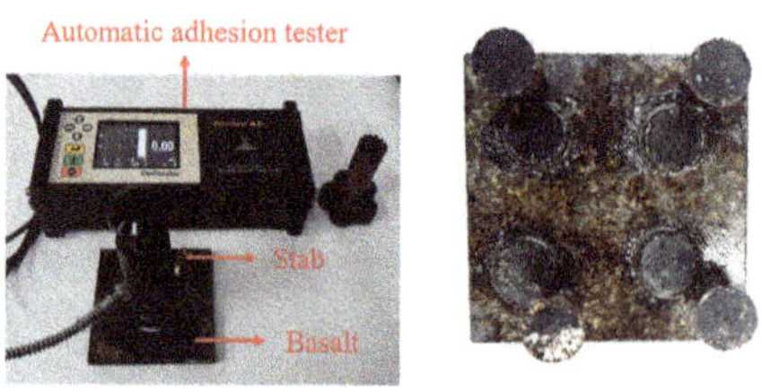

Figure 5. Automatic adhesion tester and specimen.

The steps for the pull-off test are as follows:

(1) The pull-off specimen was positioned within the sleeve of the pull-off apparatus to ensure stable installation, preventing any potential displacement during the testing procedure.
(2) The pull-off rate was set to 0.6 MPa/s. The testing apparatus was then to be activated, which would trigger the automatic commencement of the test, applying the specified pull-off force to the specimen at the predetermined rate.
(3) Upon completion of the test, the values displayed on the screen of the testing apparatus were recorded. Three parallel tests of each group were conducted to ensure statistical validity.

Following the pull-off test, the specimen is expected to fracture, potentially exhibiting one of four distinct fracture types: (a) adhesive failure at the interface of metal stab and FAM, where the fracture path follows the interface between the metal stab and FAM; (b) adhesive failure at the coarse aggregate–FAM, with the fracture path along the interface between the basalt aggregate and the FAM layer; (c) cohesive failure within the FAM, characterized by the fracture path traversing through the FAM layer; and (d) mixed failure, which incorporates a combination of the aforementioned failure types. A digital image processing technique was used to analyze the fracture surfaces, distinguishing between adhesive and cohesive failures [32,33]. Statistical analysis and calculations of the different fracture situations will be performed based on existing methodologies. The adhesive strength ($\sigma_{\text{FAM-aggregate}}$) and cohesive strength (σ_{FAM}) will be obtained from multiple tests for each specimen, and the average values will be calculated to serve as fracture parameters for the finite element model.

The strength parameters of the material were tested and calculated under different salt solution concentrations and soaking durations. Additionally, conditions at −10 °C and 0 °C were tested. To avoid the interference of freezing and thawing effects on the results, before conducting the pull-out test and SCB test at −10 °C and 0 °C environments, the specimens were soaked in a 0 °C salt solution. The obtained parameters were used in the subsequent FE models to simulate the behavior of the pavement materials. Each set of specimens consists of three groups, and the average values were computed. The results are shown in Figure 6.

As illustrated in Figure 6, both cohesive and adhesive strength values declined with increasing salt concentration and immersion time. At −10 °C, the cohesive strength after 48 h of treatment decreased by 26% to 32.5% compared to the control group, while the adhesive strength decreased by 33.5% to 63.8%. Additionally, with the same treatment duration, the cohesive strength decreased by 8% to 26%, and the adhesive strength decreased by 22.2% to 45.6%. Therefore, the treatment duration has a more significant effect on both cohesive and adhesive strength. At 0 °C, the cohesive strength after 48 h of treatment decreased by 24.9% to 44.0% compared to the control group, while the adhesive strength decreased by 37.9% to 71.6%. Additionally, with the same treatment duration, the cohesive strength decreased by 5.4% to 26.5%, and the adhesive strength decreased by 14.9% to 54.2%. It is important to emphasize that even in the absence of salt, wet treatment (immersion in a 0% salt concentration solution) leads to a reduction in the strength of the FAM–aggregate and FAM–FAM interfaces, which is consistent with previous results [34].

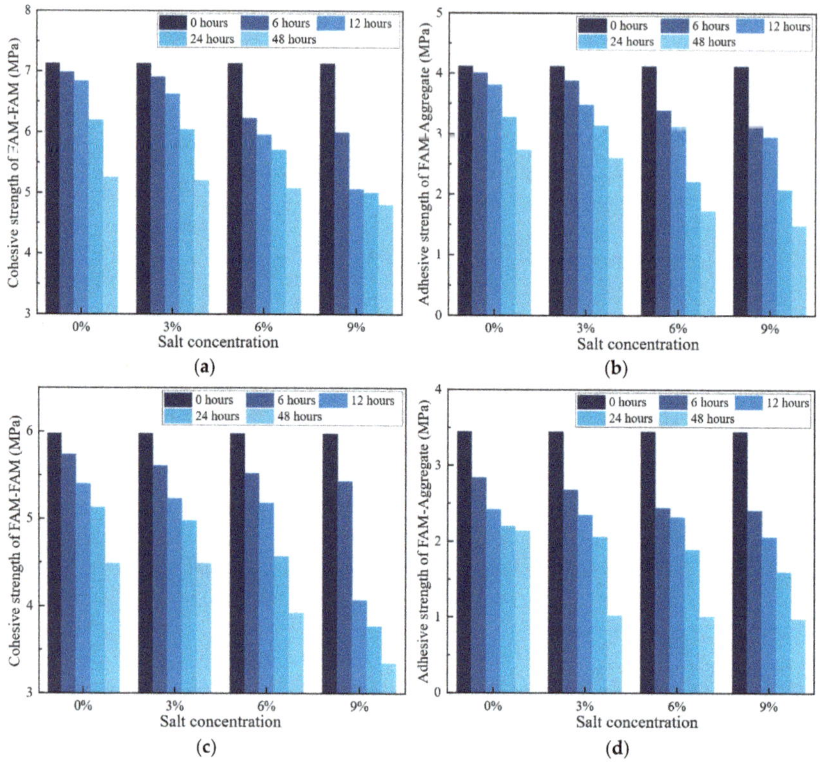

Figure 6. Cohesive strength and adhesive strength under different temperatures and salt concentrations. (**a**) cohesive strength (FAM-FAM) at −10 °C; (**b**) adhesive strength (FAM-aggregate) at −10 °C; (**c**) cohesive strength (FAM-FAM) at 0 °C; (**d**) adhesive strength (FAM-aggregate) at 0 °C.

These reductions in cohesive and adhesive strengths suggest that prolonged exposure to salt solutions at low temperatures may compromise the internal bonding within the FAM. This decline could potentially lead to increased susceptibility to cracking and structural failure. Furthermore, the diminished adhesive strength at the FAM-aggregate interface could result in reduced overall durability and performance of the pavement, increasing the likelihood of surface damage and necessitating more frequent maintenance.

3.4. SCB Test

A series of SCB specimens 150 mm in diameter, 75 mm in height, and 30 mm in thickness were prepared for obtaining the fracture parameters of FAM and asphalt concrete. These samples were cut from the cylindrical specimens of FAM and asphalt concrete 150 mm in diameter and 150 mm in height, which were prepared with the Superpave Gyratory Compactor (SGC). Additionally, an incision with 2 mm in width and 15 mm in length was made at the bottom center of each semi-circular specimen. To simulate the deicing salt environment, SCB specimens were placed in a 0 °C environment and soaked in deicing salt solutions with concentrations of 0, 3, 6, and 9%. The soaking durations were 0, 6, 12, 24, and 48 h. After treatment, the surface moisture of the specimens was wiped off, and SCB tests were conducted at −10 °C or 0 °C. Each test condition included three parallel specimens to ensure the reliability of the results.

In the experiment, a universal testing machine (UTM-250, IPC Global, Boronia, VIC, Australia) equipped with an environmental chamber was used to control the temperature, as shown in Figure 7. The specimens were placed on two metal supports in the environmental chamber without any constraints; the distance between the two bottom supports was 120 mm, and the vertical loading rate was set at 5 mm/min. A waiting period of over 4 h was required to ensure that the temperature within the specimens was uniform. Force-displacement curves were then obtained to determine the fracture properties of the FAM and asphalt concrete.

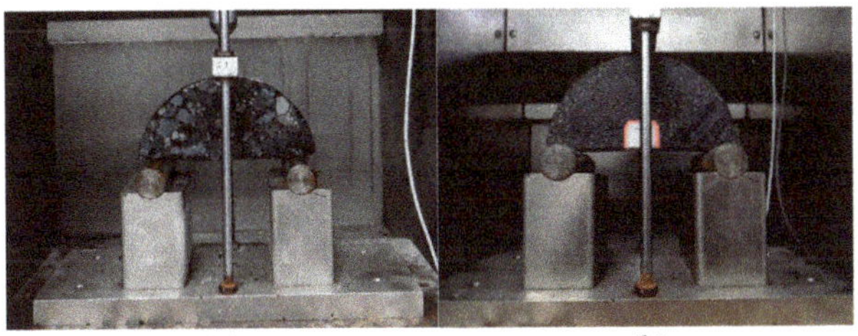

Figure 7. Test equipment and samples: (**a**) Three-point bending test of asphalt concrete, (**b**) SCB test of FAM.

In this study, the FAM was considered as a continuous matrix within the asphalt concrete, which could influence the overall fracture performance. Previous research has demonstrated that fracture energy (G_f) is a crucial indicator for evaluating fracture performance [5]. The specific calculation methodology is presented as Equation (7) [35]:

$$G_f = \frac{W_f}{(r-a)\times t} \tag{7}$$

where W_f is the fracture work, defined as the area enclosed by the load–displacement curve and the coordinate axes, mJ; r is the radius of the specimen of SCB, mm; a is the length of the notch, mm; t is the thickness of the specimen, mm. The fracture energy values of the FAM under different salt concentrations were calculated using Equation (7), and the results are presented in Figure 8.

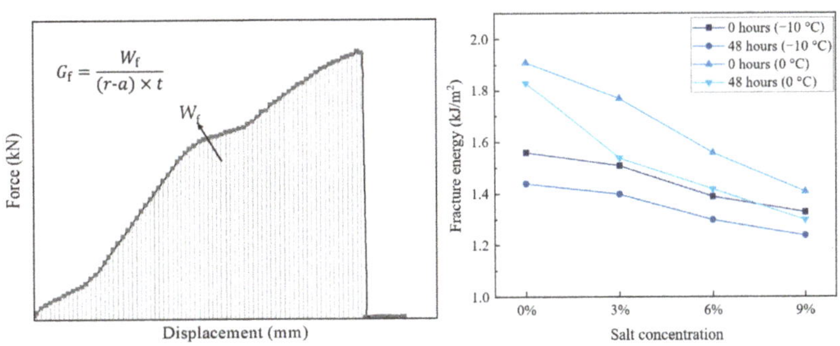

Figure 8. Fracture energy of FAM.

The lines in Figure 8 showed that the fracture energy was affected by both the salt concentration, temperature, and treatment time. The fracture energy at −10 °C is lower than that at 0 °C. In addition, the fracture energy that decreases at 0 °C is larger than that at −10 °C, with values of 4.2~13%. This reduction in fracture energy at 0 °C may suggest an increased risk of crack propagation in asphalt concrete under these conditions. Existing research has demonstrated that lower fracture energy corresponds to a higher rate of crack growth, indicating that asphalt pavements exposed to similar low-temperature environments could experience accelerated damage over time [36,37].

4. Model Development

Two types of models were developed in this study to investigate the damage of asphalt concrete pavement under coupled salt–thermal–mechanical effects. One is the heterogeneous SCB model to study the fracture mechanism of asphalt concrete and to validate the accuracy of the heterogeneous structural model, the other is the multiscale pavement model. The multi-scale pavement model includes heterogeneous and homogeneous pavement layers, homogeneous UBG (unbound granular) layers, and base layers. By analyzing meso-scale damage and macro-scale heat transfer, the damage mechanisms of asphalt pavements under salt–thermal–mechanical coupling were investigated. At the macro-scale, the vertical transfer of pavement temperature was considered. The mechanical model at the meso-scale consists of three components: salt transport, the deteriorative effects of salt on the $\sigma_{\text{FAM-aggregate}}$ and σ_{FAM}, and damage under the combined action of salt and mechanical loads. In this study, cracks can occur within the FAM (cohesive damage) or along the interface between the aggregate and FAM (adhesive damage).

4.1. Heterogeneous SCB Model

In this study, a heterogeneous asphalt concrete model was developed using digital image processing (DIP) techniques and the random aggregate method. Aggregates with a particle size greater than 2.36 mm were classified as coarse aggregates, while the Fine Asphalt Mixture (FAM) was defined as a combination of fine aggregates (particle size $\leq$ 2.36 mm), mineral powder, and asphalt. The Aggregate Image System (AIMS) was used to obtain 2D images of the coarse aggregates [38], and the gradation of coarse aggregates and the size of the models were consistent with those of the specimens in the SCB experiments. Based on the random aggregate method, all coarse aggregates were randomly placed in a designated area. Then, the FAM phase was determined through Boolean operations. Additionally, three rigid bodies were developed to simulate the SCB test setup. The horizontal distance between the centers of the two bottom rigid bodies under the model was 120 mm, consistent with the dimensions of the SCB specimens in the experimental tests. The schematic of the SCB model is shown in Figure 9.

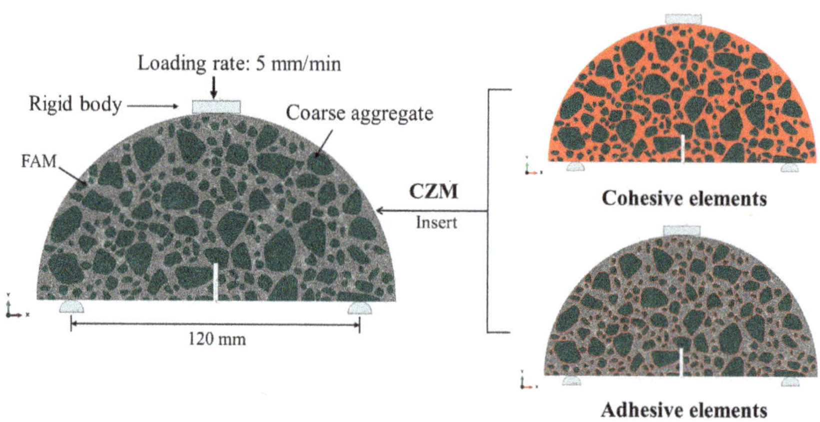

Figure 9. The schematic of the asphalt concrete FE model.

In the heterogeneous SCB models, both the coarse aggregates and the FAM were meshed using 3-node linear elements. Cohesive elements with zero thickness were inserted at the FAM–aggregate interface and within the FAM in the meshed meso-scale model to represent damage evolution and crack propagation. Specifically, the cohesive elements at the FAM–aggregate interface were employed to replicate interfacial adhesive behavior, while those within the FAM represented internal cohesive behavior [31,39], as shown in Figure 8. A vertical displacement of 5 mm/min was applied to the top rigid body to simulate the SCB test process.

The composition of the asphalt concrete was established based on findings from previous studies, which also provided the basis for deriving the dynamic modulus used in the models. The basic viscoelastic material parameters, including the elastic modulus and Poisson's ratio, were consistent with those of the FAM [40]. The detailed parameters of the models are presented in Tables 3 and 4.

Table 3. Parameters for materials in the heterogeneous model [40].

Type of Material	Elastic Modulus (MPa)	Poisson's Ratio
Aggregate	80,000	0.20
FAM	805	0.25
CZM	805	0.25

Table 4. Parameters for the Prony series of FAM and asphalt concrete [26].

Temperature (°C)	i	Relaxation Time, ρ_i (s)	Relaxation Strength, E_i (MPa) FAM	Relaxation Strength, E_i (MPa) Asphalt Concrete	T_0 (°C)	C_1	C_2
20	1	1×10^{-7}	5218.14	2240.85	20 (FAM)	25.4 (FAM)	233.2 (FAM)
	2	1×10^{-6}	5381.72	3212.28	20 (AC)	25.7 (AC)	218.9 (AC)
	3	1×10^{-5}	5499.81	3913.62	-	-	-
	4	1×10^{-4}	5570.35	4897.11	-	-	-
	5	1×10^{-3}	5112.13	5645.90	-	-	-
	6	1×10^{-2}	3608.49	5520.75	-	-	-
	7	1×10^{-1}	1489.17	3509.47	-	-	-
	8	1×10^{0}	512.63	1405.91	-	-	-
	9	1×10^{1}	154.93	531.44	-	-	-
	10	1×10^{2}	55.72	171.11	-	-	-
	11	1×10^{3}	13.97	68.48	-	-	-
	12	1×10^{4}	6.79	18.48	-	-	-
Equilibrium modulus, E_e	-	-	34.20	197.25	-	-	-

4.2. Multi-Scale FE Pavement Model

In this study, a multi-scale asphalt pavement model with a length of 3000 mm and a total thickness of 3000 mm was developed to minimize boundary effects. The model consisted of three layers with distinct materials: an asphalt concrete layer (AC-13), a cement-stabilized gravel base layer (Base layer), and a subgrade layer [41]. Except for the asphalt concrete layer, the other layers were considered homogeneous.

To reduce computational costs, the asphalt concrete layer was divided into heterogeneous and homogeneous sections. The parameters of the pavement model for the heterogeneous portion were maintained consistently with the laboratory tests. Detailed information on each layer of the pavement model is shown in Figure 10.

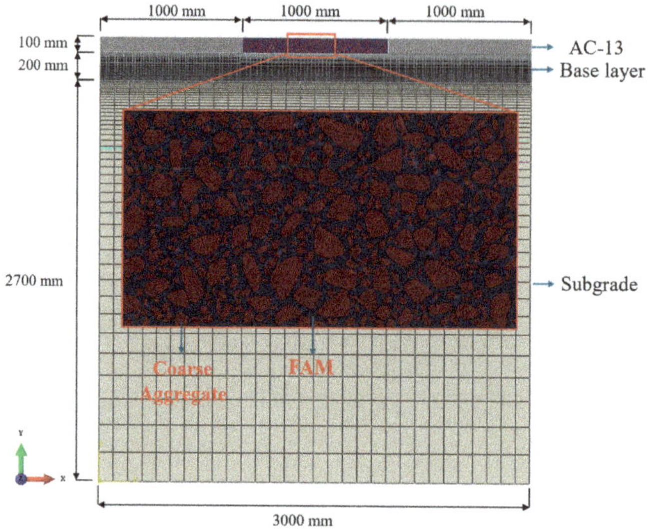

Figure 10. Multi-scale FE pavement model.

Under traffic loads, asphalt concrete exhibits time-dependent stress and strain behavior. To accurately simulate the asphalt pavement's response to dynamic loads in this multi-scale pavement model, viscoelastic parameters were assigned to both the heterogeneous FAM and the homogeneous sections of the AC-13 layer. In addition, the other layers were considered as linear elastic materials in this study. The detailed parameters of the models are shown in Tables 4 and 5 [5].

Table 5. Parameters of the models.

Material	Young's Modulus (MPa)	Poisson's Ratio	Coefficient of Thermal Expansion (10^{-5}/°C)	Thermal Conductivity [(J/m·h·°C)]	Specific Heat [(J/kg·°C)]	Density (kg/m^3)
Aggregate	80,000	0.20	1.20	5472	920	2600
FAM	805	0.25	10.52	4680	923	2200
AC-13	3592	0.35	2.19	5112	916	2340
Base layer	200	0.35	2.00	5616	912	2200
Subgrade	70	0.40	2.00	5616	1040	1800

In the multi-scale pavement model, all displacements at the bottom were constrained to zero, while vertical displacements at the lateral boundaries were permitted. The heterogeneous portion of the pavement model was meshed using 3-node linear elements, whereas the remaining homogeneous sections were discretized using 4-node bilinear elements. In this study, finite element analysis was conducted using ABAQUS/Standard software (2021) with a dynamic implicit analysis method. The interface between the asphalt surface layer and base layer was modeled using a tied contact approach to ensure precise stress transmission and mechanical continuity.

To facilitate the subsequent analysis and computation of pavement temperature fields, the following fundamental assumptions were established in this study:

(1) Except for the heterogeneous pavement structural regions, all other structural layers were assumed to be homogeneous, continuous, and isotropic.
(2) The materials between the pavement layers were assumed to be tightly bonded, ensuring continuity of temperature and heat flow at the interfaces.

(3) The lateral variation in the pavement temperature field was disregarded under the assumption that heat flow occurs solely in the vertical direction of the pavement.

5. Results and Discussion

5.1. Model Validation

To evaluate the accuracy of the computational models described above, three heterogeneous SCB asphalt concrete models with different aggregate distributions were developed. Peak load and fracture energy values for the asphalt concrete models under various conditions were obtained through the simulation of SCB tests. Moreover, the parameters for these conditions were kept consistent with those used in laboratory experiments. The schematic diagram of the damaged SCB model is shown in Figure 11. Based on the provided SCB test results and finite element simulations conducted at different temperatures (0 °C and −10 °C, treatment for 48 h), the peak load values of asphalt concrete specimens were compared under varying salt solution concentrations, as shown in Figure 12.

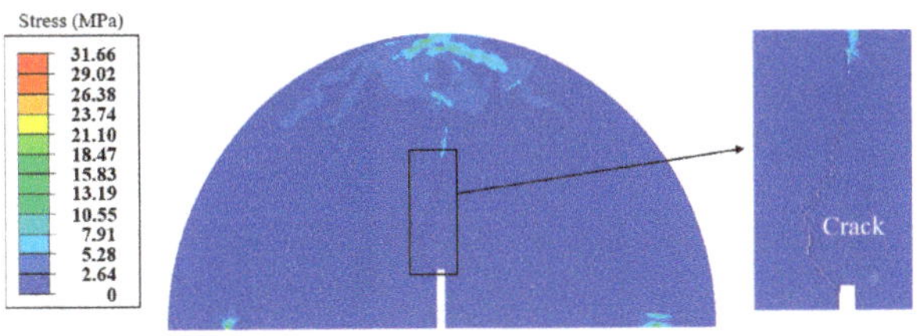

Figure 11. Schematic diagram of the damaged SCB model.

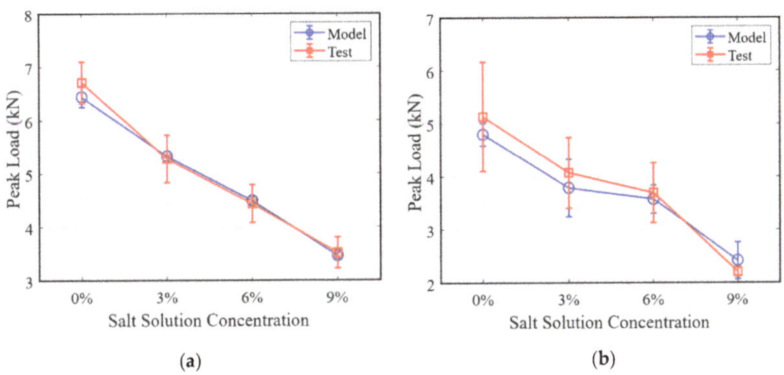

Figure 12. Comparison of the peak loads of FE models and tests at: (**a**) −10 °C, (**b**) 0 °C.

As shown in Figure 12, both experimental and simulation results indicate a consistent reduction in the peak load of asphalt concrete with increasing salt solution concentrations. Moreover, the computational values from the models generally fall within the range of the experimental results, suggesting that the overall accuracy of the model is acceptable. Specifically, at −10 °C, the peak load decreased from approximately 6.6 Kn at 0% salt concentration to around 3.5 Kn at 9% concentration. In contrast, at 0 °C, the peak load declined from around 5 Kn to 2.4 Kn as the salt concentration increased. Additionally, the peak load was 21.7% lower in the absence of salt after wet–dry cycles compared to that of the control group. Moreover, when treated with salt solutions, the peak load decreased by 37.4% compared to the control group (salt concentration = 0%).

5.2. Analysis of Salt Diffusion and Temperature Field of Pavement

In the investigation of salt damage to dense-graded asphalt pavement, salt diffusion is a critical factor. Salt ions diffuse into the pavement interior through water infiltration, yet they are difficult to reach deeper layers within a short time. Therefore, analyzing salt diffusion is essential. The salt diffusion in heterogeneous pavement structures was analyzed in this section, specifically calculating salt migration through water-mediated transport over 48 h. The computational model defined boundary conditions where the top surface had a salt concentration of 1 (representing fully saturated conditions) and the bottom surface had a concentration of 0 (indicating a completely dry state). Normalized concentration was employed to represent moisture (salt) content, with values ranging from 0 and 1. The results are shown in Figure 13.

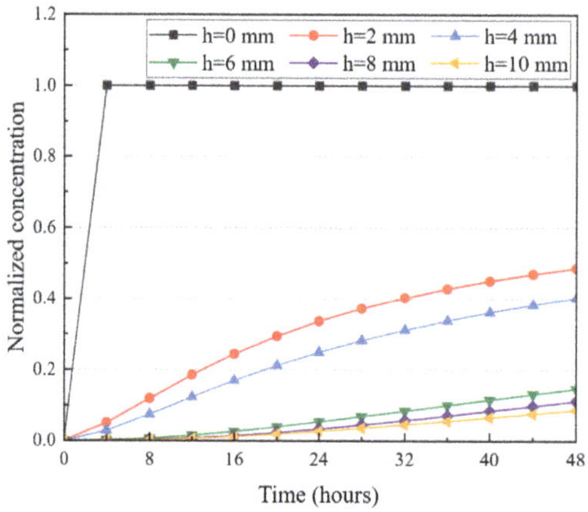

Figure 13. Changes in normalized concentration at different depths of pavement.

Figure 13 illustrates a pronounced disparity in normalized concentration changes across various depths of the asphalt pavement in 48 h. Initially, the road surface (h = 0 mm) exhibits a rapid increase in normalized concentration, indicating a high rate of salt ion penetration due to direct exposure to the saturated surface conditions. In contrast, deeper layers (h = 8 mm and h = 10 mm) demonstrate a markedly diminished rate of concentration change, suggesting a substantial reduction in the efficacy of water-mediated transport as depth increases. These findings illustrate the heterogeneity of the pavement structure, with upper layers nearing saturation while deeper layers remain relatively unsaturated.

Moreover, Figure 14 illustrates the distribution of salt within the pavement structure. The results reveal that after 48 h, salt ions approximately diffuse to a depth of 10 mm from the pavement surface. This observation aligns with findings from previous studies, where similar diffusion depths were reported, thereby supporting the reliability of the results presented in this study [42]. Moreover, due to differential diffusion coefficients between aggregate and FAM, the internal salt distribution exhibits notably non-uniform characteristics. This is because the coarse aggregates and FAM create a complex diffusion pathway due to their different diffusion coefficients. These distinct diffusion coefficients cause salt ions to interact and migrate differently within each material component, leading to localized concentration variations across the pavement structure.

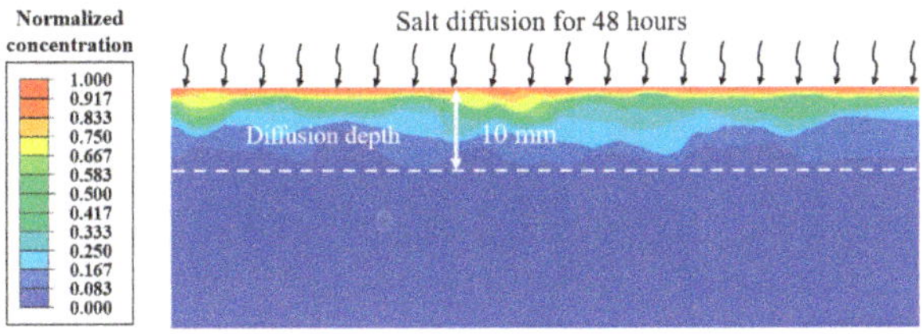

Figure 14. Salt concentration distribution of asphalt pavement at 48 h.

In addition to salt diffusion, the pavement temperature experiences periodic fluctuations due to solar radiation and diurnal cycles. Variations in solar radiation cause the atmospheric temperature to fluctuate periodically. Typically, the daily minimum temperature occurs before dawn, while the maximum temperature is reached approximately two hours after peak solar radiation. Consequently, it takes less than 10 h for the temperature to rise from the minimum to the maximum, while it takes over 14 h to drop back to the minimum. Since a single sine function cannot adequately capture this complex temperature change, a linear combination of two sine functions is employed for modeling [41].

$$T = T_a + T_m[0.96 \sin \omega(t - t_0) + 0.14 \sin 2\omega(t - t_0)] \tag{8}$$

where T_a is the daily average temperature, $T_a = 1/2(T_{max} + T_{min})$, °C; T_{max} and T_{min} are the daily maximum and minimum temperatures, respectively; T_m is the daily temperature amplitude, $T_m = 1/2(T_{max}\text{-}T_{min})$, °C; ω is the angular frequency, $\omega = 2\pi/24$, rad; the phase shift t_0 is 9, accounting for the time difference between peak solar radiation and maximum temperature. In calculations, t is expressed in hours.

Solar radiation is represented using a fitting method from the existing study [41]:

$$q(t) = q_0 \cos m\omega(t - 12), (12 - \frac{c}{2} \leq t \leq 12 + \frac{c}{2}) \tag{9}$$

where q_0 is the maximum radiation at noon, $q_0 = 0.131\ Mq$, $m = 12/c$; c is the actual effective sunshine duration, hour. Q is the total daily solar radiation, J/m^2.

Under winter de-icing conditions, pronounced temperature variations can induce significant thermal stress and potential pavement damage. Specifically, this study considers a scenario with T_{max} = 15 °C and T_{min} = −10 °C, analyzing the daily temperature fluctuations and solar radiation characteristics using the proposed fitting method. The results are illustrated in Figure 15.

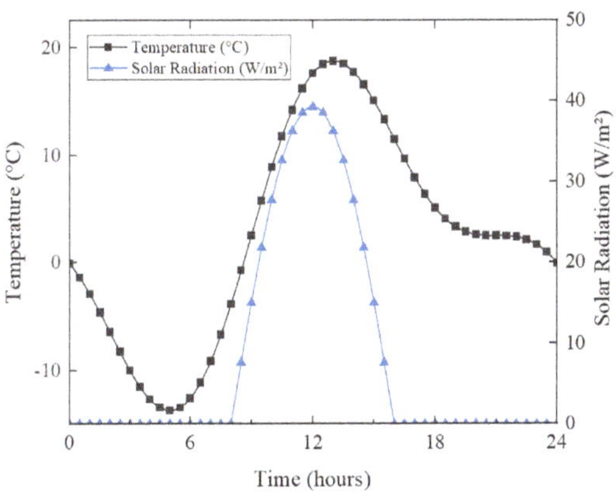

Figure 15. Temperature and solar radiation change over 24 h.

To ensure stable cyclic variations in the pavement temperature field, this study utilized a finite element model to simulate five continuous daily temperature cycles, as illustrated in Figure 15. The temperature distribution at various pavement depths during the last 48 h is shown in Figure 16.

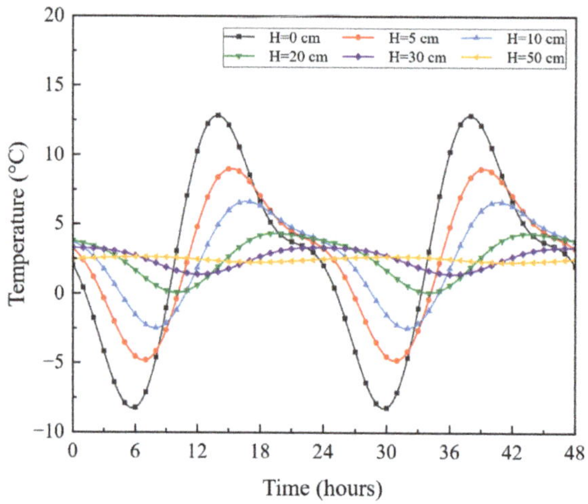

Figure 16. Pavement temperature values in 48 h.

As shown in Figure 16, the temperature fluctuation range of the pavement surface (H = 0 cm) was the largest, which means the pavement surface was affected most by the weather conditions. At the same time, with the depth increasing, the temperature was gradually stable. At a distance of 50 cm from the pavement surface, the temperature values changed slightly. Given these observations, it is recommended to consider the implementation of thermal insulation layers within the pavement structure to help mitigate temperature fluctuations at the surface. Enhancing the temperature stability of the upper layers could reduce thermal stress, thereby extending the pavement's lifespan and improving overall performance. Additionally, using materials with better thermal properties may further enhance the pavement's resilience to temperature variations, promoting a more durable infrastructure in areas with significant temperature changes.

5.3. *Responses to Coupled Salt–Thermal–Mechanical Effects*

Based on the results from Section 5.2, de-icing salt primarily diffused within the shallow surface layer in a short time, with limited influence on deeper pavement regions. Consequently, the multi-scale pavement model focuses on the structural response from the surface to a depth of 10 mm. To improve computational efficiency, the cohesive element damage parameters for this layer were set to values obtained from experimental measurements conducted at 48 h.

As shown in Figure 17, within the entire AC-13 layer ($H \leq 10$ cm), the temperature variation within one day was 20 °C. The inherent differences in thermal expansion coefficients between coarse aggregate and FAM inevitably generate internal thermal stresses. Therefore, the temperature at the 14th hour was used as the initial temperature, and the temperature field at the 7th hour of the following day was applied as the boundary condition for computational analysis. Figure 17 shows the stress distribution inside the AC-13 layer when a moving load travels to a certain position.

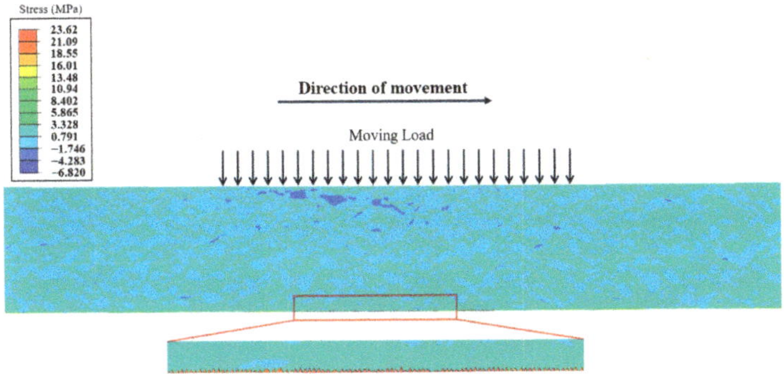

Figure 17. Distribution of stress under a moving load.

The figure illustrates that within the entire asphalt layer, the internal structure near the position of the moving load primarily exhibits a compressive and tensile characteristic, with the maximum tensile stress occurring at the bottom of the layer. Additionally, in the area near the surface, temperature stress can be observed when the moving load is not present. This is primarily due to thermal expansion and contraction resulting from environmental temperature changes affecting the surface. Since the thermal expansion coefficient of the FAM is higher than that of the coarse aggregate, tensile stresses develop within the FAM as the temperature decreases.

5.4. Effects of Salt

Based on the analysis in Section 5.2, it is evident that salt ions have difficulty diffusing deeply into the pavement in a short period (Figure 18b). To further investigate the influence of salt on the dynamic response of asphalt pavements, analyses were conducted under two conditions: one without salt (Figure 18a) and the other with salt diffusion reaching the bottom of the AC-13 layer (Figure 18c). Additionally, previous studies have shown that moisture damage affects the elastic modules of asphalt concrete; therefore, the elastic modulus of FAM was assumed to be 90% of the original elastic modulus.

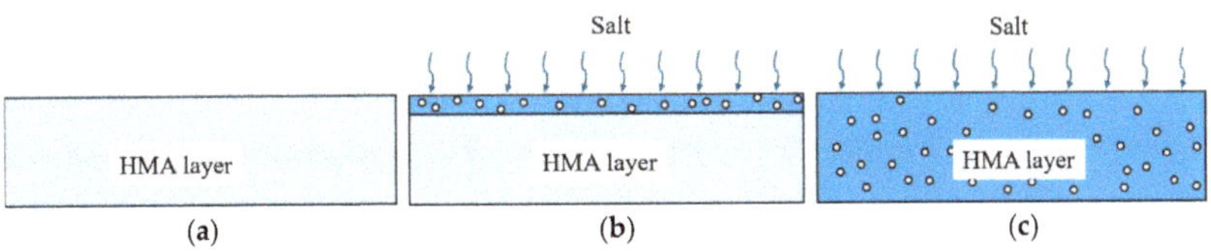

Figure 18. Stages of salt diffusion in asphalt pavement: (**a**) none; (**b**) shallow penetration; (**c**) full-depth penetration to the AC-13 base.

The pavement temperature field and moving load (40 km/h, standard load) were considered in the model, and the SDEG values for three different conditions were extracted and are presented in Figure 19. To enhance the visualization and interpretability of the model calculation results, box plots were employed, which offer a more comprehensive representation of data distribution and statistical characteristics.

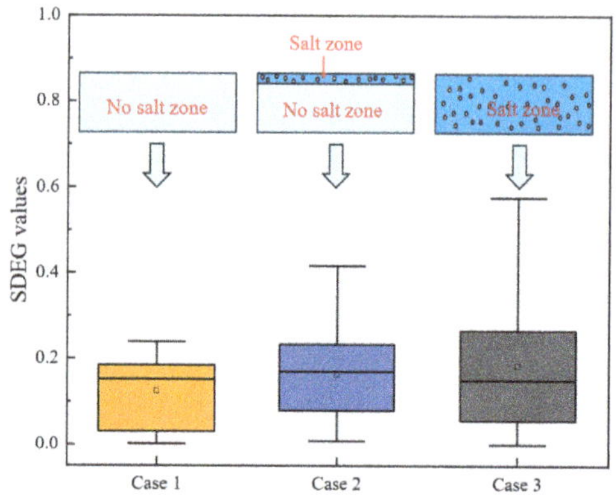

Figure 19. SDEG values under different salt diffusion conditions.

As shown in Figure 19, the SDEG values increase with greater salt diffusion depth. In the case without salt influence (Case 1), the average SDEG is 0.1250, while salt diffuses to a depth of 10 mm in the pavement (Case 2), it rises to 0.1632. When salt completely infiltrates the AC-13 layer (Case 3), the average further increases to 0.1843, although the median slightly decreases to 0.1495. This decrease in median indicates that, although more cohesive elements have sustained damage, the severity of these damages is relatively minor, resulting in a lower median value despite the increasing average SDEG.

In addition, to compare the effect of salt on the damage of elements in different parts of the AC-13 layer, the structural layers under three conditions were divided into two regions, A and B. The results are shown in Figure 20. Figure 20 illustrates the SDEG values under different salt diffusion conditions. As shown in the figure, in Case 1, the cohesive elements in Zone A exhibit almost no damage, with damage primarily concentrated in Zone B. To be more specific, the cohesive elements in Zone A exhibit minimal damage, with median SDEG values close to 0.05, while Zone B shows significantly higher damage, with median values around 0.20. This substantial difference highlights the non-uniform stress distribution within the AC-13 layer. This is because, during significant temperature drops, the coefficient of thermal contraction of the FAM is greater than that of the coarse aggregate, causing the FAM to predominantly bear tensile stress. Additionally, the AC-13 layer experiences a state of compression in the upper layer and tensile stress in the lower layer under a moving load. Under the influence of thermal stress, the tensile stress in the lower layer increases, further causing damage to the structure.

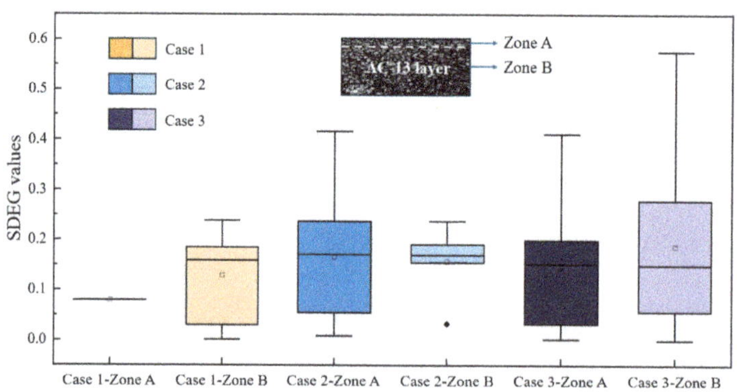

Figure 20. SDEG values under different salt diffusion conditions and different locations.

Conversely, in Case 3, the cohesive elements in Zone A also experience damage due to temperature fluctuations and moving loads, with the damage degree of the cohesive elements at the bottom being significantly higher than that of Zone A. This indicates that the presence of salt reduces the structural load-bearing capacity. In Case 2, the damage level of the cohesive elements in Zone A exceeds that of Zone B. This result is attributable not only to stress factors but also to the reduction in cohesive strength caused by salt. The results in Figure 17 indicate that the deterioration of materials due to salt action is more pronounced than the effects caused solely by moving loads.

5.5. *Effects of Moving Load*

To investigate the effects of moving loads on the damage of asphalt pavements in salt environments, two distinct vertical loads were applied to the multiscale finite element model shown in Figure 10. The first load was a standard load of 50 kN, while the second was a heavy load of 100 kN. Assuming a contact pressure of 0.9 MPa, the moving load was simplified to a line load with an equivalent circular diameter in the 2D model [27,29]. The application of the moving load was implemented using a user-defined DLOAD subroutine in ABAQUS. Initially, the starting position, length, and speed of the wheel load were defined. Subsequently, the current coordinates of the load's sides were calculated based on the given time. Finally, the subroutine determined the load value by assessing whether the current spatial coordinates of the analysis points fell within the dynamically updated load boundaries.

Figure 21 demonstrates that the SDEG values for damaged cohesive elements under both standard and heavy loads exhibit similar effects on the upper section of the AC-13 layer. However, in the lower section of the layer, where there is no influence of salt, the maximum SDEG values of cohesive elements under heavy load increase by 16% compared to those under standard load. This increase suggests that heavy loads can lead to more significant localized damage in the absence of salt, which may weaken the cohesive elements further. Interestingly, the average SDEG value under heavy load is lower than that under standard load. This anomaly occurs because more cohesive elements sustain damage under heavy loads that do not experience damage under standard loads, leading to a situation where these zero values were not considered in the average calculation.

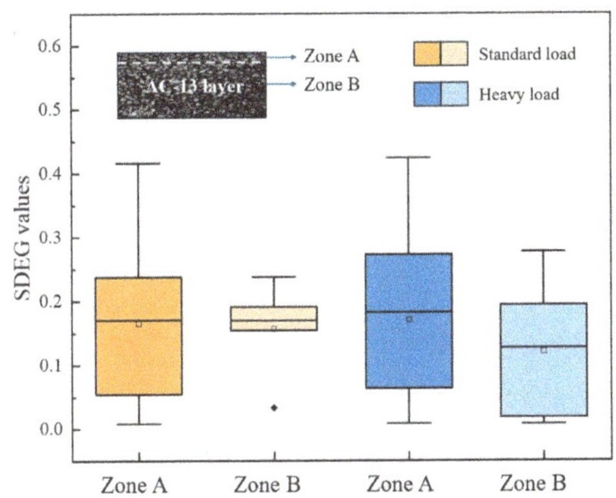

Figure 21. SDEG values under different moving loads.

Asphalt concrete is recognized as a viscoelastic material, with its dynamic response significantly influenced by the speed of the applied moving load. Therefore, two different speeds (40 km/h and 100 km/h) were considered in this study to study the effect of high and low speeds. When moving loads with different speeds acted at the same location as the model, the SDEG values of cohesive elements were obtained and are shown in Figure 22.

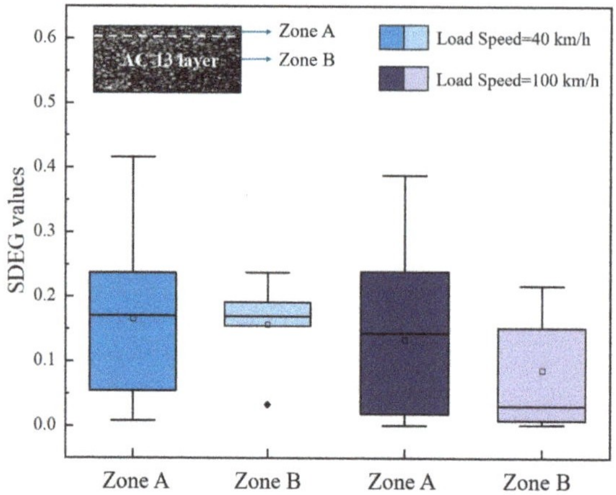

Figure 22. SDEG values under different moving load speeds.

Figure 22 illustrates that the SDEG values under a moving load of 40 km/h are significantly higher than those observed under a 100 km/h load. This discrepancy can be attributed to the longer duration of stress application that occurs at lower speeds. When vehicles move at 40 km/h, they exert stress on the pavement for an extended period, resulting in greater stress accumulation within the material. This prolonged exposure allows more time for deformation and increases the likelihood of failure, ultimately leading to more extensive damage to the cohesive elements of the pavement structure.

Moreover, the lower speed of 40 km/h correlates with a reduced frequency of load application, enabling more substantial damage to accumulate between successive load impacts. In contrast, at 100 km/h, the increased frequency of load applications reduces the time available for stress relaxation within the material. Consequently, this finding suggests that maintenance strategies should not solely concentrate on high-speed traffic conditions but also consider the cumulative damage potential associated with lower speeds.

6. Conclusions

In this study, a multi-scale pavement FE model was developed to investigate the fracture mechanism of asphalt concrete and evaluate the damage of asphalt pavement under wheel load in a salt environment. CZM was used to simulate the fracture within the asphalt concrete model, and the SDEG value was employed to evaluate raveling susceptibility. The accuracy of the model was validated by laboratory experiments. Additionally, the damage mechanisms of pavement under coupled salt–thermal–mechanical effects were investigated in this study. The main findings of this study can be summarized as follows:

(1) According to the laboratory tests, at −10 °C, cohesive strength dropped by 26% to 32.5% and adhesive strength by 33.5% to 63.8% after 48 h. At 0 °C, cohesive strength decreased by 24.9% to 44.0% and adhesive strength by 37.9% to 71.6%. Additionally, fracture energy at 0 °C decreased by 4.2% to 13% compared to −10 °C.

(2) Salt ions diffused approximately 10 mm into the pavement after 48 h. The non-uniform distribution of salt within the pavement structure was attributed to the different diffusion coefficients of the aggregate and the FAM.

(3) Maximum tensile stress occurred at the bottom of the asphalt layer, with compressive and tensile characteristics exhibited near the moving load's path. Temperature-related

stresses manifested in the surface area even without the moving load, primarily driven by thermal expansion and contraction due to environmental fluctuations.

(4) The presence of salt significantly reduces the load-bearing capacity of the upper layer, resulting in greater damage to cohesive elements compared to the lower layer. Additionally, the detrimental effects of salt on material deterioration in the upper layer are more pronounced than those caused solely by the stresses from moving loads. It is recommended to implement measures to mitigate salt exposure on the upper layer of pavements.

(5) Lower moving speeds lead to increased damage in cohesive elements due to longer stress application times and lower load frequencies, resulting in greater stress accumulation compared to higher speeds.

To further confirm the reliability and applicability of the multi-scale asphalt concrete model developed in this research, it is recommended to incorporate additional validation methods, including field tests and evaluations of other mechanical properties.

Author Contributions: Conceptualization, J.C.; Methodology, J.M. and Y.L.; Validation, M.T. and Y.L.; Formal analysis, M.T.; Investigation, M.T.; Data curation, J.M.; Writing—original draft, J.M.; Writing—review & editing, J.C. and Y.L.; Funding acquisition, J.M. and J.C. All authors have read and agreed to the published version of the manuscript.

Funding: This research was sponsored by the Guangxi Transportation Investment Group independently initiated scientific research projects (Grant No. 2022834), and the Natural Science Foundation of Hunan Province (Grant No. 2024JJ5438).

Institutional Review Board Statement: Not applicable.

Informed Consent Statement: Not applicable.

Data Availability Statement: The original contributions presented in this study are included in the article. Further inquiries can be directed to the corresponding author.

Conflicts of Interest: Author Jin Ma, Mingfeng Tang, and Yu Liu were employed by the company Guangxi Communication Investment Technology Co. Ltd. The remaining authors declare that the research was conducted in the absence of any commercial or financial relationships that could be construed as a potential conflict of interest.

References

1. Leiva-Padilla, P.; Moreno-Navarro, F.; Iglesias, G.R.; Rubio-Gamez, M.C. Thermal characterisation of electroconductive layers for anti-icing and de-snowing applications on roads. *Road Mater. Pavement Des.* **2022**, *23*, 168–183. [CrossRef]
2. Cui, Y.C.; Zhang, L.; Xing, C.; Tan, Y.Q. Anti-icing properties and application of superhydrophobic coatings on asphalt pavement. *Constr. Build. Mater.* **2024**, *419*, 135452. [CrossRef]
3. Zhang, X.M.; Chen, H.; Hoff, I. The mutual effect and reaction mechanism of bitumen and de-icing salt solution. *Constr. Build. Mater.* **2021**, *302*, 124213. [CrossRef]
4. Vega-Zamanillo, A.; Juli-Gándara, L.; Calzada-Pérez, M.A.; Teijón-López-Zuazo, E. Impact of temperature changes and freeze-thaw cycles on the behaviour of asphalt concrete submerged in water with sodium chloride. *Appl. Sci.* **2020**, *10*, 1241. [CrossRef]
5. Wu, Z.M.; Shi, C.J.; Gao, P.W.; Zhang, H.L.; Hu, X. Moisture susceptibility of asphalt mixture subjected to chloride-based deicing salt solutions under simulated environmental conditions. *J. Mater. Civ. Eng.* **2023**, *35*, 158718. [CrossRef]
6. Zhang, K.; Luo, Y.F.; Xie, W.; Wu, J.J. Evaluation of road performance and adhesive characteristic of asphalt binder in salt erosion environment. *Mater. Today Commun.* **2020**, *25*, 101593. [CrossRef]
7. Long, Z.W.; You, L.Y.; Xu, F.; Tang, X.Q.; Ding, Y.H.; Khanal, A.; Miao, Y. Nanomechanical-atomistic insights on interface interactions in asphalt mixtures with various chloride ion erosion statuses. *J. Colloid Interface Sci.* **2022**, *628*, 891–909. [CrossRef]
8. Long, Z.; Guo, N.; Xiao, Y.; You, L.; Xu, F.; Liu, Y.; Tang, X.; Ding, Y. Nanomechanical experiments and molecular dynamics explorations for the deterioration evolution of asphalt driven by chloride salt erosion. *J. Mater. Civ. Eng.* **2023**, *35*, 04023382. [CrossRef]
9. Tang, Y.; Fu, Z.; Hou, Y.; Ma, F.; Wang, W.; Zhang, T.; Yang, P.; Zhao, P. Salt damage on asphalt surfaces under repeated dissolution at molecular scale. *Colloids Surf. A-Physicochem. Eng. Asp.* **2024**, *690*, 133829. [CrossRef]

10. Hamedi, G.H.; Harehdasht, S.L.; Talebi, H. Assessing low-temperature cracking in asphalt mixtures through mix design and thermodynamic parameters. *J. Mater. Civ. Eng.* **2024**, *36*, 04024169. [CrossRef]
11. Krami, K.; Benamara, A.; Radouani, M.; Ettayeb, M. Evaluation of water sensitivity and the impact of thermal cycles on asphalt mixtures with different aggregates and hydrated lime as additive: Experimental study. *Case Stud. Constr. Mater.* **2024**, *20*, 145–152. [CrossRef]
12. Chen, J.Q.; Wang, J.; Shi, Z.; Zhang, Z.; Dan, H.C. Computational fracture analysis of steel slag asphalt mixture subjected to moisture damage. *Constr. Build. Mater.* **2024**, *433*, 136697. [CrossRef]
13. Sobieski, T.; Arámbula-Mercado, E.; Martin, A.E. A method for characterising RAP degree of binder activity. *Road Mater. Pavement Des.* **2022**, *23*, 2250–2270. [CrossRef]
14. Liang, Y.H.; Chen, J.Q.; Li, L. Computational evaluation of the fracture behavior of porous asphalt concrete exposed to moisture and salt erosion. *Materials* **2024**, *17*, 1505. [CrossRef]
15. Wang, L.; Wang, W.; Guan, H.F.; Li, Y.X.; Pei, K.; Chen, B.Y.S. Effect of salt freeze-thaw cycle on crack resistance of steel slag rubber powder modified asphalt mixture. *J. Mater. Civ. Eng.* **2024**, *36*, 04023612. [CrossRef]
16. Xiong, R.; Chu, C.; Qiao, N.; Wang, L.; Yang, F.; Sheng, Y.; Guan, B.; Niu, D.; Geng, J.; Chen, H. Performance evaluation of asphalt mixture exposed to dynamic water and chlorine salt erosion. *Constr. Build. Mater.* **2019**, *201*, 121–126. [CrossRef]
17. Chen, J.Q.; Wu, Q.L.; Wang, H.; Quan, Z.Q.; Dan, H.C. Modeling and analysis of ice condensation on bridge deck pavement surface based on heat transfer theory and finite element method. *Appl. Therm. Eng.* **2024**, *241*, 122344. [CrossRef]
18. Li, Z.; Han, D.; Chen, H.; Zhao, Y. Simulation analysis and optimization of temperature field for hot in-place asphalt pavement recycling. *J. Transp. Eng. Part B-Pavements* **2023**, *149*, 04023009. [CrossRef]
19. Liu, G.; Qian, Z.D.; Wu, X.Y.; Chen, L.L.; Liu, Y. Investigation on the compaction process of steel bridge deck pavement based on DEM-FEM coupling model. *Int. J. Pavement Eng.* **2023**, *24*, 2169443. [CrossRef]
20. Chen, J.Q.; Sun, C.Q.; Sun, X.; Dan, H.C.; Huang, X.Y. Finite difference model for predicting road surface ice formation based on heat transfer and phase transition theory. *Cold Reg. Sci. Technol.* **2023**, *207*, 103772. [CrossRef]
21. Chen, J.Q.; Wang, H.; Xie, P.Y. Pavement temperature prediction: Theoretical models and critical affecting factors. *Appl. Therm. Eng.* **2019**, *158*, 113755. [CrossRef]
22. Dan, H.C.; Zhang, Z.; Chen, J.Q.; Wang, H. Numerical simulation of an indirect tensile test for asphalt mixtures using discrete element method software. *J. Mater. Civ. Eng.* **2018**, *30*, 6–10. [CrossRef]
23. Dan, H.C.; Gao, L.S.; Wang, H.; Tang, J. Discrete-element modeling of mean texture depth and wearing behavior of asphalt mixture. *J. Mater. Civ. Eng.* **2022**, *34*, 04022027. [CrossRef]
24. Niazi, S.; Najmeddine, A.; Shakiba, M. A coupled thermo-hydro-mechanical framework for simulating the failure response of asphalt concrete under freezing conditions. *Cold Reg. Sci. Technol.* **2024**, *218*, 104073. [CrossRef]
25. Xu, J.P.; Zheng, C.F.; Song, Z.F.; Zhou, S.; Chen, W.X. Mesodamage mechanism of asphalt mixtures with different structural types under frost heaving of ice crystals. *J. Mater. Civ. Eng.* **2023**, *35*, 04022377. [CrossRef]
26. Sun, Y.R.; Zhang, Z.; Gong, H.R.; Zhou, C.J.; Chen, J.Y.; Huang, B.S. 3D Multiscale modeling of asphalt pavement responses under coupled temperature-stress fields. *J. Eng. Mech.* **2022**, *148*, 04022010. [CrossRef]
27. Yu, S.J.; Wang, H.; Chen, J.Q. Ravelling susceptibility of porous asphalt pavement in salt environments using multi-scale modelling analysis. *Int. J. Pavement Eng.* **2025**, *26*, 2467222. [CrossRef]
28. Qiu, X.; Deng, J.; Fu, Q.; Lou, Y.; Ye, Y.; Zhang, D. Investigation of chloride salt erosion on asphalt binders and mixtures: Performance evaluation and correlation analysis. *Materials* **2025**, *18*, 156. [CrossRef]
29. Manrique-Sanchez, L.; Caro, S.; Arámbula-Mercado, E. Numerical modelling of ravelling in porous friction courses (PFC). *Road Mater. Pavement Des.* **2018**, *19*, 668–689. [CrossRef]
30. Chen, S.; Yin, D.; Gao, Y.; Zhang, Y.; Ma, T.; Chen, C. Damage behavior of porous asphalt concrete under load–pore pressure effects. *J. Mater. Civ. Eng.* **2024**, *36*, 04024041. [CrossRef]
31. Wang, H.; Wang, J.; Chen, J.Q. Fracture simulation of asphalt concrete with randomly generated aggregate microstructure. *Road Mater. Pavement Des.* **2018**, *19*, 1674–1691. [CrossRef]
32. Xu, Y.Z.; Zheng, C.F.; Feng, Y.P.; Guo, X.D. Low-temperature cohesive and adhesive strength testing of contact surface between bitumen and mineral aggregates by image analysis. *Constr. Build. Mater.* **2018**, *183*, 95–101. [CrossRef]
33. Yee, T.S.; Hamzah, M.O.; Van den Bergh, W. Evaluation of moisture susceptibility of asphalt-aggregate constituents subjected to direct tensile test using imaging technique. *Constr. Build. Mater.* **2019**, *227*, 116642. [CrossRef]
34. Zhang, K.; Luo, Y.; Li, Z.; Zhao, Y.; Zhao, Y. Evaluation of performance deterioration characteristics of asphalt mixture in corrosion environment formed by snow-melting agents. *J. Mater. Civ. Eng.* **2022**, *34*, 101593. [CrossRef]
35. *TP105 A*; Standard Method of Test for Determining the Fracture Energy of Asphalt Mixtures Using the Semicircular Bend Geometry (SCB). AASHTO: Washington, DC, USA, 2013.
36. Pirmohammad, S.; Abdi, M.; Ayatollahi, M.R. Effect of support type on the fracture toughness and energy of asphalt concrete at different temperature conditions. *Eng. Fract. Mech.* **2021**, *254*, 107921. [CrossRef]

37. Kim, K.W.; El Hussein, M. Variation of fracture toughness of asphalt concrete under low temperatures. *Constr. Build. Mater.* **1997**, *11*, 403–411. [CrossRef]
38. Chen, J.; Yu, S.; Huang, W. Two-dimensional microstructure-based model for evaluating the permeability coefficient of heterogeneous construction materials. *Materials* **2023**, *16*, 5892. [CrossRef]
39. Chen, J.Q.; Ouyang, X.; Sun, X. Numerical investigation of asphalt concrete fracture based on heterogeneous structure and cohesive zone model. *Appl. Sci.* **2022**, *12*, 11150. [CrossRef]
40. Teng, G.C.; Zheng, C.F.; Chen, X.Y.; Lan, X.L.; Zhu, Y.M.; Shan, C. Numerical fracture investigation of single-edge notched asphalt concrete beam based on random heterogeneous FEM model. *Constr. Build. Mater.* **2021**, *304*, 124581. [CrossRef]
41. Du, C.; Sun, Y.; Chen, J.; Zhou, C.; Liu, P.; Wang, D.; Oeser, M. Coupled thermomechanical damage behavior analysis of asphalt pavements using a 2d mesostructure-based finite-element method. *J. Transp. Eng. Part B-Pavements* **2021**, *147*, 04021012. [CrossRef]
42. Guo, Q.; Liu, Q.; Zhang, P.; Gao, Y.; Jiao, Y.; Yang, H.; Xu, A. Temperature and pressure dependent behaviors of moisture diffusion in dense asphalt mixture. *Constr. Build. Mater.* **2020**, *246*, 118500. [CrossRef]

materials

MDPI

Article

Influence of Aging and Salt–Alkali Coupling on the Fatigue and Self-Healing Behavior of Graphene Oxide-Modified Asphalt

Ruixia Li [1], Wei Zhu [1], Hailong Chen [2], Xiao Li [1], Linhao Sun [3] and Jinchao Yue [1,*]

[1] School of Water Conservancy and Transportation, Zhengzhou University, Zhengzhou 450001, China; ruixiali031@163.com (R.L.); zw19990603@163.com (W.Z.); lixiao106118@163.com (X.L.)
[2] Shanghai Urban Construction Maintenance Management Co., Ltd., Shanghai 200032, China; chen1543372126@163.com
[3] Zhengzhou Public Utility Investment and Development Group Co., Ltd., Zhengzhou 450001, China; sunlinhao2023@163.com
* Correspondence: yuejc@zzu.edu.cn

Abstract: The harsh environments in saline–alkaline areas and high-altitude regions with intense ultraviolet radiation pose great challenges to the durability of asphalt pavements. The fatigue performance of asphalt binder significantly determines the actual service life of asphalt pavements. Existing studies have predominantly focused on the impact of individual environmental factors (e.g., aging and saline–alkaline erosion) on asphalt performance, yet there remains a notable research gap in the systematic analysis of asphalt's fatigue and self-healing behavior under coupled multi-factor interactions, particularly regarding the synergistic effects of UV aging and saline–alkaline conditions. Therefore, it is of great importance to understand the influence rules of the coupling effect of aging and salt–alkaline characteristics on the properties of asphalt materials. In this study, 70# base asphalt and GO-modified asphalt were taken as the research objects. Frequency sweep tests, linear amplitude sweep (LAS) tests, and LAS-based healing tests were conducted using a dynamic shear rheometer. The fatigue and self-healing properties of the two asphalt materials under different aging conditions and aging and salt–alkali coupling effects were analyzed based on the viscoelastic continuum damage theory. The results showed that the degree of aging can increase the stress peak of asphalt materials under small strains and also increase their stress attenuation rate. Except for short-term aging and salt–alkali effects, the aging and salt–alkali coupling effects generally further reduce the stress peaks of asphalt materials. Aging can increase the fatigue life of asphalt and increase the fatigue life attenuation rate of asphalt. The aging and salt–alkali coupling effects will reduce the fatigue life of asphalt and increase the decline rate of the asphalt fatigue life. The self-healing efficiency of asphalt is affected by the degree of aging, and the aging and salt–alkali coupling effects further reduce the self-healing efficiency of asphalt materials. This paper elucidates the influence mechanisms of intense UV irradiation and saline–alkaline environments on GO-modified asphalt, providing theoretical and practical references for its future engineering applications in harsh environmental conditions.

Keywords: GO-modified asphalt; aging; aging and salt–alkali coupling effects; viscoelastic continuum damage theory; fatigue performance; self-healing performance

1. Introduction

Asphalt materials are widely used in high-grade pavements and urban roads due to their excellent road performance. However, prolonged exposure to natural environments subjects asphalt pavements to the combined effects of water, oxygen, ultraviolet (UV) radiation, and saline–alkaline conditions. These factors induce distresses such as cracks and potholes during service, significantly reducing pavement lifespan [1]. In saline–alkaline regions, the unique climatic and geological conditions exacerbate these challenges. Rainwater dissolves salts from the soil, forming saline–alkaline solutions that erode asphalt pavements [2]. Furthermore, high temperatures and intense radiation in these areas induce photo-oxidative aging and the salt-induced degradation of asphalt materials [3].

Extensive studies have explored the impact of saline–alkaline environments on asphalt performance. Li et al. [4] investigated the effects of water, acid, alkaline, and salt solutions on asphalt through immersion tests, revealing changes in the complex modulus and phase angle, which provided foundational insights into asphalt–saline interactions. Xu et al. [5] demonstrated that salt solutions could enhance the rheological and mechanical properties of asphalt under specific conditions, complementing Zhang's findings and highlighting the complexity of saline–alkaline effects. Xiong et al. [6] further reported that an increased salt concentration reduces low-temperature rheological properties and weakens adhesion, though it may induce hardening under certain conditions. However, existing studies primarily focused on single-factor impacts and failed to simulate real-world multi-factorial coupling effects such as simultaneous salt erosion, UV exposure, and thermal aging.

Fatigue damage, caused by vehicular loads and environmental stressors, critically governs asphalt pavement durability. In harsh environments such as in saline–alkaline and high-altitude regions, fatigue resistance becomes even more critical. As a viscoelastic material, asphalt exhibits self-healing behavior during load-free intervals, which significantly mitigates fatigue damage. However, saline–alkaline environments may impede molecular diffusion, thereby suppressing self-healing efficiency [7]. Additionally, thixotropy, steric hardening, and nonlinear viscoelasticity further complicate fatigue performance [8]. Thus, integrating self-healing capability into fatigue life prediction models is essential for accurate pavement design.

To enhance asphalt performance, modifiers such as styrene–butadiene–styrene (SBS), styrene–butadiene rubber (SBR), and polyethylene (PE) are commonly employed. Recently, nanomaterials like graphene oxide (GO) have gained attention due to their high compatibility with asphalt, attributed to abundant carboxyl, hydroxyl, and epoxy groups [9]. Studies by Habib et al. [10] demonstrated that GO improves both high- and low-temperature elasticity. Huang et al. [11] found that GO enhances the anti-deformation and oxidation resistance of asphalt mixtures after aging. Wu et al. [12] further revealed that GO inhibits the volatilization of saturated components and forms aromatic stacking with resins, improving thermal stability. However, existing studies on GO-modified asphalt have primarily focused on investigating its rheological properties under high/low-temperature conditions while research on its fatigue and self-healing performance in complex natural environments—particularly under saline–alkaline exposure and combined high-temperature, high-humidity, and high-radiation conditions—remains insufficiently explored, lacking systematic and in-depth investigations.

In summary, although some scholars have studied the influence of single environmental factors (such as saline–alkaline erosion or ultraviolet aging) on asphalt performance, the action mechanism of multi-factor coupling effects (e.g., the synergistic effects of saline–alkaline, aging, and ultraviolet effects) on the fatigue and self-healing behavior of GO-modified asphalt in the actual service environment has not been systematically revealed. The existing literature has mostly focused on optimizing the high- and low-temperature performance of GO-modified asphalt, yet it has lacked an in-depth analysis of its damage evolution and self-healing characteristics under extreme multi-factor coupling conditions. This paper takes GO-modified asphalt and 70# base asphalt as the research objects. By simulating the complex environment of saline–alkaline solution erosion (DW, NaCl solution, Na_2SO_4 solution, and Na_2CO_3 solution), UV aging, and thermal-oxygen aging, combined with the viscoelastic continuum damage theory (VECD theory), the linear amplitude sweep (LAS) test is adopted to study the variation law of fatigue and self-healing performance of GO-modified asphalt in complex environments. This provides a key theoretical basis for the application of GO-modified asphalt in road engineering in regions with complex environments.

2. Materials and Methods

2.1. Materials

The experimental program utilized 70# base asphalt, a widely adopted pavement binder in Chinese highway engineering. Following the Test Specifications for Highway Engineering Asphalt and Asphalt Mixtures (JTG E20-2011) [13], comprehensive characterization of fundamental asphalt properties was conducted, with the obtained data systematically compiled in Table 1. Graphene oxide (GO) materials were supplied by Suzhou Danfeng Graphene Technology Co., Ltd. (Suzhou, China), and their essential physical parameters were rigorously evaluated in compliance with the international Standard ISO TS 21356:2021 "Nanotechnology-Graphene Oxide" [14]. The quantified material characteristics are presented in Table 2.

Table 1. Fundamental properties of 70# base asphalt.

Property	Unit	Test Result	Specification Requirement	Test Method (JTG E20-2011)
Penetration (25 °C)	0.1 mm	64.3	60–80	T0604-2011
Softening point	°C	47.5	≥45	T0606-2011
Ductility	cm	34	≥25	T0605-2011

Table 2. Fundamental properties of graphene oxide (GO).

Material	Technical Parameter	Measured Value
Graphene Oxide	Purity (%)	>96
	Sheet Diameter	10–50
	Thickness (μm)	3.5–7.0
	Number of Layers	6–10
	Specific Surface Area (m^2/g)	100–300
	Appearance	Black Powder Particles

2.2. Sample Preparation

In this paper, GO-modified asphalt was prepared using a high-speed shearer and 70# base asphalt. GO-modified asphalts with doping amounts of 0.4%, 0.6%, 0.8%, 1.0%, and 1.2% were prepared. The 70# base asphalt was preheated in an oven at 150 °C, and then, while maintaining this temperature, a certain mass fraction of GO was added. It was sheared by a high-speed shearer at a rotation speed of 5000 rad/min for 60 min. The sheared GO-modified asphalt was placed in an oven at 120 °C for 1 h of curing and development. The basic technical indicators of the GO-modified asphalt and 70# base asphalt are shown in Table 3. Considering the penetration, softening point, and ductility of GO-modified asphalts with different doping amounts, the GO-modified asphalt with a doping amount of 1% was selected for subsequent experiments.

Table 3. Basic technical indicators of GO-modified asphalt and 70# base asphalt.

Asphalt Type	Penetration (25 °C)/0.1 mm	Softening Point/°C	Ductility/cm (10 °C)
70#	64.3	47.5	34
0.4% GO	61.3	48.1	23.2
0.6% GO	60.9	49.5	21.1
0.8% GO	59.2	50.2	20.3
1.0% GO	58.7	50.8	19.4
1.2% GO	58.4	51.0	18.9

To simulate the complex conditions occurring in actual road surfaces, the GO-modified asphalt and 70# base asphalt were subjected to short-term aging, long-term aging, and ultraviolet aging treatments. For the short-term aging test, a rotary thin-film oven of model SYD-0609 was used. The set temperature was 163 °C, the air flow was 4000 ± 200 mL/min, the rotation speed of the turntable was 15 ± 0.2 r/min, and the aging time was 85 min. For the long-term aging, an asphalt pressure aging test chamber provided by Tianjin Gangyuan Test Instrument Factory (Tianjin, China) was employed. The set temperature was 100 °C, the pressure was 2.1 MPa, and the time was 20 h. For the ultraviolet aging test, an ultraviolet aging irradiation test chamber produced by Shaoxing Shangyu Huanke Testing Equipment Co., Ltd. (Shaoxing, China) was used, with an ultraviolet light intensity of 120 W/m^2 in the sample area.

After the aging treatment, the samples were subjected to coupling test treatments with salt–alkali solutions (DW, NaCl solution, Na_2SO_4 solution, and Na_2CO_3 solution), and GO-modified asphalt and 70# base asphalt treated by schemes such as short-term aging and salt–alkali coupling effects, long-term aging and salt–alkali coupling effects, and ultraviolet aging and salt–alkali coupling effects were obtained.

Dynamic shear rheometer (DSR) tests were conducted using asphalt specimens with a diameter of 8 mm and a thickness of 2 mm. The asphalt, heated to a molten state, was poured into the mold and trimmed to ensure a smooth surface. After cooling, the specimens were stored in a constant-temperature chamber maintained below 5 °C. Three parallel tests were performed for each group to ensure data reliability.

2.3. Experimental Methods

2.3.1. Frequency Sweep Test

A Dynamic Shear Rheometer (DSR) of the model Discover HR-1, which is produced by TA Instruments Inc. in the United States, was used to conduct Frequency sweep (FS) tests in accordance with AASHTO T 315 standard to characterize the viscoelastic properties of asphalt. This non-destructive testing method enables precise measurement of rheological parameters without inducing microstructural damage to the specimens [15]. The tests were conducted at 15 °C, 25 °C, and 35 °C. A carrier with a diameter of 8 mm was selected, the strain amplitude was 0.1%, and the scanning frequency range was 0.1–100 rad/s. The dynamic shear modulus, phase angle, and other parameters measured by this experiment were prepared for the subsequent fatigue tests. The DSR test procedure is shown in Figure 1.

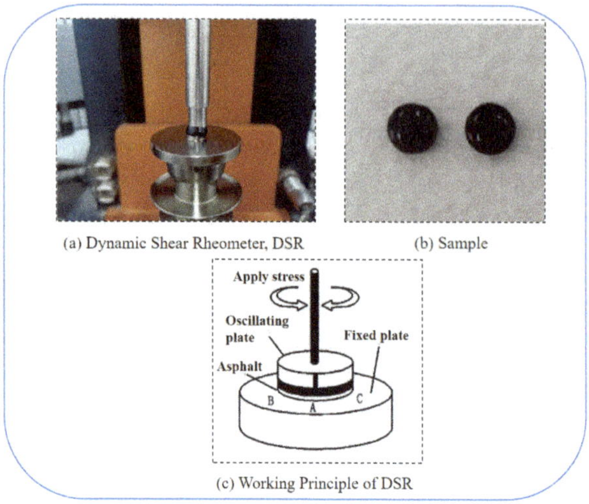

Figure 1. DSR test procedure.

2.3.2. Fatigue Test

The fatigue test (LAS test) for GO-modified asphalt and 70# base asphalt was conducted using a dynamic shear rheometer (DSR). The test was carried out according to the AASHTO T 391 standard, adopting 8 mm diameter parallel plates. The test temperature was set at 25 °C, with the strain range selected as 0.1–30%, the strain increment as 0.1%, and the loading time chosen as 300 s [16].

The VECD theory is used to describe the variation patterns of material behavioral characteristics under dynamic loads. Since it was further developed through studies of numerous scholars, it has now become a popular research method. It integrates the continuum damage theory and viscoelastic theory to characterize the strength, stiffness of materials under high strain, and material damage conditions. Precisely for this reason, the VECD theory is widely applied to evaluate the fatigue performance of asphalt materials. Through micromechanics and continuum mechanics, the internal state changes of asphalt materials under loading are defined as damage variables, and the damage degree of asphalt materials is assessed by quantifying these variables. In this study, the VECD theory used the work potential theory to relate the modulus and damage, thereby quantifying the damage parameter D. Based on the stress–strain response data obtained from the LAS test and combined with the VECD theory for nonlinear data fitting [17], the fatigue damage characteristic curve (DCC) of the asphalt material was obtained. The damage degree D of asphalt was calculated according to Formula (1).

$$D \cong \sum_{i=1}^{N} \left[\pi\gamma_0^2(|G^*|\sin\delta_{i-1} - |G^*|\sin\delta_i)\right]^{\frac{\alpha}{1+\alpha}} (t_i - t_{i-1})^{\frac{1}{1+\alpha}} \tag{1}$$

Here, γ_0 is the strain amplitude, %; $|G^*|$ is the complex dynamic shear modulus, Pa; δ is the phase angle, °; and α is the rheological parameter in the non-damaged state [18].

The DCC curve was fitted according to Formula (2).

$$C = C_0 - C_1 D^{C_2} \tag{2}$$

Here, C_0 is 1, and C_1 and C_2 are model fitting parameters.

The parameters C_1 and C_2 were obtained, and the fatigue life of the asphalt material could be calculated according to Formulas (3)–(5) and combined with the C value corresponding to the peak stress in the stress–strain curve in AASHTO TP 101-14 as the fatigue failure point D_f.

$$D_f = \left(\frac{C_0 - C_{peak}}{C_1}\right)^{\frac{1}{C_2}} \tag{3}$$

$$A = \frac{f(D_f)^k}{k(\pi C_1 C_2)^{\alpha}}, B = 2\alpha \tag{4}$$

$$N_f = A(\gamma_{\max})^{-B} \tag{5}$$

Here, A and B are fatigue life equation parameters, f = 10 Hz, $K = 1 + (1 - C_2)\,\alpha$, C_{peak} is the C value corresponding to the peak stress, and N_f is the fatigue life of the asphalt material. Through the above formulas, the fatigue life of an asphalt material at any strain can be predicted.

2.3.3. Self-Healing Test

Based on the VECD theory model, this study used the self-healing test based on LAS (LASH) first proposed by scholar Xie [18] in 2017. This test can greatly reduce the duration of traditional healing tests and simplify the test process. In this study, a 60 min intermittent period was introduced during the standard LAS test process. Combined with the VECD theory, the self-healing performance of GO-modified asphalt and 70# base asphalt was evaluated through the change in the damage variable parameter D before and after healing. The key to the LASH healing test is to determine the damage degree and select an appropriate healing time. Tests were carried out on GO-modified asphalt and 70# base asphalt under different aging and aging and salt–alkali coupling effects. The test process was as follows. First, a standard LAS test was carried out on the asphalt sample, and the damage parameter D_f defining the fatigue failure of the asphalt based on the peak stress was calculated. In this study, 50% of the damage degree was selected as the damage degree at the start of healing, and the corresponding strain amplitude was calculated accordingly. Then, an asphalt sample treated in the same way was selected for the LAS test before healing. When the strain amplitude reached the strain amplitude corresponding to the damage degree at the start of healing, the test was stopped, and the test sample was left to heal for 1 h. Immediately after the end of the intermittent time, the standard LAS test was continued until the strain reached 30%, and the loading rate and the loaded strain amplitude were kept the same as those before healing. The loading process is shown in Figure 2a. The damage characteristic curve was obtained by using the LASH-based healing test, and its form is shown in Figure 2b.

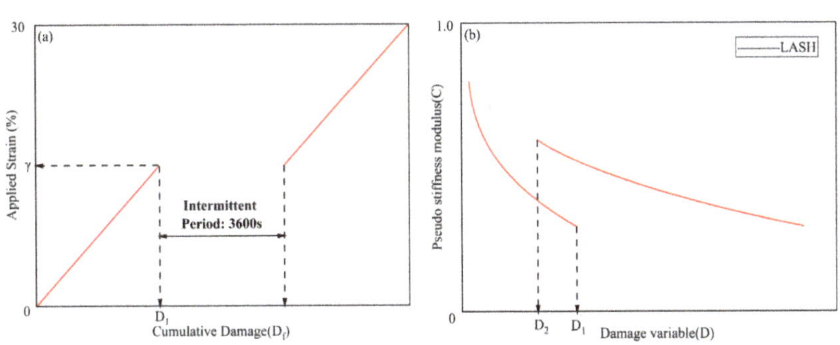

Figure 2. Schematic diagram of the LASH test loading principle. (**a**) Relationship between applied strain (γ) and cumulative damage (D_f), with an intermittent period of 3600s. (**b**) Relationship between pseudo-stiffness modulus (C) and damage variable (D). The curve represents LASH.

The healing index $\% H_s$ was calculated through the change in fatigue damage D before and after the intermittent period, as shown in Formula (6), where the calculation of D_2 needs to subtract the healing time [19].

$$\%H_S = \frac{D_1 - D_2}{D_1} \times 100\% \tag{6}$$

3. Results and Discussion

3.1. Fatigue Performance Analysis

3.1.1. Stress–Strain Response Analysis

According to the AASHTO TP101 specification, the peak of the stress–strain curve was taken as the fatigue failure point in the LAS fatigue test [20]. Based on the VECD theory, the LAS results at 25 °C were analyzed to obtain the stress–strain curves of GO-modified asphalt in complex environments, as shown in Figure 3.

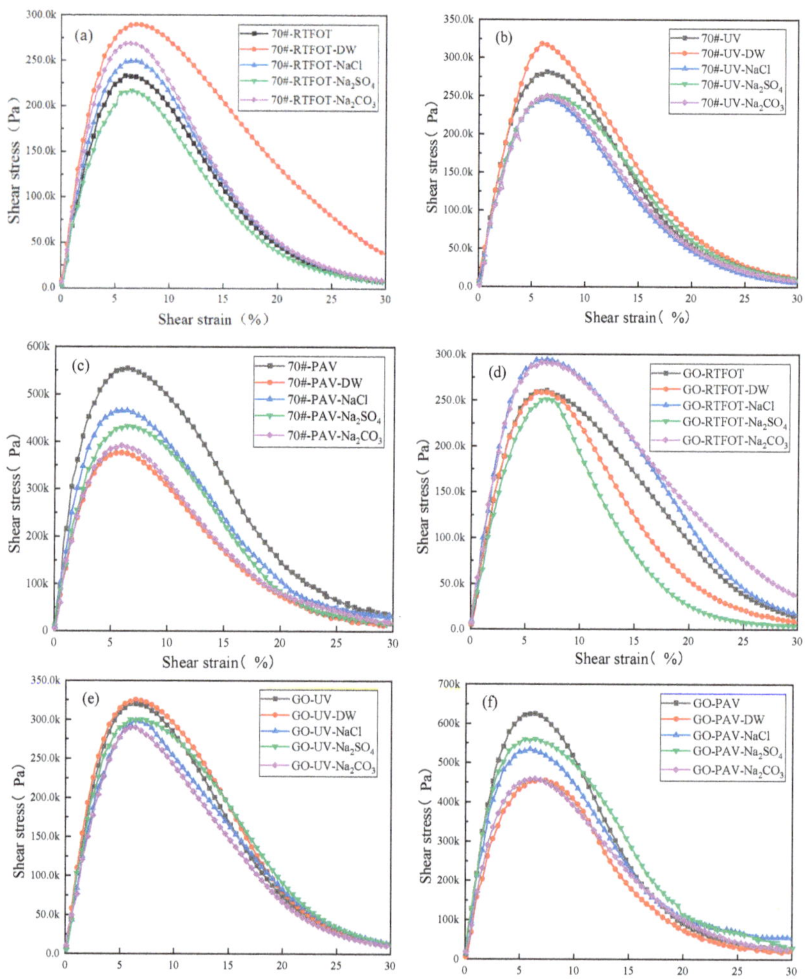

Figure 3. Stress–strain curves of 70# base asphalt and GO-modified asphalt.

Figure 3 shows that both GO-modified asphalt and 70# base asphalt exhibited a peak stress. As the strain continuously increased, the shear stress gradually increased to the peak stress and then began to decrease, indicating that fatigue damage occurred in GO-modified asphalt and base asphalt. With the deepening of the aging degree, the peak stress of both GO-modified asphalt and base asphalt gradually increased, and the influencing trend was PAV > UV > RTFOT > original sample.

Under the short-term aging and salt–alkali coupling effects, for GO-modified asphalt, sodium chloride and sodium sulfate increased the stress peak of GO-modified asphalt, having a "hardening" effect. Sodium carbonate, however, reduced the stress peak of GO-modified asphalt, weakening its deformation resistance, indicating that the short-term aged GO-modified asphalt was most sensitive to the alkaline salt environment and has a certain anti-erosion ability to sodium chloride and sodium sulfate. For base asphalt, sodium chloride and sodium carbonate reduced the strain dependence of the asphalt, with sodium carbonate having a more significant effect, while sodium sulfate reduced the stress–strain level of the base asphalt and increased its strain dependence. Thus, it can be seen that the deteriorating effect of sodium sulfate was the most obvious, accelerating the aging of the base asphalt to a certain extent.

After the ultraviolet aging and salt-alkali solution coupling effect, the saline–alkaline solutions all reduced the peak stress of GO-modified asphalt and base asphalt, indicating that the erosion of the saline–alkaline solutions reduced the deformation resistance of the asphalt and increased its strain dependence. The difference between the two was that the peak stress of GO-modified asphalt was greater than those of base asphalt, indicating that GO-modified asphalt had a greater ability to resist loads at the same strain level and a smaller strain dependence than base asphalt. Overall, for areas with ultraviolet irradiation, saline–alkaline solutions had an adverse impact on the road performance of asphalt pavements.

The GO-modified asphalt after long-term salt–alkaline coupling had greatly reduced its stress peak. The GO-modified asphalt under the sodium carbonate solution showed the worst performance, indicating that sodium carbonate had the most significant deteriorating effect on it, seriously affecting the deformation resistance of GO-modified asphalt. The stress–strain curve of the base asphalt under the long-term-salt–alkaline coupling effect was basically the same as that of GO-modified asphalt. The saline–alkaline solutions all reduced its stress peak, and the sodium carbonate solution had the most serious impact.

3.1.2. Fatigue Damage Characteristic Curve Analysis

Based on the VECD theory and the fatigue failure criterion, the fatigue failure point was determined. Through the LAS experimental data, the DCC curve could be drawn, which characterized the quantitative change law between the pseudo-stiffness modulus and the cumulative damage variable. The results are shown in Figure 4.

It can be seen from Figure 4 that the pseudo-stiffness modulus gradually decreased with the accumulation of the damage variable and finally reached the end-point. It was the fatigue failure point determined according to the peak theory, representing the basic fatigue damage characteristics of asphalt materials under different environmental effects. Due to the changes in the physical and chemical structures of asphalt materials caused by aging and salt–alkaline effects, different end-points occurred. The end-point value of the original GO-modified asphalt was the smallest, indicating that the cumulative damage variable was the smallest when it reached the fatigue failure point.

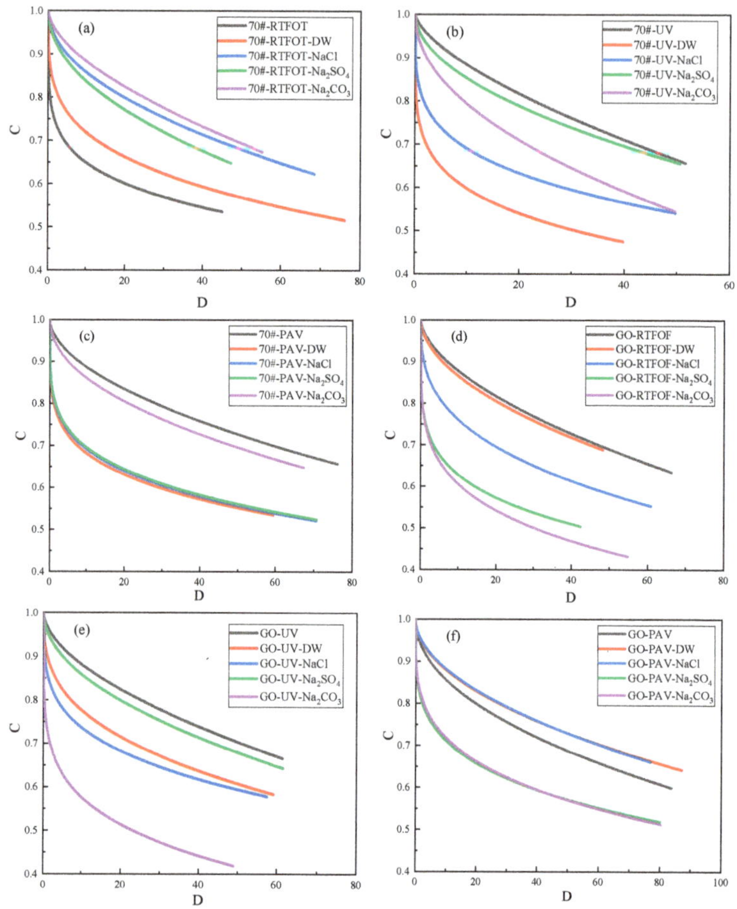

Figure 4. C-D curves.

For GO-modified asphalt, after the short-term aging and salt–alkali coupling effects, all DCC curves were below that of the short-term aged GO-modified asphalt, and their end-points had decreased to different degrees. This indicates that the salt–alkaline environment weakened the deformation resistance of GO-modified asphalt. For base asphalt, after the short-term aging and salt–alkali coupling effects, all DCC curves were above that of the short-term aged asphalt, which meant that the salt–alkaline solution could enhance the deformation resistance of the short-term aged base asphalt, and the sodium carbonate solution had the most significant enhancement effect.

Under the ultraviolet aging and salt–alkali coupling effects, the DCC curves of GO-modified asphalt and 70# base asphalt after the action of the salt–alkaline solution were all below those under the single ultraviolet aging effect. This showed that under the same damage degree, the salt–alkaline solution accelerated the decline rate of the asphalt pseudo-stiffness modulus, indicating that the erosion of the salt–alkaline solution reduced the fatigue performance of asphalt to a certain extent. However, the sodium carbonate solution had the most serious impact on GO-modified asphalt while 70# base asphalt was most sensitive to the DW solution.

Under the long-term aging and salt–alkali coupling effects, all DCC curves were below that of the 70# base asphalt after long-term aging, indicating that the salt–alkaline environment weakened the deformation resistance of asphalt to a certain extent. The DW solution had the most significant erosion effect on 70# base asphalt. For GO-modified asphalt, the sodium chloride solution and the DW solution were above it, indicating that these two could improve the deformation resistance of asphalt to a certain extent. At the same time, the sodium carbonate solution had the most significant deteriorating effect on the long-term aged GO-modified asphalt. Preliminary analysis showed that it may have been because it was easy for the sodium carbonate solution to form voids inside the asphalt, thus reducing its fatigue resistance.

3.1.3. Fatigue Life Analysis

Generally, asphalt materials under different environments have different DCC trends, and each DCC curve has different C_f and D_f values. It is usually considered that the larger the D_f value is, the better the durability and fatigue resistance of the asphalt material will be [21]. However, the fatigue damage rates of each asphalt material under load also vary, making it difficult to conduct a quantitative analysis of the asphalt fatigue life using the DCC curve. Therefore, it is necessary to analyze through the fatigue life equation to accurately evaluate the differences in the fatigue lives of modified asphalts under different environments. The fatigue life equations of GO-modified asphalt and base asphalt are shown in Figure 5.

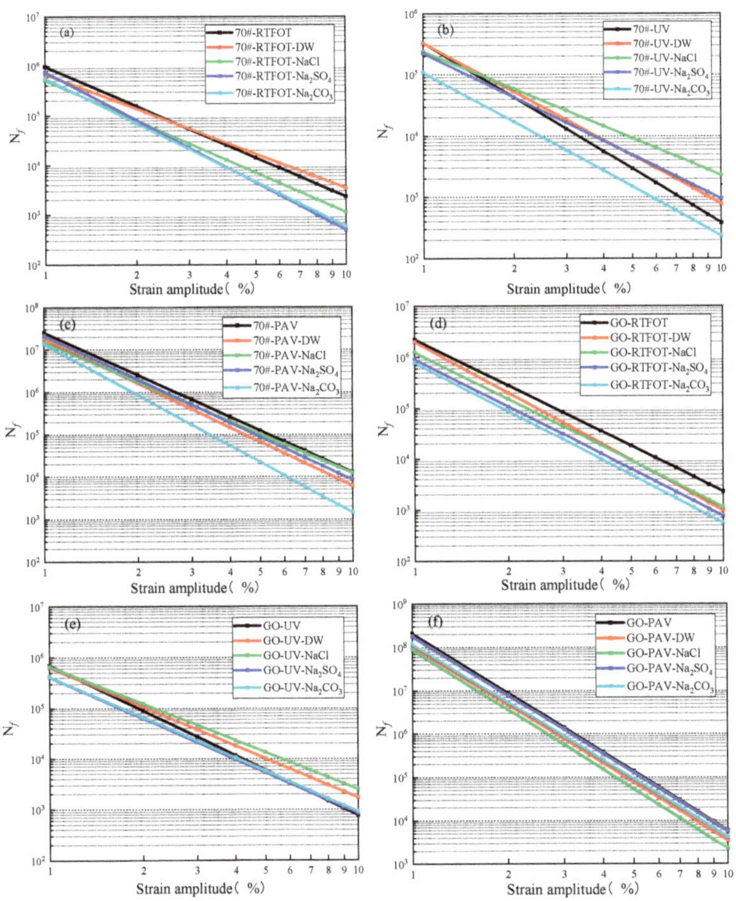

Figure 5. Fatigue life equations of 70# base asphalt and GO-modified asphalt.

Figure 5 shows that after aging, the fatigue lives of the GO-modified asphalt and the base asphalt alike increased to a certain extent. Meanwhile, the fatigue life decreased with the increase in strain, and the degree of aging could boost the material's fatigue-resistant property to some extent. By comparing the six sets of data of the GO-modified asphalt and the base asphalt, the overall influencing trend was long-term aging > short-term aging > ultraviolet aging. However, aging also accelerated the decline rate of the fatigue life of asphalt materials. The aging process actually involved the aggregation of asphaltene micelles and the condensation of aromatic fractions, which raised the asphalt's molecular weight [22]. In the initial stage of aging, the increase in molecular weight and the formation of more complex molecular structures boosted the material's fatigue-resistant property to a certain extent. But as the degree of aging deepened, the brittleness of the asphalt increased. At high strain levels, the material was more likely to develop cracks and fail, thus leading to a faster decline in fatigue life.

For GO-modified asphalt, the fatigue life of GO-modified asphalt under the short-term aging/long-term aging salt–alkaline coupling effect decreased. The reason may have been that the erosion of the salt–alkaline solution accelerated the fatigue damage of GO-modified asphalt, and the ions in it destroyed the molecular structure of GO-modified asphalt, resulting in a decrease in the fatigue life. Under the ultraviolet aging and salt–alkali coupling effects, it could be clearly seen that the fatigue life of GO-modified asphalt soaked in the salt–alkaline solution increased to a certain extent under large-strain levels, and the sodium chloride solution had the most obvious improvement effect, which may have been related to the direct irradiation of ultraviolet rays.

For 70# base asphalt, under the coupling effect of short-term aging and long-term aging with salt–alkaline solutions, due to the erosion of the salt–alkaline solutions on the base asphalt, the fatigue life of the base asphalt decreased under strain levels, and the sodium carbonate solution had the greatest impact, greatly reducing the fatigue life of the base asphalt. The reason may have been that the ions in the salt–alkaline solution caused an oxidation reaction of asphalt molecules, resulting in the breakage of molecular bonds inside and the destruction of the balance of its colloidal structure, leading to a decline in the fatigue life of asphalt. Under the ultraviolet aging and salt–alkali solution coupling effect, it could be found that the sodium carbonate solution had the most significant impact on the fatigue life of the base asphalt, reducing its fatigue life, while the sodium chloride and sodium sulfate solutions could increase the fatigue life of the base asphalt under large-strain levels to a certain extent.

3.2. *Healing Performance Analysis*

3.2.1. Healing Index Analysis

Based on the VECD theory, the healing performance of base asphalt and GO-modified asphalt was evaluated by combining the LASH healing test. The healing performance was evaluated by calculating the healing index based on the damage variables before and after healing. The test results are shown in Figure 6.

The healing index quantifies the healing ability of asphalt by analyzing the relationship between the pseudo-stiffness modulus and the damage degree. The larger the value is, the stronger the healing ability of the asphalt is. It can be seen from Figure 6 that the original base asphalt and the original GO-modified asphalt had the highest self-healing rates. After the asphalt underwent short-term aging, ultraviolet aging, and long-term aging, its self-healing decreased. And the influencing effect was long-term aging > ultraviolet aging > short-term aging. The reason was that compared with short-term aging and ultraviolet aging, under long-term aging, oxygen reacted with the double bonds in asphalt molecules, generating more oxidation products, changing the molecular structure of the asphalt, thus affecting the diffusion rate of the asphalt and reducing its self-healing efficiency.

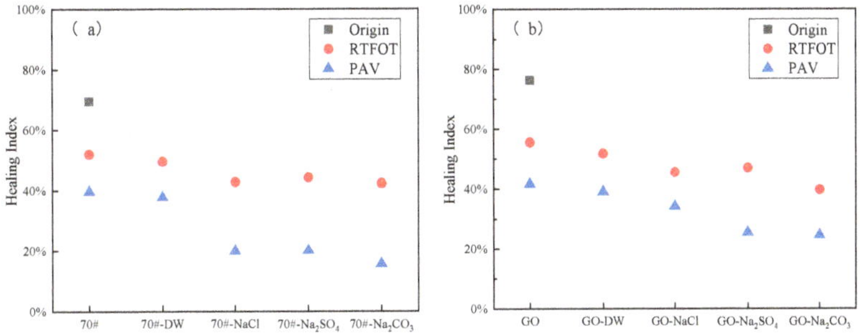

Figure 6. Self-healing indices of GO-modified asphalt and 70# base asphalt under different aging conditions.

It can be seen from Figure 6 that the self-healing efficiency of GO-modified asphalt was basically better than that of base asphalt in various complex environments, indicating that GO had a positive impact on the self-healing performance of asphalt, manifested as an increase in the self-healing index of asphalt. The addition of GO molecules could form hydrogen bonds with asphalt molecules and generate intermolecular van der Waals forces, and form aromatic ring stacking with aromatics and resins, enabling GO to be stably dispersed in asphalt and form stable connections, thus improving the road performance of asphalt. At the same time, it could affect the transformation between components and enhance the self-healing diffusion movement of asphalt molecules, thereby increasing the self-healing index of asphalt. Currently, there has been no unified conclusion on the modification mechanism of GO, and further research is needed.

When the salt–alkaline and aging were coupled, the self-healing rates of both base asphalt and GO-modified asphalt declined to a certain extent. The erosion of the DW had a relatively small impact on reducing the self-healing efficiency of base asphalt and GO-modified asphalt, indicating that the DW had had little impact on the self-healing of asphalt. For base asphalt, the self-healing efficiency was the lowest under the coupling of the sodium carbonate solution and aging, indicating that the base asphalt was most sensitive to alkaline salts, and the alkaline salts had the most serious deteriorating effect on the base asphalt. For GO-modified asphalt, the sodium sulfate solution had the most serious impact on GO-modified asphalt. It was speculated that the reason was that SO_4^{2-} affected the connection between GO molecules and asphalt molecules, thus reducing its self-healing efficiency.

3.2.2. Fatigue Life Analysis Before and After Healing

With the introduction of the load intermittent period, we could clearly observe the recovery of the asphalt modulus in the LASH test, which promoted the recovery of the asphalt fatigue life. Therefore, it is necessary to analyze this. For this section, the fatigue life of the self-healed asphalt was calculated based on the VECD theory. At the same time, according to the empirical formula proposed by Mannan [23], the fatigue life coefficient after healing was calculated. A fatigue life conversion coefficient μ was introduced to quantitatively analyze the improvement effect of the fatigue life after self-healing. The calculation formula of this coefficient is shown in Formula (7).

$$\mu = \frac{N_f(with - rest)}{N'_f(without - rest)} \tag{7}$$

Here, N_f and N'_f are the fatigue lives of asphalt calculated by the LAS test and the LASH healing test, respectively, according to the peak stress failure criterion. The calculation results of the fatigue life conversion coefficient μ are shown in Table 4.

Table 4. Fatigue life conversion coefficient μ of asphalt under different aging and salt–alkali coupling effects.

Asphalt Type	Original	RTFOT	UV	PAV
70#	1.407	1.349	1.312	1.299
70#-DW	-	1.321	1.308	1.298
70#-NaCl	-	1.303	1.276	1.261
70#-Na_2SO_4	-	1.282	1.208	1.255
70#-Na_2CO_3	-	1.207	1.199	1.174
GO	1.469	1.402	1.349	1.248
GO-DW	-	1.389	1.338	1.238
GO-NaCl	-	1.325	1.321	1.185
GO-Na_2SO_4	-	1.363	1.261	1.102
GO-Na_2CO_3	-	1.299	1.272	1.141

It can be seen from Table 4 that the μ value of the original asphalt was the highest, and the μ value decreased after aging, indicating that aging hindered the recovery of the asphalt fatigue life. With the deepening of aging, the μ value showed a downward trend, and the influencing trend was long-term aging > ultraviolet aging > short-term aging. After the aging and salt–alkali coupling effects, it could be found by comparison that the μ values all decreased, indicating that the salt–alkaline environment hindered the recovery of the asphalt fatigue life. For more intuitive display, and according to the research of domestic and foreign scholars [24], when the strain level $\gamma = 2.5\%$ and 5%, they corresponded to the actual strain levels of thick-layer and thin-layer asphalt pavements, respectively, and when the strain level was 5%, the fatigue life of asphalt had a high correlation with the fatigue life of the mixture. Figure 7 shows the comparison of the fatigue lives of the two asphalts before and after healing at a strain of 5%.

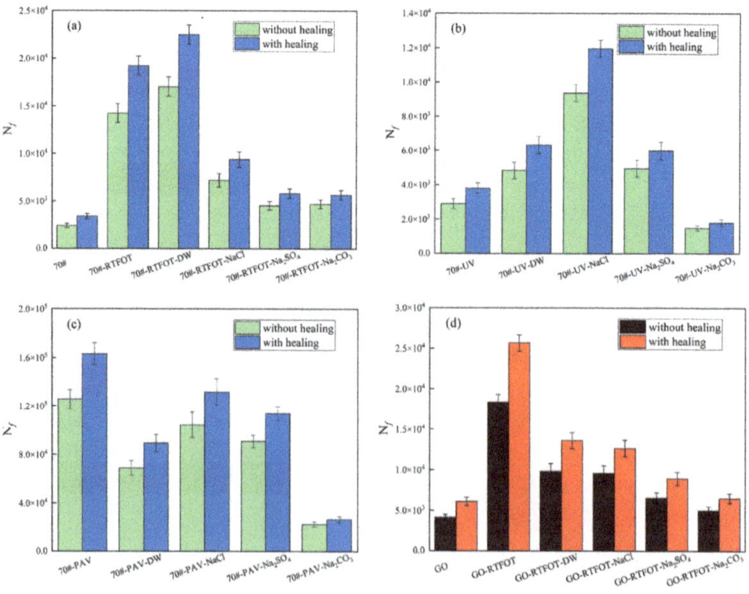

Figure 7. *Cont.*

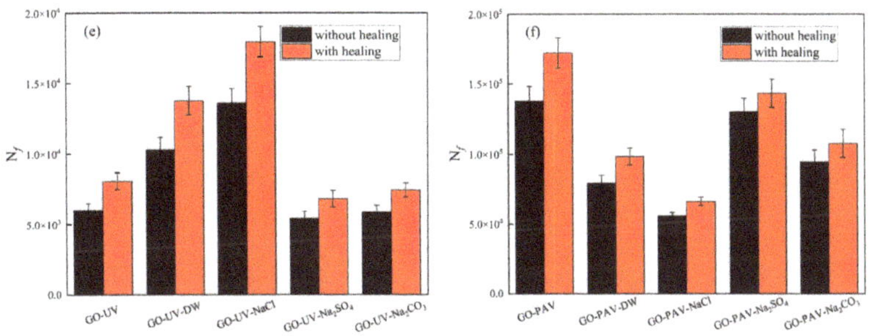

Figure 7. Fatigue lives before and after healing.

It can be seen from Figure 7 that due to the introduction of the intermittent period, the self-healing performance of asphalt had a certain growth effect on its fatigue life. In the 60 min intermittent period, it could be seen that the fatigue lives of all asphalt samples increased to different degrees. The growth efficiency of the original sample was the highest while the growth rates of the asphalt after short-term aging, ultraviolet aging, and long-term aging decreased. The long-term aged asphalt had the worst recovery effect. The reason was that in a high-temperature and high-pressure environment, the light components in the asphalt volatilized, and the increases in hard components hindered the flow of the asphalt, suppressing the healing of micro-cracks in the asphalt, thus reducing the ability of the asphalt to withstand the secondary load.

After the aging and salt–alkali coupling effects, it could be found that the growth effect of the asphalt fatigue life was further reduced, indicating that the salt–alkaline solution also inhibited the fatigue and self-healing performance of asphalt. For the asphalt under the short-term aging and salt–alkali coupling effects, it could be seen that the fatigue life growth effect of the asphalt under the action of the sodium carbonate solution was the worst. Under the ultraviolet aging and salt–alkali coupling effects and the long-term aging and salt–alkali coupling effects, for the base asphalt, the fatigue life growth effect of the asphalt under the action of the sodium carbonate solution was the worst, while GO-modified asphalt was more sensitive to the sodium sulfate solution, resulting in a poor fatigue life growth effect.

In addition, by comparing the fatigue lives of 70# base asphalt and GO-modified asphalt after healing, it could also be found that the fatigue life growth effect of GO-modified asphalt was generally higher than that of base asphalt, indicating that GO molecules could improve the self-healing performance of base asphalt. This benefited from the compatibility of GO molecules with asphalt, enabling them to form a more stable molecular structure with asphalt molecules, thus improving the self-healing performance of asphalt and enhancing its fatigue life growth effect.

4. Conclusions

This paper studied the influence of different aging and salt–alkali coupling effects on the fatigue and self-healing of asphalt and analyzed it, combined with the viscoelastic continuum damage theory, expounding the change rules of the fatigue and self-healing performance of asphalt under different aging and salt–alkali coupling effects. The main conclusions are as follows:

(1) Aging increased the peak stress but accelerated the stress decline rate. Under the short-term aging and salt–alkali coupling effects, the sodium carbonate solution could significantly increase the peak stress of asphalt while the sodium sulfate solution would significantly reduce them. Under the ultraviolet aging and salt–alkali coupling effects and

long-term aging and salt–alkali coupling effects, the salt–alkaline solutions had an adverse impact on asphalt, especially the sodium carbonate solution.

(2) The fatigue lives of GO-modified asphalt and base asphalt increased with the increase in the aging degree, but with the increase in the strain, the decline rate of the fatigue life of asphalt materials also increased. Under the salt–alkaline-aging coupling effect, the presence of salt–alkaline solutions reduced the fatigue life of asphalt, with the alkaline solution having the most significant impact. At the same time, it also increased the decline rate of the fatigue life of asphalt materials.

(3) Saline–alkaline solutions degraded the self-healing performance of both GO-modified and base asphalt. Under different aging and salt–alkali coupling effects, the salt–alkaline effect reduced the self-healing performance of the two kinds of asphalt, making the growth amplitude of their fatigue life decrease. The growth was the smallest under the long-term aging and salt–alkali coupling effects solution, and the asphalt life growth under the erosion of the sodium carbonate solution was the smallest, indicating that asphalt was most sensitive to the alkaline environment.

In this study, a large number of laboratory tests were carried out to investigate the fatigue and self-healing properties of GO-modified asphalt and 70# base asphalt under different aging effects and aging and salt–alkali coupling effects. However, there are still some issues that need further research. First, this paper has only focused on the research on asphalt binders. Future studies can explore the fatigue and self-healing properties of asphalt mixtures when subjected to different aging scenarios, with a particular focus on the coupling of aging and salt–alkali effects. This is of great significance for the research on the road-use performance of asphalt pavements in saline-alkali areas and high-altitude areas with strong ultraviolet radiation. Second, regarding the research on the self-healing properties of asphalt, it is necessary to further consider the influence of factors such as temperature, thixotropy, and intermittent time on the self-healing properties of GO-modified asphalt. This is conducive to improving the prediction accuracy of the fatigue life of asphalt.

Author Contributions: R.L.: funding acquisition, methodology, supervision, writing—review and editing; W.Z.: validation, investigation, writing—original draft; H.C.: formal analysis, resources; X.L.: resources, visualization; L.S.: resources, project administration.; J.Y.: supervision, project administration, funding acquisition. All authors have read and agreed to the published version of the manuscript.

Funding: The authors thank the National Natural Science Foundation of China (Grant No. 51708513), the General Program of Henan Provincial Natural Science Foundation (Grant No. 242300420003), the Henan Provincial Science and Technology Research Project (Grant No. 252102241014) from the Henan Provincial Department of Education, and the Henan Provincial Department of Housing and Urban-Rural Development (grant number HNJS-2024-K28) for their financial support.

Institutional Review Board Statement: Not applicable.

Informed Consent Statement: Not applicable.

Data Availability Statement: The original contributions presented in the study are included in the article; further inquiries can be directed to the corresponding author.

Conflicts of Interest: Author Hailong Chen was employed by the company Shanghai Urban Construction Maintenance Management Co., Ltd. Author Linhao Sun was employed by the company Zhengzhou Public Utility Investment and Development Group Co., Ltd. The remaining authors declare that the research was conducted in the absence of any commercial or financial relationships that could be construed as a potential conflict of interest.

References

1. Agzenai, Y.; Pozuelo, J.; Sanz, J.; Perez, I.; Baselga, J. Advanced self-healing asphalt composites in the pavement performance field: Mechanisms at the Nano level and new repairing methodologies. *Recent Pat. Nanotechnol.* **2015**, *9*, 43. [CrossRef] [PubMed]
2. Xiong, R.; Jiang, W.; Yang, F.; Li, K.; Guan, B.; Zhao, H. Investigation of Voids Characteristics in an Asphalt Mixture Exposed to Salt Erosion Based on CT Images. *Materials* **2019**, *12*, 3774. [CrossRef] [PubMed]
3. Sun, G.H. Experiments and Research on Diseases of Asphalt Concrete Pavements in Saline Soil Areas. Master's Thesis, Jilin University, Changchun, China, 2013.
4. Tang, Z.; Wu, X.; Guo, B.; Zhang, L.; Jia, D. Preparation of butadiene–styrene–vinyl pyridine rubber–graphene oxide hybrids through co-coagulation process and in situ interface tailoring. *J. Mater. Chem.* **2012**, *22*, 7492–7501. [CrossRef]
5. Habib, N.Z.; Aun, N.C.; Zoorob, S.E.; Lee, P.I. Use of Graphene Oxide as a Bitumen Modifier: An Innovative Process Optimization Study. *Adv. Mater. Res.* **2015**, *1105*, 365–369. [CrossRef]
6. Huang, J.; Ma, Q.; Huang, L. Rheological characteristics and thermal-oxidative aging resistance of graphene oxide compounded with polyphosphoric acid modified asphalt. *J. Chongqing Jiaotong Univ. (Nat. Sci. Ed.)* **2023**, *42*, 71–79.
7. Zeng, Q.; Liu, Y.; Liu, Q.; Liu, P.; He, Y.; Zeng, Y. Preparation and modification mechanism analysis of graphene oxide modified asphalts. *Constr. Build. Mater.* **2020**, *238*, 117706. [CrossRef]
8. Li, Z. Research on the Influence of Water, Acid, and Salt Solutions on Asphalt Aging. Master's Thesis, Changsha University of Science and Technology, Changsha, China, 2018.
9. Xu, R. Research on the Influence of Sodium Salt Erosion on the Self-Healing and Fatigue Characteristics of Asphalt. Master's Thesis, Guangxi University, Nanning, China, 2020.
10. Xiong, R.; Feng, B.; Qiao, N.; Ji, K.; Wang, H. Evaluation of asphalt performance under sulfate freeze-thaw cycles. *J. Chongqing Jiaotong Univ. (Nat. Sci. Ed.)* **2022**, *41*, 70–75.
11. An, X.; Wang, R.; Kang, X.; Yue, J. A more accurate fatigue characterization of GO-modified asphalt binder considering non-linear viscoelastic behavior and UV exposure effects. *Int. J. Fatigue* **2023**, *168*, 107396. [CrossRef]
12. Shan, L.; Tan, Y.; Underwood, S.; Kim, Y.R. Application of thixotropic to analyze fatigue and healing characteristics of asphalt binder. *Trans. Res. Rec.* **2010**, *2179*, 85–92. [CrossRef]
13. *JTG E20-2011*; Research Institute of Highway Ministry of Transport. Standard Test Methods of Bitumen and Bituminous Mixtures for Highway Engineering. China Communications Press: Beijing, China, 2011.
14. *ISO/TS 21356-1:2021*; Nanotechnologies-Structural Characterization of Graphene-Part 1: Graphene from Powders and Dispersions [Technical Specification]. First Edition 2021-03; International Organization for Standardization: Geneva, Switzerland, 2021.
15. Johnson, C.M. Estimating Asphalt Binder Fatigue Resistance Using an Accelerated Test Method. Ph.D. Thesis, University of Wisconsin Madison, Madison, WT, USA, 2010.
16. Hintz, C.; Velasquez, R.A.; Johnson, C.M.; Bahia, H.U. Modification and validation of the linear amplitude sweep test for binder fatigue specification. *Transp. Res. Rec.* **2011**, *2207*, 99–106. [CrossRef]
17. Lee, H.J.; Kim, Y.R. A uniaxial viscoelastic constitutive model for asphalt concrete under cyclic loading. *J. Eng. Mech.* **1998**, *124*, 32–40. [CrossRef]
18. Xie, W.; Castorena, C.; Wang, C.; Kim, Y.R. A framework to characterize the healing potential of asphalt binder using the linear amplitude sweep test. *Constr. Build. Mater.* **2017**, *154*, 771–779. [CrossRef]
19. Zhou, S.-W.; Shi, J.-T. Research on the Fatigue Property of Asphalt Based on Linear Amplitude Sweep Test. *Pet. Asph.* **2018**, *32*, 7.
20. Hintz, C.; Velasquez, R.; Li, Z.; Bahia, H. Effect of Oxidative Aging on Binder Fatigue Performance. *J. Assoc. Asph. Paving Technol.* **2011**, *80*, 527–548.
21. Jiang, J.; Lu, F.; Guo, X.; Hao, P.; Wang, W.; Yu, J. Effect of SARA Fractions on Fatigue Properties of Hard Asphalt. *Materials* **2024**, *17*, 2713. [CrossRef] [PubMed]
22. Werkovits, S.; Bacher, M.; Mirwald, J.; Rosenau, T.; Hofko, B.; Grothe, H. Photo-induced ageing processes in bitumen. *Constr. Build. Mater.* **2024**, *438*, 137186. [CrossRef]
23. Mannan, U.A.; Tarefder, R.A. Investigating different fatigue failure criteria of asphalt binder with the consideration of healing. *Int. J. Fatigue* **2018**, *114*, 198–205. [CrossRef]
24. Wang, C. Rheological Characterization on Paving Performance of Asphalt Binder. Ph.D. Thesis, Beijing University of Technology, Beijing, China, 2015.

materials

MDPI

Article

Thermal-Ultraviolet-Humidness Coupling Ageing and Regeneration Properties and Mechanisms of SBS-Modified Asphalt Under Hot–Wet Environment Conditions

Shuo Zhou [1], Dengfeng Wang [2], Liuxing Wu [3], Alimire Maimaitisidike [1], Zhiqing Wang [1], Hongbo Zhao [1] and Jiaolong Ren [1,*]

[1] School of Civil Engineering and Geomatics, Shandong University of Technology, Zibo 255000, China; 22110902175@stumail.sdut.edu.cn (S.Z.); 22120908119@stumail.sdut.edu.cn (A.M.); 23507030878@stumail.sdut.edu.cn (Z.W.); hbzhao@sdut.edu.cn (H.Z.)

[2] Zhumadian Transportation Construction Technology Center, Zhumadian 463000, China; zs2318976106@163.com

[3] Zhumadian Highway Engineering Development Co., Ltd., Zhumadian 463000, China; 19546325096@163.com

* Correspondence: worjl@sdut.edu.cn

Abstract: Styrene-butadiene-styrene (SBS)-modified asphalt, a widely utilised binder in pavement engineering, is susceptible to ageing due to the coupling effects of thermo-oxidation, ultraviolet radiation, and humidness. Due to the limited availability of high-quality asphalt resources, recycling aged asphalt has emerged as a vital strategy for addressing resource shortages and reducing environmental pollution. This study investigated the effects of thermal-ultraviolet-humidness coupled ageing on the pavement performance of SBS-modified asphalt, with a specific focus on the hot–wet climates of Guangzhou and Chengdu. Beijing's standard climate serves as a reference for this study. Additionally, industrial animal oil was chosen as a rejuvenator for aged SBS-modified asphalt. The mechanisms underlying hot–wet coupling ageing and regeneration of SBS-modified asphalt were analysed using Fourier Transform Infrared Spectroscopy (FTIR) and Fluorescence Microscopy (FM). The findings indicate that thermal-oxidation and humidness accelerate sulphide formation, resulting in a marked increase in sulfoxide groups and facilitating the migration of lighter components, ultimately leading to asphalt hardening. Under high-temperature and humidness conditions, the butadiene index (BI) of asphalt decreased by 5.96% in Chengdu and 15.78% in Guangzhou compared to Beijing. The sulfoxide index (SI) and aromaticity index (CI) increased by 3.74% and 3.89% in Chengdu, and by 9.39% and 8.54% in Guangzhou, respectively, confirming the exacerbating effect of humidness on ageing. During the regeneration process, industrial animal oil effectively diluted polar molecules in aged asphalt, resulting in reductions in SI by 38.88%, 36.74%, and 37.74%, and in CI by 63.77%, 62.54%, and 63.11% under ageing conditions in Beijing, Guangzhou, and Chengdu, respectively. Rejuvenation is achieved by replenishing lighter components, thereby promoting the aggregation and swelling of the degraded SBS chains.

Keywords: SBS-modified asphalt; coupling ageing; regeneration; thermal-ultraviolet-humidness; industrial animal oil

1. Introduction

SBS-modified asphalt is extensively used in highway and urban road construction due to its outstanding high-temperature stability, low-temperature crack resistance, and fatigue resistance [1,2]. However, during service, SBS-modified asphalt pavements undergo ageing reactions induced by the coupling effects of environmental factors, including thermal oxidation, rainfall, and ultraviolet radiation [3–5]. These reactions result in a substantial decline in performance, ultimately leading to the material failing to meet service requirements [6,7]. Consequently, examining the ageing characteristics and mechanisms of SBS-modified asphalt under these coupled environmental conditions is essential for the effective construction and maintenance of road traffic infrastructure.

Liu et al. [8] demonstrated that the coupling effects of ultraviolet radiation and temperature accelerate asphalt oxidation. Similarly, Wang et al. [9] and Li et al. [10], through complex modulus and phase angle analyses, found that the interaction between ultraviolet radiation and high temperature degrades the rheological properties of asphalt and shortens its fatigue life. Li et al. [11] observed that the coupling effect of ultraviolet radiation and temperature increased the asphaltene content while decreasing the aromatic content of SBS-modified asphalt. Zhu et al. [12] elucidated the ageing mechanism of SBS-modified asphalt under the coupled influence of ultraviolet radiation and high temperatures using Fourier Transform Infrared Spectroscopy (FTIR). These studies indicate that the ageing effect resulting from the interaction of ultraviolet radiation and temperature is significantly more pronounced than those observed under individual ageing conditions. This highlights the importance of understanding the coupling effects in the real-world service environment of asphalt. Therefore, the influence of humidness, as another critical factor, should also be prioritised, particularly in regions with humid climates [13].

Wei et al. [14] and Zhang et al. [15] found that humidness exacerbates ultraviolet-induced surface cracking in SBS-modified asphalt. Yuan et al. [16] employed microscopy to observe that water infiltrates the inner layers of asphalt through cracks, which weakening the material's performance, when coupling with high temperatures. Ma et al. [17] confirmed that the humidness-thermal coupling effect disrupts the "bee-like structure" of asphalt, thereby diminishing its bonding properties. Existing research indicates that the ageing process of SBS-modified asphalt is accelerated and intensified under the coupling influence of humidness and high temperatures, suggesting that SBS-modified asphalt pavements are particularly vulnerable to ageing in hot–wet environments [18,19].

However, during its service life, asphalt is influenced not only by dual-factor coupling ageing but also by the coupling effects of thermal-ultraviolet-humidness, which are more likely to accelerate the degradation of asphalt performance. Zou et al. [20] demonstrated that the high-temperature rutting resistance and low-temperature crack resistance of asphalt are significantly diminished after ageing due to thermal oxidation, humidness, and ultraviolet radiation. However, these tests only simulated short-term ageing conditions, and ageing factors were not applied simultaneously. Yang et al. [21] found that the coupling effects of thermal-ultraviolet-humidness significantly intensified microcracks on the surface of SBS-modified asphalt, resulting in performance degradation. Sun et al. [22] demonstrated the mitigating effect of humidness on asphalt ageing, revealing that the water film formed on the asphalt surface reduced the overall impact of the coupled effect of thermal-ultraviolet-humidness. These findings suggest that the coupling effects of thermal-ultraviolet-humidness result in severe asphalt ageing.

However, as a by-product of petroleum [23,24], the availability of high-quality asphalt is decreasing due to the depletion of petroleum resources [25]. Recycling aged SBS-modified asphalt presents an effective and cost-efficient solution to meet the growing demand for asphalt in infrastructure construction. Furthermore, if not properly managed, large quantities of waste SBS-modified asphalt can contribute to environmental pollution and resource depletion [26]. Recycling aged asphalt not only reduces carbon emissions and alleviates environmental pressures but also meets the demand for SBS-modified asphalt in transportation engineering, thereby promoting sustainable resource utilisation [27].

Numerous studies have examined the effects of various recycling technologies on asphalt ageing. Zhang et al. [28] demonstrated that rejuvenators rich in aromatic compounds can effectively restore the performance of aged asphalt by compensating for the loss of lighter components. Wu et al. [29] revealed that rubber hydrocarbon rejuvenators improve the low-temperature crack resistance of ageing SBS-modified asphalt. Zhang et al. [30] demonstrated that rejuvenators rich in lighter components significantly enhanced the high-temperature performance of SBS-modified asphalt. Yao et al. [31] utilised FTIR to elucidate the regeneration mechanism of unaged SBS-modified asphalt when applied to ageing asphalt. Furthermore, bio-oil has been extensively used in asphalt recycling due to its cost-effectiveness and environmental benefits [32–34]. Yang et al. [35] validated the feasibility of bio-oil rejuvenators through computational simulations. Li et al. [36] employed FTIR and scanning electron microscopy to elucidate the regeneration mechanism of soybean bio-asphalt. These results indicate that both soybean bio-asphalt and SBS-modified asphalt contain numerous functional groups, which can promote the expansion of SBS to form a network structure. However, existing studies have primarily focused on regeneration under single ageing conditions while neglecting regeneration following thermal-ultraviolet-humidness coupling ageing conditions. Given the substantial impact of multifactor coupling ageing, which causes more severe damage in SBS-modified asphalt, investigating the regeneration properties and mechanisms of SBS-modified asphalt after coupling ageing is of critical importance.

Given that the asphalt performance is significantly influenced by the combined effects of various factors during traffic operations, the objectives of this study are as follows:

- To investigate the thermal-ultraviolet-humidness coupling ageing properties and mechanisms of SBS-modified asphalt under hot–wet environmental conditions.
- To elucidate the regeneration effects and mechanisms of an industrial animal oil rejuvenator on SBS-modified asphalt subjected to coupling ageing.

Therefore, the hot–wet conditions of Guangzhou and Chengdu were selected as research subjects, while the climatic conditions of Beijing served as a control. The thermal-ultraviolet-humidness coupling ageing properties and mechanisms of SBS-modified asphalt were examined under these environmental conditions. Moreover, the coupling aged SBS-modified asphalt was rejuvenated using industrial animal oil, highlighting the effectiveness and underlying mechanisms of the regeneration process. This study not only contributes to the advancement of anti-ageing technologies for SBS-modified asphalt but also promotes the reuse of multifactor coupling ageing.

2. Materials and Methods

2.1. Materials

The SBS-modified asphalt produced by Qilu Petrochemical Company (Zibo, China) is manufactured by blending AH-70# base asphalt with 4.5% SBS content relative to the base asphalt, and all key parameters met the specifications "Technical Specifications for Construction of Highway Asphalt Pavements (JTG F40-2004)" [37]. An industrial animal oil produced by Guangzhou Fufei Chemical Technology Co., Ltd. (Guangzhou, China) was selected as the rejuvenator. Tables 1 and 2 present the key technical parameters of the SBS-modified asphalt and the industrial animal oil, respectively. The functional groups of industrial animal oils were analysed by Fourier transform infrared spectroscopy (FTIR), as illustrated in Figure 1.

Table 1. Performance indicators of SBS-modified asphalt.

Technical Indicators	Unit	Prescribed Standards	Experimental Results
Penetration (25 °C, 100 g, 5 s)	mm	40~60	57
Softening Point	°C	>75	82.5
Ductility at 5 °C	cm	>20	61.4
Elastic Recovery (25 °C)	%	≥75	92
Kinematic Viscosity at 135 °C	Pa·S	<3	2.710
Flash Point (Open Cup)	°C	>230	329
Solubility in Trichloroethylene	%	>99	99.74
Density at 15 °C	g/cm^3	Actual measurement	1.033

Table 2. Performance indicators of industrial animal oil.

Technical Indicators	Indicators
Appearance	Light yellow
Kinematic Viscosity at 40 °C, mm^2/s	30–50
Open Flash Point, °C	260
Density at 20 °C	0.95

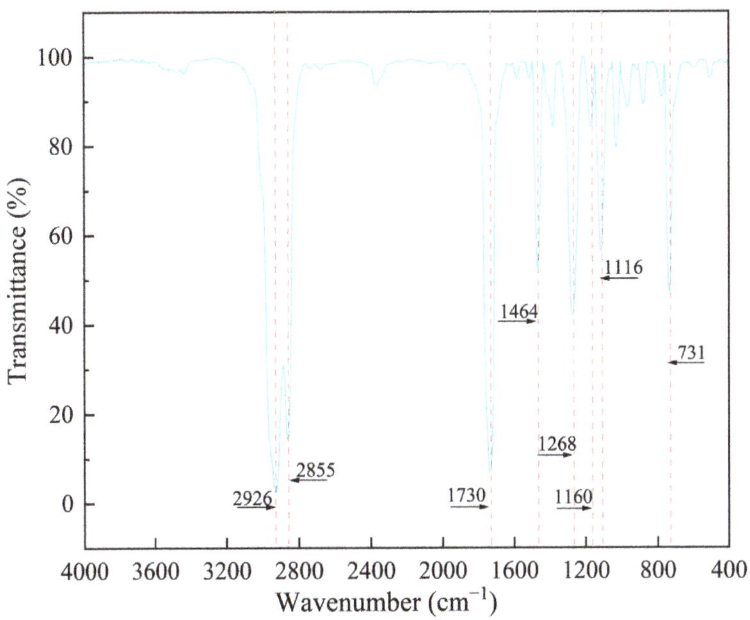

Figure 1. Functional groups in industrial animal oil.

As illustrated in the figure, the absorption peaks observed at 2926 cm^{-1}, 2855 cm^{-1}, and 1464 cm^{-1} correspond to the asymmetric stretching vibration, symmetric stretching vibration, and bending vibration of methylene (CH_2) groups, respectively. The prominent absorption peak at 1730 cm^{-1} is attributed to the stretching vibration of the carbonyl group (C=O), indicating a substantial presence of carbonyl compounds in the industrial animal oil. The absorption peak at 1268 cm^{-1} is attributed to the phenolic hydroxyl group, while the peak at 1116 cm^{-1} corresponds to the stretching vibration of ether linkages (C-O-C) within lipid chains. Additionally, the prominent absorption peak at 731 cm^{-1} is associated with the out-of-plane bending vibration of aromatic carbon-hydrogen (C-H) groups in benzene rings, indicating the presence of aromatic compounds in industrial animal oil.

2.2. Experimental Methods

2.2.1. Ageing Process

According to the Chinese test standard "Standard Test Methods of Bitumen and Bituminous Mixtures for Highway Engineering (JTG E20-2011) [38]", the Rolling Thin Film Oven Test (RTFOT) is utilised to conduct thermal-oxidative ageing on asphalt binders. Each test specimen is prepared with a mass of 35 ± 0.5 g and subjected to thermal ageing at 163 °C ± 0.5 °C for a duration of 85 min. Subsequently, the thermal-oxidative ageing asphalt is placed in a coupling ageing test chamber, where it undergoes the coupling action of thermal oxidation, ultraviolet radiation, and humidness. The parameters for each ageing condition are detailed in Section 2.2.2. The aged asphalt is then mixed with varying contents of industrial animal oil (0%, 3%, 6%, and 9%) for a continuous duration of 25 min at 1500 r/min in a shearing unit, ensuring thorough integration of the industrial animal oil with the aged asphalt. The detailed process is illustrated in Figure 2.

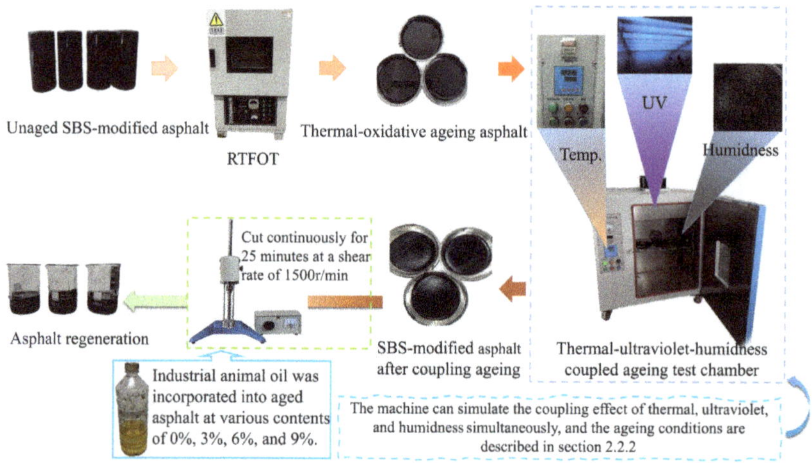

Figure 2. Experimental flow chart.

2.2.2. Ageing Conditions

To accurately simulate the asphalt ageing environments in Beijing, Guangzhou, and Chengdu, the specific experimental conditions present in each city were analysed. Given that asphalt ageing is most pronounced during the summer, the summer climatic characteristics of these three cities were used as benchmarks to design experimental groups incorporating distinct ageing factors. A thermal-ultraviolet-humidness coupling ageing chamber was employed to simulate the ageing processes of asphalt under controlled laboratory conditions, effectively replicating real-world ageing phenomena.

(1) Based on meteorological data from the China Meteorological Data Network [39], a statistical analysis was conducted on the total summer rainfall in Beijing, Guangzhou, and Chengdu from 2001 to 2020, and the average daily rainfall was calculated accordingly. Detailed data are presented in Table 3.

(2) According to the "Technical Specifications for Construction of Highway Asphalt Pavements (JTG F40-2004)" [37], Beijing is classified as a hot summer and cold winter climatic zone, while Chengdu is categorised as a warm-winter climatic zone. In all three cities,

the average maximum temperature during the hottest month exceeds 30 °C, with temperatures of 35 °C and above being more frequent in Guangzhou. Consequently, the average summer temperatures in Beijing, Guangzhou, and Chengdu are 30 °C, 35 °C, and 30 °C, respectively. Generally, the surface temperature of asphalt pavement is approximately 20 to 35 °C higher than the ambient air temperature. Therefore, experimental temperatures of 65 °C, 70 °C, and 65 °C were chosen as the ageing environmental conditions for these three cities.

(3) This study utilised a thermal-ultraviolet-humidness coupling ageing chamber equipped with six 40 W ultraviolet lamps. The summer solar radiation levels in Beijing, Guangzhou, and Chengdu were determined using solar radiation data from the China Meteorological Data Network, with ultraviolet radiation accounting for approximately 5% to 7% of the total solar radiation. Based on the energy equivalence, the ultraviolet radiation exposure time for the three cities were calculated using Equation (1). Specific data are presented in Table 4.

$$T = \frac{Y}{3600 \cdot X} \quad (1)$$

where T represents the indoor ultraviolet irradiation duration (h); X denotes the irradiation intensity of the ultraviolet lamps (W/m^2); and Y corresponds to the average summer solar ultraviolet irradiation intensity (MJ/m^2).

Table 3. Urban Rainfall Amount.

City	Total Summer Rainfall (mm)	Average Daily Rainfall (mm)
Beijing	6862.80	3.8
Guangzhou	17,556.03	9.7
Chengdu	13,773.7	7.6

Table 4. Ultraviolet irradiation time.

City	The Average Total Solar Radiation in Summer (MJ/m^2)	Ultraviolet Radiation Intensity (MJ/m^2)
Beijing	1755.6–2048.2	95.095
Guangzhou	1254–1504.8	68.97
Chengdu	1003.2–1254	56.43

Based on the aforementioned findings, the experimental coupled ageing conditions were categorised as follows: Beijing served as the benchmark environmental conditions (BEC), Guangzhou represented a high-temperature and high-humidness environment conditions (HEC), and Chengdu was classified as a slightly high-temperature and high-humidness environmental conditions (SHEC). Detailed parameters are illustrated in Table 5.

Table 5. Coupling ageing conditions.

City	Humidness (mm)	Temperature (°C)	Ultraviolet Duration (h)
Beijing	3.8	65	110
Guangzhou	9.7	70	80
Chengdu	7.6	65	65

2.2.3. Test Methods

The high and low temperature properties of asphalt after ageing and regeneration were evaluated using the softening point test and the 5 °C ductility test, respectively. According to the Chinese test standard "Standard Test Methods of Bitumen and Bituminous Mixtures for Highway Engineering (JTG E20-2011) [38]", two specimens are measured for the softening point, and three specimens are used for the 5 °C ductility.

A Nicolet 5700 Fourier transform infrared spectrometer (Waltham, MA, USA) and an LW100 FT/B fluorescence microscope (Cossim, Beijing, China) were employed to analyse the microstructure of the asphalt, investigate the chemical functional groups present in aged and recycled asphalt, and evaluate the size and dispersion of the SBS. The infrared spectrum of the asphalt material ranged from 400 cm^{-1} to 4000 cm^{-1}, with a resolution of 0.4 cm^{-1}. The fluorescence microscope provided a magnification of 400×.

3. Pavement Properties

3.1. Pavement Properties After Ageing

The softening point and ductility of SBS-modified asphalt under different ageing conditions were tested, and the results are illustrated in Table 6 and Figure 3.

Table 6. The performance of SBS-modified asphalt pavement after ageing.

Ageing Conditions	Softening Point (°C)	Ductility (cm)
Beijing	65.9	20.9
Guangzhou	67.7	16.5
Chengdu	66.4	19.8
Thermal-oxidative ageing	62.4	43.9

Under coupled ageing conditions, the performance of the SBS-modified asphalt undergoes significant changes. As illustrated in Figure 3, the softening point and ductility of the SBS-modified asphalt exhibited varying degradation patterns under three distinct environmental conditions: Guangzhou, Chengdu, and Beijing. Thermal-oxidative ageing results in a 24.36% reduction in the softening point and a 28.50% decrease in ductility, indicating a significant deterioration in high-temperature performance due to the coupled effects of thermal-oxidative ageing. Following thermal-ultraviolet-humidness coupling ageing, the softening points under the BEC, HEC, and SHEC increased by 5.61%, 8.49%, and 6.41%, respectively, compared to those observed after thermal-oxidative ageing. In contrast, the ductility values decreased by 52.39%, 62.41%, and 54.90% under the BEC, HEC, and SHEC, respectively. The degree of ageing among the three cities can be ranked as follows: HEC > SHEC > BEC. Specifically, compared to BEC in Beijing, the HEC in Guangzhou exhibited a 2.73% increase in the softening point, but a 21.05% greater reduction in ductility. Similarly, the SHEC in Chengdu showed a 0.76% increase in softening point, accompanied by a 5.26% decrease in ductility. This phenomenon can be attributed to variations in the environmental parameters. The humidness and ultraviolet radiation levels in the HEC are 2.55 times and 0.72 times higher, respectively, than those in the BEC, with a temperature that is 5 °C higher. The SHEC in Chengdu exhibits humidness and ultraviolet radiation levels that are 2.0 times and 0.59 times higher, respectively, than those recorded at BEC under comparable temperature conditions. Notably, although the BEC in Beijing experiences the most intense ultraviolet radiation, the severity of ageing is lower than that observed under HEC. This finding that asphalt ageing results from the synergistic interactions among thermal, ultraviolet, and humidness factors. Moreover, the coupled high-temperature and high-humidness conditions in the HEC environment induce significantly more severe ageing effects compared to the environmental conditions in BEC. This observation further confirms that ultraviolet radiation does not play a dominant role in the ageing process of SBS-modified asphalt. Furthermore, although the high-temperature and humidness environment conditions in Chengdu present lower ultraviolet conditions compared to the benchmark conditions in Beijing, their increased humidness suggests that the coupling of humidness and ultraviolet radiation amplifies the synergistic ageing mechanism involving thermal, ultraviolet exposure, and humidness.

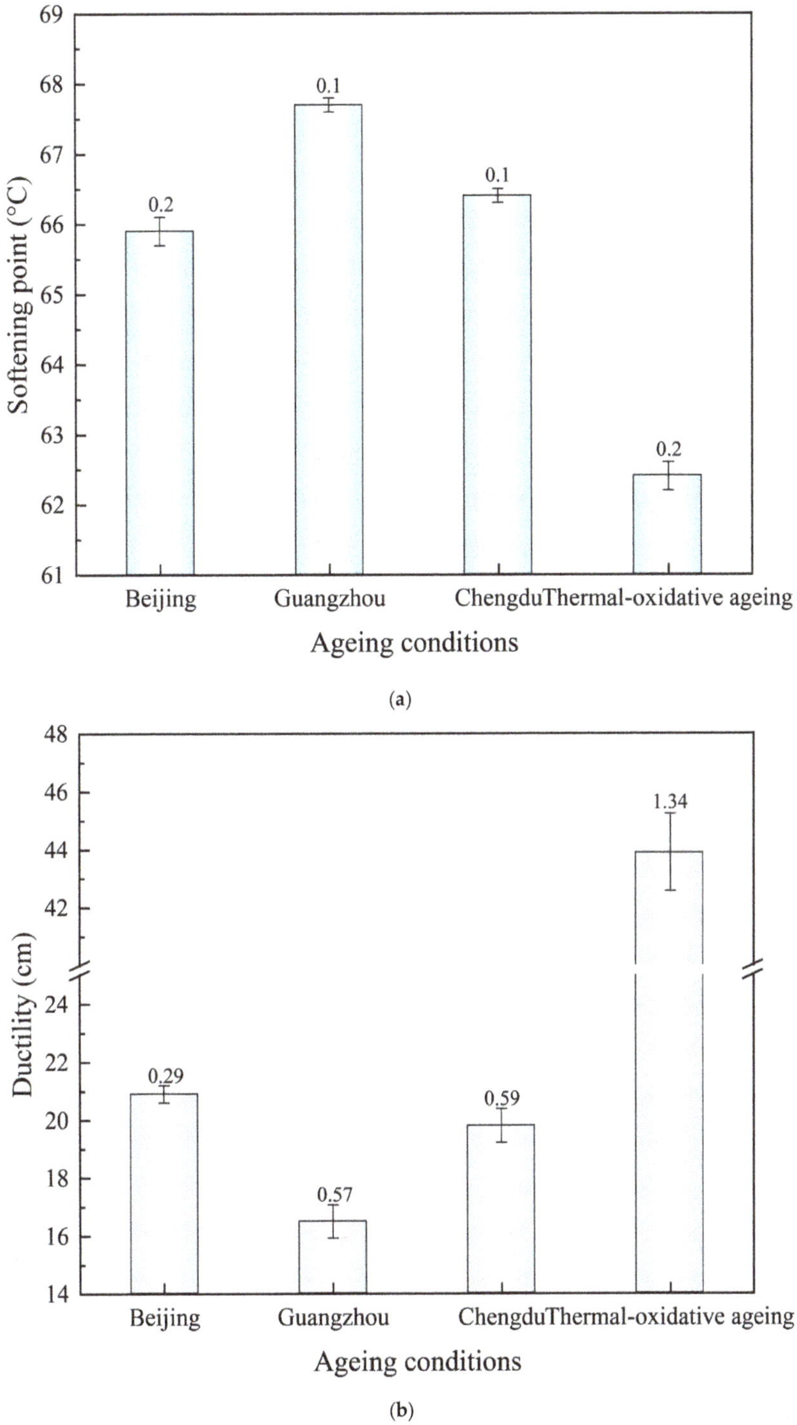

Figure 3. The performance of SBS-modified asphalt pavement after ageing. (**a**) Softening point. (**b**) Ductility.

3.2. Pavement Properties After Regeneration

Softening point and ductility tests were conducted on recycled asphalt with varying additive contents of 0%, 3%, 6%, and 9%. The results are presented in Table 7 and Figure 4.

Table 7. Road performance of SBS-modified asphalt at varying modifier contents.

Ageing Conditions	Softening Point (°C)			Ductility (cm)		
	3%	6%	9%	3%	6%	9%
Beijing	61.4	57.3	54.8	42.3	57.3	63.1
Guangzhou	63.9	60.0	57.3	36.1	48.3	54.6
Chengdu	62.3	58.5	55.8	40.7	54.9	61.7
Thermal-oxidative ageing	57.4	53.5	51.3	58.1	71.4	81.5

As illustrated in Figure 4, the softening point of the SBS-modified asphalt decreases notably with increasing concentrations of industrial animal oil, while its ductility is significant enhanced. This observation indicates that the high-temperature performance of the SBS-modified asphalt does not improve during the regeneration process; however, its stress-bearing capacity under low-temperature conditions is markedly enhanced. When 3% industrial animal oil was added, the ductility of the BEC-, HEC-, SHEC-aged, and thermal-oxidative SBS-modified asphalts increased by 102.39%, 118.79%, 105.55%, and 32.35%, respectively, compared to the aged SBS- modified asphalt. Conversely, the softening points decrease by 6.82%, 5.61%, 6.17%, and 8.01%, respectively. The three types of coupled aged asphalt exhibited significant improvements in ductility, with the thermal-oxidative aged asphalt showing the smallest increase in ductility. In terms of the reduction in the softening point, the thermal-oxidative aged asphalt experienced the most pronounced decrease, whereas HEC-aged asphalt exhibited the least reduction. These findings indicate that the addition of a low content of industrial animal oil effectively replenishes the components lost by the asphalt during coupled ageing, thereby significantly enhancing its low-temperature performance. When the industrial animal oil content reached 6%, the ductility of the thermal-oxidative aged asphalt was 1.16 times greater than that of the unaged asphalt. In this scenario, the ductility of the BEC and SHEC-aged asphalts recovered to 93.32% and 89.41% of their unaged states, respectively, while the HEC-aged asphalts recovered only 78.66% of its unaged state. When the industrial animal oil content was further increased to 9%, the ductility of the BEC and SHEC-aged asphalt surpassed that of the unaged asphalt, whereas the HEC-aged asphalt recovered only 88.93% of the ductility of the unaged asphalt. Concurrently, the softening points were reduced to 66.42%, 67.64%, and 69.45% of the unaged asphalt. The recovery performance of the HEC-aged indicates that thermal-ultraviolet-humidness coupling effects under high-temperature and high-humidness conditions induce significant ageing damage, leading to significant degradation in low-temperature performance, which is challenging to restore. In contrast, the decline in the high-temperature performance was partially mitigated by the increased hardness of the asphalt.

In order to further investigate the regenerative effects of industrial animal oil at varying contents, we calculated the changes in softening point and ductility under different ageing conditions. The results are presented in Figure 5.

As illustrated in Figure 5, the addition of industrial animal oil to aged asphalt results in both increases and decreases in the softening point and ductility. Initially, these changes increase and then decrease, with the maximum change occurring at an industrial animal oil content of approximately 4%. This indicates that the restorative effect of industrial animal oil on aged asphalt initially enhances its properties before gradually diminishing. The addition of industrial animal oil can supplement essential components in asphalt, thereby improving the performance of aged asphalt and facilitating its restoration. However, as the integration of industrial animal oil approaches saturation, the restorative effect also declines. Notably, as the content of industrial animal oil increases, the absolute value of the softening point shows a downward trend, while the absolute value of ductility exhibits an upward trend.

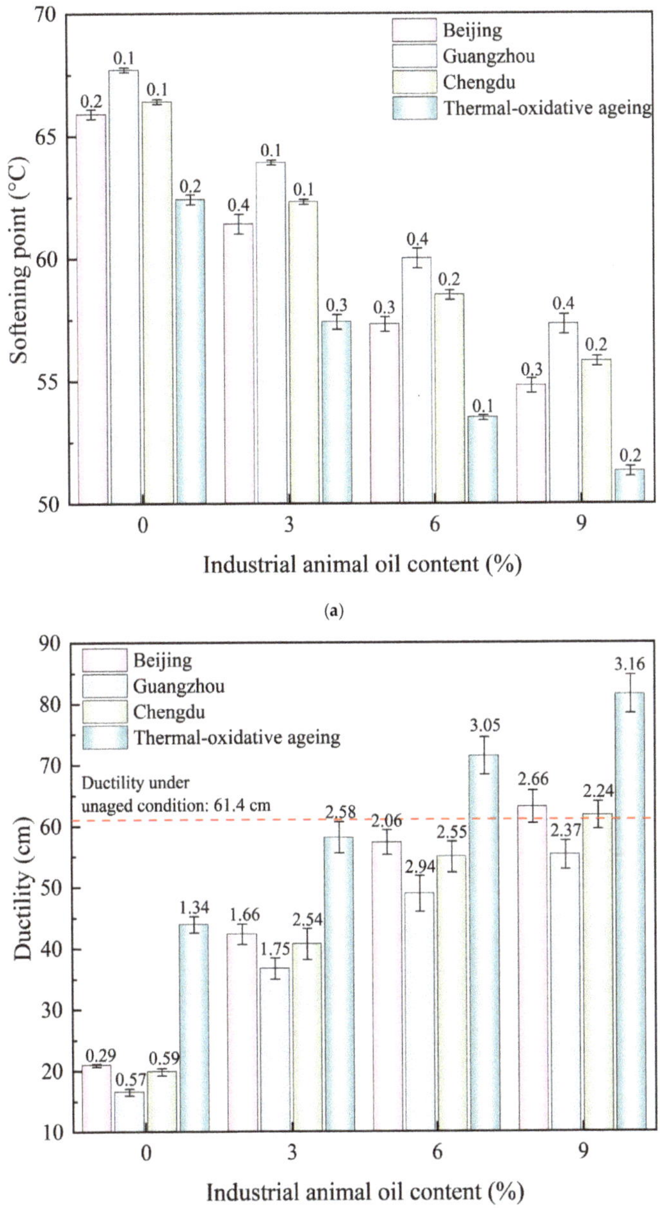

Figure 4. Road performance of SBS-modified asphalt at varying modifier contents. (**a**) Softening point. (**b**) Ductility.

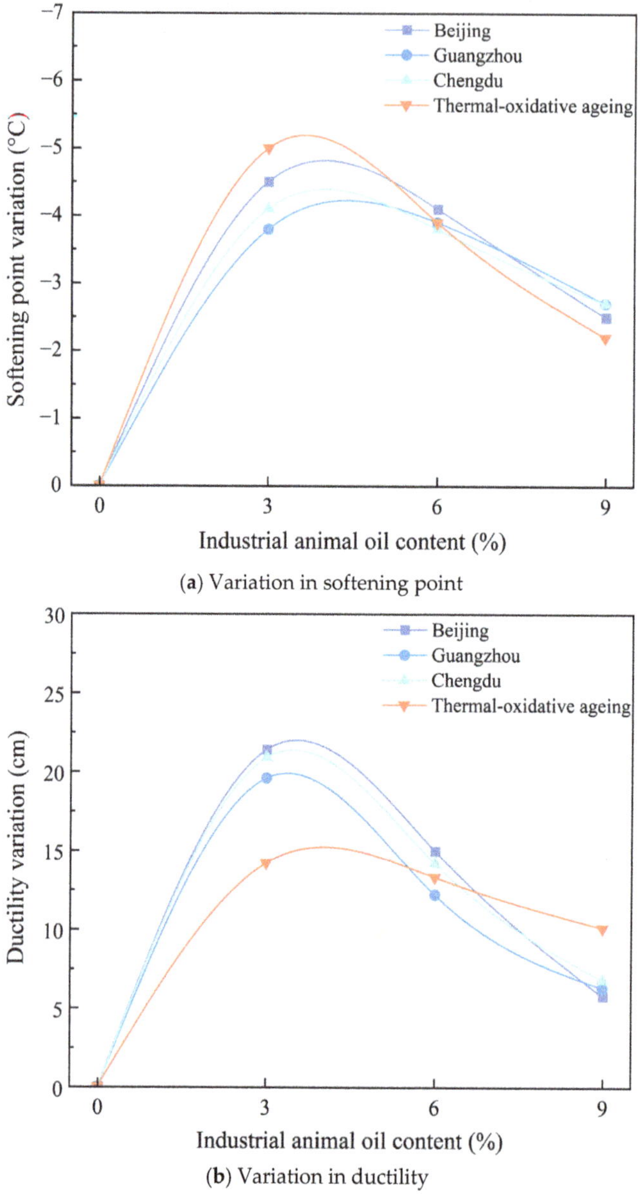

Figure 5. Variation in softening point (**a**) and ductility (**b**) with varying amounts of industrial animal oil.

In addition, as illustrated in Figure 5a, when the content of industrial animal oil is less than 6%, the reduction in the softening point of asphalt under thermal-oxidative ageing conditions is greatest, while the reduction observed in Guangzhou is the smallest. However, as the content of industrial animal oil continues to increase, the softening point of aged asphalt under thermal-oxidative ageing conditions decreases more rapidly, falling below the levels observed under coupled ageing conditions. As illustrated in Figure 5b, when the content of industrial animal oil is less than 6%, the increase in the ductility of asphalt under thermal-oxidative ageing conditions is minimal and significantly lower than that of asphalt subjected to coupled ageing conditions. However, as the content of industrial animal oil continues to increase, the ductility of aged asphalt under thermal-oxidative ageing conditions improves more significantly than under coupled ageing conditions. This indicates that, when the content of industrial animal oil reaches a certain level, it becomes easier for the asphalt to reach saturation under coupled ageing conditions.

4. Ageing and Regeneration Mechanisms

4.1. Functional Group

Figure 6 illustrates the evolution of functional groups in SBS-modified asphalt under various ageing conditions, as analysed by FTIR spectroscopy.

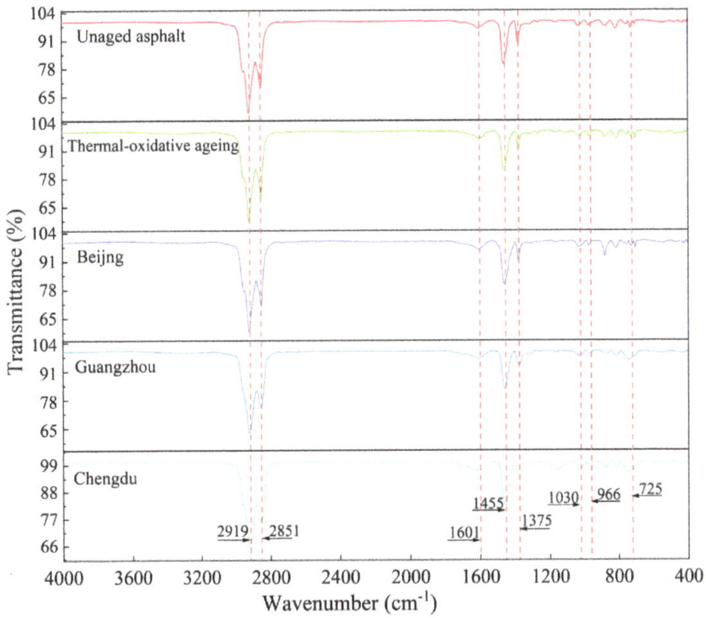

Figure 6. Functional groups of SBS-modified asphalt under different ageing conditions.

As illustrated in Figure 6, distinct absorption peaks are evident in the FTIR spectra of asphalt subjected to various ageing conditions at 2919 cm^{-1}, 2851 cm^{-1}, 1601 cm^{-1}, 1455 cm^{-1}, 1375 cm^{-1}, 1030 cm^{-1}, 966 cm^{-1}, and 725 cm^{-1}. The strong peaks at 2919 cm^{-1} and 2851 cm^{-1} are attributed to the asymmetric and symmetric stretching vibrations of C-H bonds in methylene ($-CH_2-$) groups, respectively. Additionally, the absorption peaks detected at 1601 cm^{-1}, 1455 cm^{-1}, 1375 cm^{-1}, and 1030 cm^{-1} are assigned to the stretching vibrations of C=C bonds with in condensed aromatic rings, the asymmetric bending vibrations of methyl ($-CH_3$) groups, the symmetric bending vibrations of methyl ($-CH_3$) groups, and the stretching vibrations of sulfoxide (S=O) functional groups, respectively. Notably, the peaks at 966 cm^{-1} and 725 cm^{-1} are attributed to the bending vibrations of C=C bonds in polybutadiene and the wagging vibrations of methylene ($-CH_2-$) groups in the aromatic rings of polystyrene, respectively, serving as characteristic absorption peaks for SBS modifiers. The intensity of the absorption peaks varied, while their positions remained relatively stable. This indicates that the types of functional groups in the asphalt did not change, only their concentrations did. Under different ageing conditions, the decrease in intensity at 966 cm^{-1} indicated the scission or crosslinking of butadiene within the asphalt. Meanwhile, changes in peak intensity at 1030 cm^{-1} and 1600 cm^{-1} indicate an increased degree of oxidation and enhanced aromatic conjugation, respectively. Previous studies have established a correlation between the areas of these absorption peaks and the concentrations of the respective functional groups [40]. Given the relatively small absorption peak area of carbonyl groups, which can introduce significant measurement errors, this study employs three indices to characterise the ageing and regeneration mechanisms of SBS-modified asphalt: the Butadiene Index ($BI = A_{966}/A_{1375}$), which quantifies SBS content; the Sulfoxide Index ($SI = A_{1030}/A_{1375}$), which measures S=O groups; and the Aromaticity Index ($CI = A_{1600}/A_{1375}$), which assesses C=C bonds. A_{966}, A_{1030}, A_{1600}, and A_{1375} represent the absorption peak areas at 966 cm^{-1}, 1030 cm^{-1}, 1600 cm^{-1}, and 1375 cm^{-1}, respectively. The calculated values of BI, SI, and CI under various ageing conditions are presented in Table 8 and Figure 7.

Table 8. BI, SI, and CI under various ageing conditions.

Ageing Conditions	Butadiene Index	Sulfoxide Index	Aromaticity Index
Beijing	0.1658	0.5077	1.6991
Guangzhou	0.1396	0.5554	1.8442
Chengdu	0.1559	0.5267	1.7651
Thermal-oxidative ageing	0.2227	0.4261	1.3830
Not aged	0.2952	0.3605	1.2453

As illustrated in Figure 7, the asphalt subjected to various ageing conditions exhibited notable changes in its chemical properties compared to the unaged asphalt. Specifically, the butadiene index (BI) of the aged asphalt decreased significantly, while the sulfoxide index (SI) and aromaticity index (CI) showed increasing trends. These findings suggest a reduction in the concentration of butadiene functional groups in the SBS-modified asphalt during the ageing process, accompanied by an increase in sulfoxide and oxygen-containing groups, as well as aromatic components. This observation highlights the molecular degradation of the SBS polymers and the progression of oxidation reactions during asphalt ageing. Thermal-oxidative ageing resulted in a 24.56% decrease in the BI, an 18.18% increase in the SI, and an 11.06% increase in the CI. These changes provide clear evidence of the pronounced oxidation reactions occurring in asphalt under thermal-oxidative conditions. The substantial decrease in the BI was accompanied by a marked reduction in the softening point, as discussed in Section 3.1. The migration of lighter components during this process is insufficient to mitigate the deterioration of asphalt's high-temperature performance, resulting in a pronounced decline in its high-temperature characteristics. Following thermal-ultraviolet-humidness coupling ageing treatment, BEC, HEC, and SHEC exhibited additional BI reductions of 25.57%, 37.31%, and 30.00%, respectively, compared to thermal-oxidative ageing. Furthermore, increases in SI of 19.17%, 30.35%, and 23.62% were observed, along with increases in cohesion CI of 22.85%, 33.35%, and 27.63%, respectively. The severity of butadiene degradation and the increase in sulfoxide and aromatic groups followed the order: high temperature and humidness environment conditions > slightly high temperature and humidness environment conditions > benchmark environment conditions. This trend aligns with the ageing intensity hierarchy established in Section 3.1. These findings indicate that thermal-ultraviolet-humidness coupling ageing predominantly accelerates the thermal-oxidative degradation of butadiene, leading to the disruption of the cross-linked network structure within the asphalt. Concurrently, this process promotes the oxidation of thioether compounds into sulfoxide groups, enhancing intermolecular interactions and inducing the migration of lighter components, As a result, the hardness of the asphalt increases. Consequently, these chemical transformations lead to elevated softening points and reduced ductility in asphalt performance metrics.

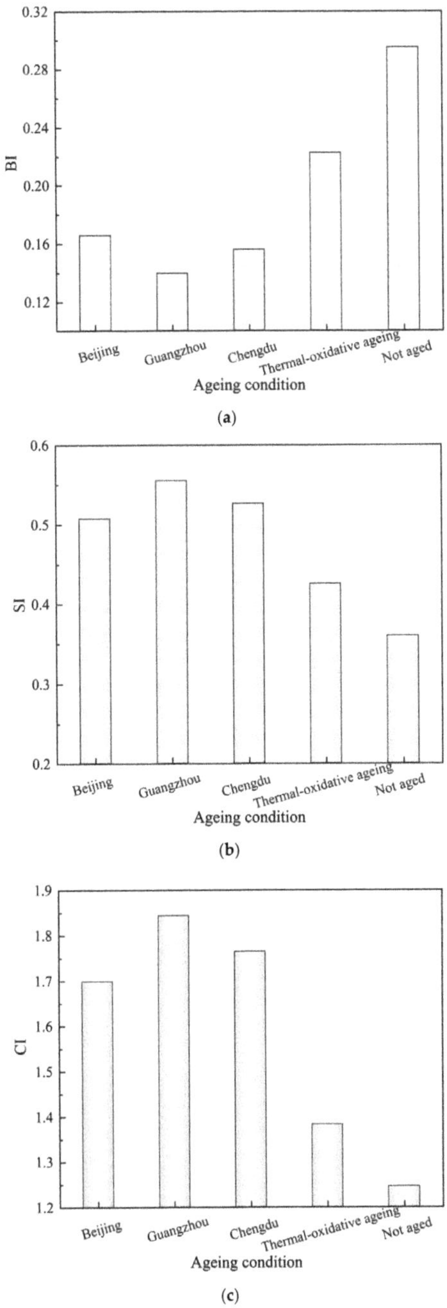

Figure 7. BI, SI, and CI under various ageing conditions. (**a**) BI after ageing. (**b**) SI after ageing. (**c**) CI after ageing.

A comparative analysis between HEC and BEC revealed that HEC exhibited a 15.78% lower BI, 9.39% higher SI, and 8.54% higher CI. These results indicate that the combined effects of Guangzhou have a greater impact on asphalt ageing than Beijing, which are characterised by high ultraviolet radiation. A high-temperature and high-humidness environment accelerates the degradation of butadiene and the oxidation reactions within the asphalt matrix. Compared to the findings of Ma et al. [41], who demonstrated the prompting effects of a coupled humidness and thermal environment on asphalt ageing, the above results indicate that a similarly detrimental effect is also exerted by highly humidness and thermal conditions present in coupled thermal-ultraviolet-humidness ageing. A comparison between SHEC and BEC revealed that, SHEC exhibited a 5.96% decrease in BI, accompanied by increases of 3.74% and 3.89% in SI and CI, respectively. These findings indicate that Chengdu exert stronger coupling effects than Beijing, thereby confirming the role of humidness in accelerating the degradation and oxidation of butadiene. Mechanistically, under high-humidness conditions, the temperature-induced formation of abundant polar groups within the asphalt leads to the formation of hydrogen bonds with water molecules. This interaction facilitates molecular migration within the asphalt and intensifies oxidation reactions. Notably, the ageing of asphalt in Beijing was constrained by the surface hardening of the aged materials, which impeded the penetration of ultraviolet radiation into the deeper layers. However, this constraint had a limited effect on the coupling interactions between temperature and humidness, as liquid water or temperature-induced water vapour could still diffuse into the interior of the asphalt, disrupting its colloidal structures. Collectively, these findings underscore that the coupled influence of temperature and humidness exerts a more substantial effect on butadiene degradation and oxidation than ultraviolet radiation, with elevated humidness levels significantly enhancing the ageing mechanisms.

FTIR analysis of rejuvenated asphalt was conducted to calculate the corresponding BI, SI, and CI indices, as illustrated in Figure 8, Table 9, and Figure 9, respectively.

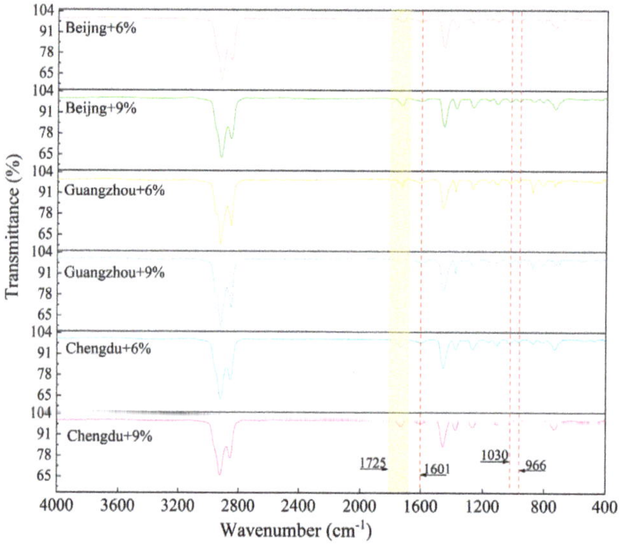

Figure 8. Regenerated functional groups of aged asphalt with 6% and 9% industrial animal oil contents.

Table 9. BI, SI, and CI values for the conditions of Beijing, Guangzhou, and Chengdu with 6% and 9% industrial animal oil contents.

Ageing Conditions	Butadiene Index		Sulfoxide Index		Aromaticity Index	
	6%	9%	6%	9%	6%	9%
Beijing	0.1834	0.1912	0.3103	0.2936	0.6155	0.5347
Guangzhou	0.1503	0.1538	0.3513	0.3386	0.6908	0.6353
Chengdu	0.1714	0.1770	0.3279	0.3122	0.6512	0.5784

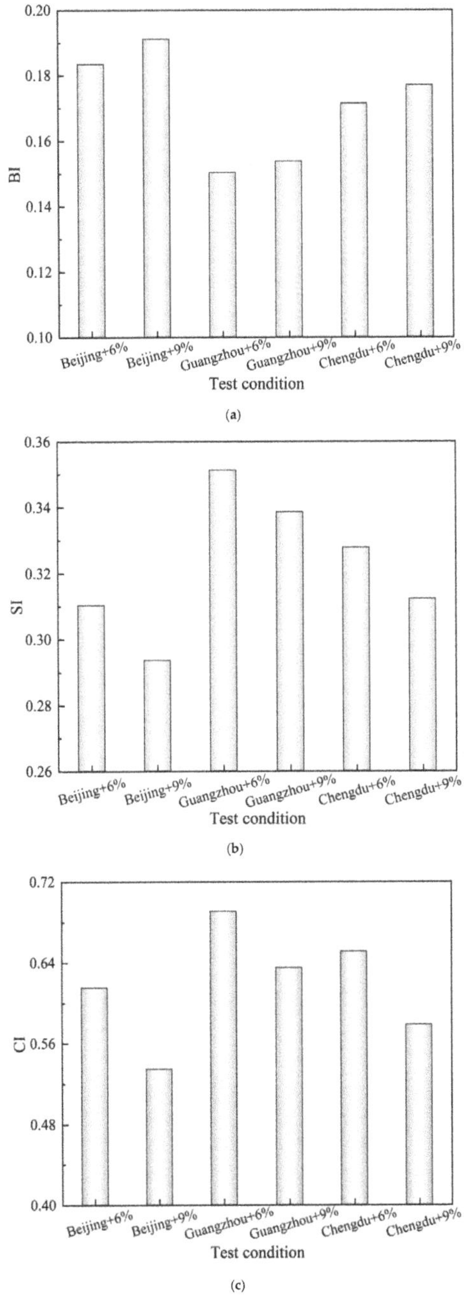

Figure 9. BI, SI, and CI values for samples from Beijing, Guangzhou, and Chengdu with 6% and 9% industrial animal oil contents. (**a**) BI after regeneration. (**b**) SI after regeneration. (**c**) CI after regeneration.

As illustrated in Figure 8, a new peak emerges at 1725 cm^{-1}, indicating chemical interactions between industrial animal oil and aged asphalt. This peak was attributed to the stretching vibration of the carbonyl groups (C=O) present in the fatty acids of industrial animal oils. Figure 9 illustrates that, with increasing industrial animal oil content, all groups exhibited significant upward trends in BI, accompanied by notable reductions in SI and CI, consistent with the observed variations, aligning in pavement properties. This indicates that the incorporation of industrial animal oil not only increases the butadiene content but also decreases the concentrations of both polar and non-polar functional groups, further validating its effectiveness in rejuvenation. Compared to asphalt subjected to coupling ageing, the incorporation of a 6% content of industrial animal oil resulted in increases in the BI of 10.64%, 7.66%, and 9.96% for BEC, HEC, and SHEC, respectively. This modification was accompanied by reductions in the SI by 38.88%, 36.74%, and 37.74%, and decreases in the CI by 63.77%, 62.54%, and 63.11%, respectively. The relatively modest improvements in BI suggest that the restorative effects of industrial animal oil on SBS polymers are limited. In contrast, the significant reductions in SI and CI indicate that the primary rejuvenation mechanism involves the dissolution and dilution of both polar and non-polar molecules generated during the ageing process, along with the replenishment of lighter components lost over time. When the concentration of industrial animal oil was increased to 9%, the BI, SI, and CI values initially exhibited slight increases before subsequently decreasing under all ageing conditions. Although these indices approach the levels observed in the unaged asphalt, they remain below the benchmarks established for unaged materials. This observation suggests that, while higher contents of industrial animal oil can further enhance asphalt performance, the overall efficacy of the rejuvenation process remains limited. Specifically, the lower index levels observed in SHEC and HEC at a 9% content indicate an intensified interaction between humidness and other ageing factors under their respective environmental conditions. The persistent performance gap between regenerated asphalt and unaged asphalt underscores the inherent limitations of industrial animal oil in fully restoring aged asphalt to its original state, particularly in response to complex ageing mechanisms driven by interrelated environmental factors.

4.2. Microscopic Morphology

The distribution characteristics and variations in SBS in the asphalt binders are illustrated in Figures 10 and 11, respectively. These observations were made under different ageing conditions and following industrial animal oil regeneration, using a microscope with 400× magnification.

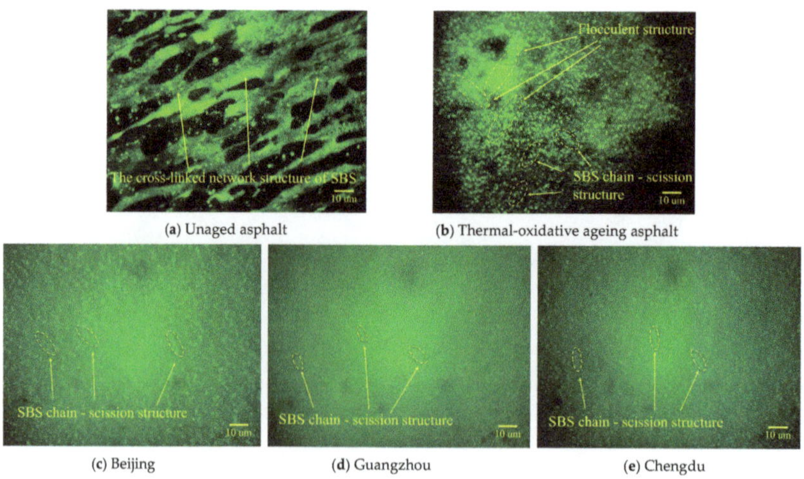

Figure 10. Morphology and distribution of SBS-modifiers in: (**a**) unaged asphalt, (**b**) thermal-oxidative ageing asphalt, and coupling-aged asphalts from (**c**) Beijing, (**d**) Guangzhou, and (**e**) Chengdu.

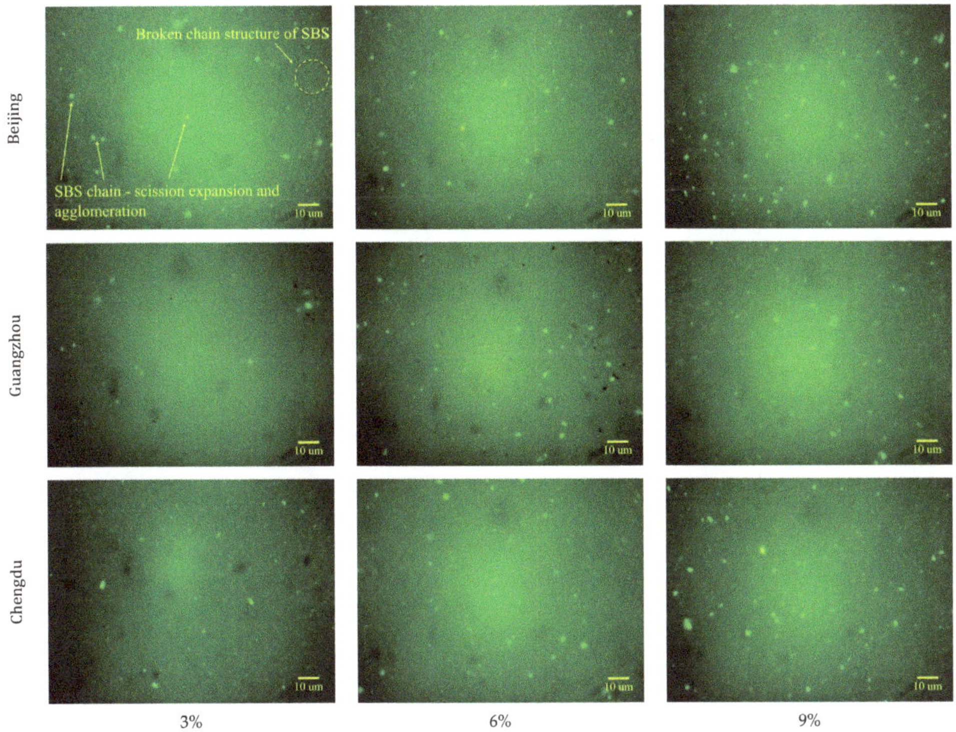

Figure 11. Morphology and distribution of SBS modifiers in samples from Beijing, Guangzhou, and Chengdu under varying contents of industrial animal oil (3%, 6%, and 9%).

As illustrated in Figure 10, unaged asphalt Figure 10a displays a continuous cross-linked network structure of SBS polymers, which substantially enhances the viscosity and cohesion of the asphalt. This structure configuration serves as a fundamental mechanism for maintaining stability and ensuring superior performance. However, following thermal-oxidative ageing Figure 10b, a significant portion of the cross-linked network undergoes thermal-oxidative degradation, resulting in the formation of flocculent and fragmented chain structures. This structural evolution results from two primary mechanisms: (i) the volatilisation of light components (e.g., saturates and aromatics), which reduces the compatibility of asphalt with SBS and weakens structural support, and (ii) the thermal-oxidative degradation of butadiene segments, which disrupts the molecular architecture of SBS. This observation aligns with the decline in the butadiene index reported in Section 4.1. Notably, the substantial reduction in the softening point of thermal-oxidative aged asphalt, as documented in Section 3.1, further underscores the critical importance of cross-linked network integrity in maintaining high-temperature performance.

Under coupled environmental ageing conditions, BEC (Figure 10c), HEC (Figure 10d), and SHEC (Figure 10e) exhibited distinct patterns of structural degradation. As depicted in panels (Figure 10c–e), the aged asphalt exhibit a complete disintegration of the original cross-linked network into smaller, dispersed, and fragmented chains, accompanied by a noticeable reduction in SBS fluorescence intensity. Comparative analysis revealed the following: (i) high-temperature and humidness environment conditions (Guangzhou) resulted in the most dispersed SBS fragments, characterised by the smallest particle dimensions and the weakest fluorescence intensity; (ii) benchmark environment conditions (Beijing) exhibited the largest particle dimensions and the most pronounced fluorescence intensity among the fragmented chains; and (iii) the moderately high-temperature and humidness environment conditions (Chengdu) displayed intermediate characteristics. These findings indicate that high-temperature and humid environments exert the most significant degradation effects on SBS. The humid conditions in the SHEC promote chain scission decomposition, while the SBS degradation caused by the high ultraviolet radiation environment in the BEC is comparatively less pronounced. Under ultraviolet radiation, water undergoes photolysis, generating abundant hydroxyl radicals that act as strong oxidants. These radicals attack the C=C double bonds in the butadiene segments of SBS oxidation reactions. The decomposition process leads to the formation of numerous polar groups, ultimately resulting in the breakdown of the SBS structure. Due to the lower humidness conditions in the BEC, this process was leading to relatively improved sizes and connectivity of SBS chain scission structures. The extent of the aforementioned degradation effects was correlated with the changes in the butadiene index described in Section 4.1. This correlation that the combined effects of high temperature and humidness have a more pronounced impact on the structural degradation of SBS than the effects of increased ultraviolet radiation.

The incorporation of industrial animal oil significantly influenced the microstructural and performance characteristics of aged asphalt. As illustrated in Figure 11, the fragmented SBS chains become miscible with the light components introduced by the industrial animal oil, leading to a gradual attenuation of the fluorescence intensity. Simultaneously, the residual SBS fragments swell and coalesce upon absorbing light components, forming prominent fluorescent light spots. These microstructural transformations were macroscopically manifested as reduced softening points and enhanced ductility, indicating improved material plasticity. With increasing industrial animal oil content, a significant proliferation of fluorescent, enlarged SBS aggregates was observed. This phenomenon reflects two primary mechanisms: (i) the replenishment of light components, which enhances asphalt-SBS compatibility, and (ii) the facilitation of SBS fragment coalescence and swelling. These mechanisms are consistent with the trend of BI elevation documented in Section 4.1.

Comparative analysis of the SBS distribution under various ageing conditions revealed sparse and unevenly distributed fluorescent aggregates under Guangzhou, whereas Beijing produced the most abundant and uniformly distributed fluorescent aggregates. Notably, industrial animal oil was unable to reconstruct fractured SBS molecular chains or restore the cross-linked network architecture. As a result, rejuvenated asphalt exhibits significant performance recovery disparities between high- and low-temperature properties. While low-temperature performance improves, the addition of industrial animal oil fails to compensate for losses in high-temperature performance and, in some cases, even accelerates their decline. These findings underscore the necessity of maintaining strict control over the high-temperature stability of industrial animal oil-rejuvenated asphalt in practical pavement engineering applications to ensure satisfactory road service performance.

5. Conclusions

The ageing and regeneration properties, along with the mechanisms of SBS-modified asphalt under the coupled effects of thermal, ultraviolet, and humid conditions in a hot–wet environment, are investigated in this study. The main conclusions are as follows:

- Thermal-oxidative ageing results in the deterioration of asphalt performance, and the severity of ageing is further amplified under thermal-ultraviolet-humidness coupling conditions due to increased humidness and temperature. This is evidenced by the observation that the extent of ageing under high-temperature and humidness environmental conditions is greater than that observed under slightly high-temperature and humidness conditions, both of which are more severe than those observed under benchmark environmental conditions. A comparison between the experimental conditions of slightly high-temperature and humidness environmental conditions (Chengdu) and the benchmark environmental conditions (Beijing) revealed that humidness intensifies the effects of coupled ageing.

- The types of functional groups present in the SBS-modified asphalt remained unchanged under coupled ageing conditions. However, the intensity of the absorption peak associated with butadiene fracture or crosslinking (966 cm^{-1}) decreased, while the intensities of the absorption peaks corresponding to oxidation (1030 cm^{-1}) and aromatic conjugation (1600 cm^{-1}) increased. Following thermal-oxidative ageing, the butadiene index of the asphalt decreased, whereas the sulfoxide and aromaticity indices increased. This process is accompanied by the breakdown of the cross-linked network, a structural feature critical for maintaining performance retention. Thermal-ultraviolet-humidness coupling ageing intensifies the thermal-oxidative degradation of butadiene, leading to the complete conversion of the cross-linked network into fragmented chain structures. Simultaneously, thermal oxidation induces oxidation reactions in sulphide compounds, resulting in the formation of additional sulfoxide groups that strengthen intermolecular forces. Humidness further amplifies this process, facilitating the transfer of lighter components, ultimately leading to asphalt hardening.
- As the proportion of industrial animal oil increased, the low-temperature performance of the SBS-modified asphalt improved; however, the high-temperature performance did not show a corresponding enhancement. At a 9% concentration, the elongation of the aged asphalt under high-temperature and humidness environmental conditions (Guangzhou) remained below the levels observed in the unaged asphalt. Moreover, the most significant changes in the softening point and ductility occurred at a 4% content, after which the extent of change gradually diminished.
- FTIR analysis confirms the chemical interactions between industrial animal oil and aged asphalt, as evidenced by the appearance of a new carbonyl (C=O) peak at 1725 cm^{-1}. Simultaneously, the butadiene index increased while the sulfoxide index decreased. The primary function of industrial animal oil is to dissolve and dilute the polar molecules generated during the ageing process while replenishing the light components lost over time. The light components introduced by industrial animal oil facilitate the aggregation and expansion of SBS chain fragments; however, they are ineffective in restoring broken crosslinked chain structure of SBS.
- Under benchmark environmental conditions (Beijing), the dissolution and dilution effects on the oxidation products in asphalt are most pronounced, resulting in a noticeable recovery of broken chains. However, as the degree of oxidation in asphalt increases under slightly high-temperature and humidness environmental conditions (Chengdu) and high-temperature and humidness environmental conditions (Guangzhou), the ability of industrial animal oil to mitigate oxidation products is diminished, further exacerbating SBS chain breakage.

Author Contributions: Conceptualization, S.Z. and J.R.; methodology, D.W. and J.R.; validation, J.R.; formal analysis, J.R.; investigation, S.Z., D.W., L.W. and A.M.; resources, L.W.; data curation, S.Z., A.M., Z.W. and J.R.; writing—original draft preparation, S.Z., D.W., L.W. and J.R.; writing—review and editing, H.Z. and J.R.; visualisation, A.M., Z.W. and J.R.; supervision, A.M. and J.R.; project administration, D.W., L.W., H.Z. and J.R.; funding acquisition, H.Z. and J.R. All authors have read and agreed to the published version of the manuscript.

Funding: This study is sponsored in part by the National Natural Science Foundation of China under grant 51808326 and 42377174, and the Natural Science Foundation of Shandong Province under grant ZR2022ME198, to which the authors are very grateful.

Institutional Review Board Statement: Not applicable.

Informed Consent Statement: Not applicable.

Data Availability Statement: The original contributions presented in this study are included in the article. Further inquiries can be directed to the corresponding author.

Conflicts of Interest: Author Dengfeng Wang was employed by the Zhumadian Transportation Construction Technology Center. Author Liuxing Wu was employed by the Zhumadian Highway Engineering Development Co., Ltd. The remaining authors declare that the research was conducted in the absence of any commercial or financial relationships that could be construed as a potential conflict of interest.

References

1. Zhang, C.; Wang, H.; You, Z.; Gao, J.; Irfan, M. Performance test on styrene-butadiene-styrene (SBS) modified asphalt based on the different evaluation methods. *Appl. Sci.* **2019**, *9*, 467. [CrossRef]
2. Liang, B.; Shi, K.; Niu, Y.; Liu, Z.; Zheng, J. Probing the modification mechanism of and customized processing design for SBS-modified asphalts mediated by potentiometric titration. *Constr. Build. Mater.* **2020**, *234*, 117385. [CrossRef]
3. Qin, Y.; Meng, Y.; Lei, J.; Fu, T.; Xu, R.; Yang, X.; Zhao, Q.; Xi, C. Study on the microscopic characteristics and rheological properties of thermal-oxidative aged and virgin-old recycled asphalts. *Eur. Polym. J.* **2021**, *154*, 110499. [CrossRef]
4. Huang, Y.; Bird, R.; Heidrich, O. Development of a life cycle assessment tool for construction and maintenance of asphalt pavements. *J. Clean. Prod.* **2009**, *17*, 283–296. [CrossRef]
5. Hou, X.D.; Xiao, F.P.; Wang, J.G.; Amirkhanian, S. Identification of asphalt aging characterization by spectrophotometry technique. *Fuel* **2018**, *226*, 230–239. [CrossRef]
6. Zhang, Y.J.; Sun, C.; Peng, X.; Pei, Q.; Zhang, Y.; Kang, A.H.; Li, Y.Q.; Cao, Z.D. Study on the synergistic regeneration effect of tall oil and crumb rubber on aged SBS modified asphalt. *Constr. Build. Mater.* **2024**, *444*, 137870. [CrossRef]
7. Guan, Y.; Zhang, Y.; Sang, T.; Ding, Y.; Yan, Z.; Kang, A. Evaluating impact of thermo-oxidative and ultraviolet aging on performance of hot in-place recycled asphalt mixtures. *Materials* **2024**, *17*, 5813. [CrossRef]
8. Liu, X.; Wu, S.; Liu, G.; Li, L. Effect of ultraviolet aging on rheology and chemistry of LDH-modified bitumen. *Materials* **2015**, *8*, 5238–5249. [CrossRef]
9. Wang, F.; Zhang, L.; Zhang, X.; Li, H.; Wu, S. Aging mechanism and rejuvenating possibility of SBS copolymers in asphalt binders. *Polymers* **2020**, *12*, 92. [CrossRef]
10. Li, R.; Wang, Y.; Zhu, W.; Chen, Y.; Yue, J. Research on the rheological performance of fast-melting SBS-modified asphalt under complex environmental factors. *Coatings* **2024**, *14*, 1241. [CrossRef]
11. Li, L.; Guo, Z.; Ran, L.; Zhang, J. Study on low-temperature cracking performance of asphalt under heat and light together conditions. *Materials* **2020**, *13*, 1541. [CrossRef] [PubMed]
12. Zhu, C.; Li, D.; Zhang, H.; Wang, Z.; Li, J. Synergistic effect of surface modified nano-zinc oxide and organic vermiculite on long-term ultraviolet and thermal-oxidative aging resistance of different asphalt binders. *Constr. Build. Mater.* **2023**, *409*, 133832. [CrossRef]
13. Wang, H.; Guo, Y.; Shen, A.; Yang, X.; Li, P. Effect of nanoclays on moisture susceptibility of SBS-modified asphalt binder. *Adv. Mater. Sci. Eng.* **2020**, *2020*, 2074232. [CrossRef]
14. Wei, H.; Bai, X.; Qian, G.; Wang, F.; Li, Z.; Jin, J.; Zhang, Y. Aging mechanism and properties of SBS modified bitumen under complex environmental conditions. *Materials* **2019**, *12*, 1189. [CrossRef]
15. Zhang, Y.; Wei, H.; Dai, Y. Influence of different aging environments on rheological behavior and structural properties of rubber asphalt. *Materials* **2020**, *13*, 3376. [CrossRef]
16. Yuan, J.; Li, H.; Li, T.; Huang, J.; Sun, W. Evolution of asphalt performance under the coupled aging of ultraviolet, high temperature and water. *Case Stud. Constr. Mater.* **2024**, *20*, e03220. [CrossRef]
17. Ma, X.; Wang, D.J.; Liu, S.; Jiang, J.W.; Kan, J.Q.; Tu, M.Y. Analysis of asphalt microscopic and force curves under water-temperature coupling with AFM. *Case Stud. Constr. Mater.* **2024**, *20*, e03071. [CrossRef]
18. Ma, L.; Varveri, A.; Jing, R.; Erkens, S. Comprehensive review on the transport and reaction of oxygen and moisture towards coupled oxidative ageing and moisture damage of bitumen. *Constr. Build. Mater.* **2021**, *283*, 122632. [CrossRef]
19. Yan, G.; Xu, J.; Zhang, R.; Yan, E.H. Effect of water diffusion and thermal coupling condition on SBS modified asphalts' surface micro properties. *Constr. Build. Mater.* **2020**, *273*, 121758.
20. Zou, Y.; Pang, L.; Xu, S.; Wu, S.; Yuan, M.; Amirkhanian, S.; Xu, H.; Lv, Y.; Gao, X. Investigation of the rheological properties and chemical structure of asphalt under multiple aging conditions of heat, UV and aqueous solution. *Materials* **2022**, *15*, 5711. [CrossRef]

21. Yang, B.; Li, H.; Sun, Y.; Zhang, H.; Liu, J.; Yang, J.; Zhang, X. Chemo-rheological, mechanical, morphology evolution and environmental impact of aged asphalt binder coupling thermal oxidation, ultraviolet radiation and water. *J. Clean. Prod.* **2023**, *388*, 135866. [CrossRef]
22. Sun, Y.; Li, H.; Yang, B.; Han, Y.; Zou, Z. Investigation on rheological properties and aging mechanism of asphalt under multiple environmental conditions. *Constr. Build. Mater.* **2024**, *443*, 137713. [CrossRef]
23. Ren, J.L.; Xu, Y.S.; Zhao, Z.D.; Chen, J.C.; Cheng, Y.Y.; Huang, J.D.; Yang, C.X.; Wang, J. Fatigue prediction of semi-flexible composite mixture based on damage evolution. *Constr. Build. Mater.* **2022**, *318*, 126004. [CrossRef]
24. Ren, J.; Zhang, X.; Zhao, H.; Zhang, L.; Yang, J.; Tang, G.; Chen, J.; Wang, J.; Li, M. Data-driven model for fourier transform infrared spectrum characteristics of the aged modified bio-asphalt binder. *Measurement* **2022**, *202*, 111879. [CrossRef]
25. Pahlavan, F.; Amirul, R.; Deng, S.G.; Peter, L.; Elham, H.F. Investigation of balanced feedstocks of lipids and proteins to synthesize highly effective rejuvenators for oxidized asphalt. *ACS Sustain. Chem. Eng.* **2020**, *8*, 7656–7667. [CrossRef]
26. Ren, J.; Zhang, L.; Zhao, H.; Zhao, Z.; Wang, S. Determination of the fatigue equation for the cement-stabilized cold recycled mixtures with road construction waste materials based on data-driven. *Int. J. Fatigue* **2022**, *158*, 106765. [CrossRef]
27. Deng, Y.; Shi, X.; Zhang, Y. Determination of composite modulus of concrete containing reclaimed asphalt pavement using micromechanical modelling and soft computing techniques. *J. Clean. Prod.* **2021**, *327*, 129486. [CrossRef]
28. Zhang, S.; Cui, Y.; Du, C.; Liu, L.; Chen, Q. Low-temperature performance and micro-structure of warm mix recycled composite aged asphalt. *Constr. Build. Mater.* **2024**, *440*, 137443. [CrossRef]
29. Jing, W.; Ding, S.; Wang, L.; Lu, W.; Ge, D. Performance evaluation of styreneic methyl copolymer regenerated SBS-modified asphalt and its mixture with high content RAP. *Case Stud. Constr. Mater.* **2023**, *19*, e02657. [CrossRef]
30. Zhang, N.T.; Fan, G.P.; Lv, S.T.; He, F.W.; Fan, X.P.; Peng, X.H.; Liu, T.J.; Liu, H.L. The influence of SBS modification on the rheological property of asphalt before and after regeneration. *Constr. Build. Mater.* **2021**, *310*, 125239. [CrossRef]
31. Yao, X.G.; Wang, Y.; Xu, T. Development on recycling, aging simulation and regeneration methods of reclaimed styrene-butadiene-styrene modified asphalt. *J. Clean Prod.* **2021**, *312*, 127767. [CrossRef]
32. Feng, P.; Li, J.; Wang, J.; Wang, H.; Xu, Z. Xu, Effect of bio-oil species on rheological behaviors and gasification characteristics of coal bio-oil slurry fuels. *Processes* **2020**, *8*, 1045. [CrossRef]
33. Gao, J.; Wang, H.; You, Z.; Hasan, M.R.M.; Lei, Y.; Irfan, M. Rheological behavior and sensitivity of wood-derived bio-oil modified asphalt binders. *Appl. Sci.* **2018**, *8*, 919. [CrossRef]
34. Zhang, R.; You, Z.; Ji, J.; Shi, Q.; Suo, Z. A review of characteristics of bio-oils and their utilization as additives of asphalts. *Molecules* **2021**, *26*, 5049. [CrossRef]
35. Yang, T.; Chen, M.; Zhou, X.; Xie, J. Evaluation of thermal-mechanical properties of bio-oil regenerated aged asphalt. *Materials* **2018**, *11*, 2224. [CrossRef]
36. Li, J.; Zhang, F.; Liu, Y.; Muhammad, Y.; Su, Z.; Meng, F.; Chen, X. Preparation and properties of soybean bio-asphalt/SBS modified petroleum asphalt. *Constr. Build. Mater.* **2019**, *201*, 268–277. [CrossRef]
37. *JTG F40-2004*; Technical Specifications for Construction of Highway Asphalt Pavements. Research Institute of Highway Ministry of Transport: Beijing, China, 2004.
38. *JTG E20-2011*; Standard Test Methods of Bitumen and Bituminous Mixtures for Highway Engineering. Ministry of Transport of the People's Public of China: Beijing, China, 2011.
39. Available online: https://data.cma.cn/ (accessed on 1 July 2024).
40. Ren, J.; Zhang, X.; Peng, C.; Wang, Y.; Wang, Y.; Zhao, H.; Xu, X.; Xia, L.; Wang, C.; Li, G.; et al. Short-term aging characteristics and mechanisms of SBS-modified bio-asphalt binder considering time-dependent effect. *Constr. Build. Mater.* **2022**, *352*, 129048. [CrossRef]
41. Ma, X.; Zhang, C.; Li, R.; Zheng, L.; Xu, J. Effect of temperature and humidity environment on the microstructure and physicochemical properties of SBS modified asphalt. *Case Stud. Constr. Mater.* **2024**, *22*, e04192. [CrossRef]

materials

MDPI

Article

The Effect of Ionic Soil Stabilizer on Cement and Cement-Stabilized Iron Tailings Soil: Hydration Difference and Mechanical Properties

Hongtu Li [1,2], Jian Jia [1,2,*], Xiaolei Lu [1,2], Xin Cheng [1,2,*], Jiang Zhu [1,2], Lina Zhang [1,2], Peipei Guo [3] and Gongning Zhai [3]

1 Shandon Provincial Key Laboratory of Preparation and Measurement of Building Materials, University of Jinan, Jinan 250022, China; hongtu-li@foxmail.com (H.L.); mse_luxl@ujn.edu.cn (X.L.); mse_zhuj@ujn.edu.cn (J.Z.); mse_zhangln@ujn.edu.cn (L.Z.)

2 School of Materials Science and Engineering, University of Jinan, Jinan 250022, China

3 Qingdao Highway Development Center, Qingdao 266075, China; jiajianxml@foxmail.com (P.G.); qdgljhc@qd.shandong.cn (G.Z.)

* Correspondence: chm_xuml@ujn.edu.cn (J.J.); chengxin@ujn.edu.cn (X.C.)

Abstract: The ionic soil stabilizer (ISS) can synergistically enhance the mechanical properties and improve the engineering characteristics of iron tailings soil in conjunction with cementitious materials such as cement. In this paper, the influence of ISS on the cement hydration process and the charge repulsion between iron tailings soil particles was studied. By means of Isothermal calorimetry, X-ray diffraction (XRD), Scanning electron microscope (SEM), and Low-field nuclear magnetic resonance microscopic analysis methods such as (LF-NMR), X-ray photoelectron spectroscopy (XPS), Non-evaporable water content and Zeta potential were used to clarify the mechanism of ISS-enhanced cement stabilization of the mechanical properties of iron tailings soil. The results show that in the cement system, ISS weakens the mechanical properties of cement mortar. When ISS content is 1.67%, the 7 d compressive strength of cement mortar decreases by 59.8% compared with the reference group. This retardation arises due to carboxyl in ISS forming complexes with Ca^{2+}, creating a barrier on cement particle surfaces, hindering the hydration reaction of the cement. In the cement-stabilized iron tailings soil system, ISS has a positive modification effect. At 0.33% ISS, compared with the reference group, the maximum dry density of the samples increased by 6.5%, the 7 d unconfined compressive strength increased by 35.3%, and the porosity decreased from 13.58% to 11.85%. This is because ISS reduces the double electric layer structure on the surface of iron tailings soil particles, reduces the electrostatic repulsion between particles, and increases the compactness of cement-stabilized iron tailings soil. In addition, the contact area between cement particles increases, the reaction energy barrier height decreases, the formation of $Ca(COOH)_2$ reduces, and the retarding effect on hydration weakens. Consequently, ISS exerts a beneficial effect on augmenting the mechanical performance of cement-stabilized iron tailings soil.

Keywords: ionic soil stabilizer; cement hydration; cement-stabilized iron tailings soil; mechanical properties

1. Introduction

Traditional calcium-based stabilizers are the most widely used for soil stabilization [1–3]. However, the use of lime and cement to treat soil containing sulfate will lead to an increase

 https://doi.org/10.3390/ma18071444

in volume change, and water stability cannot meet the needs of engineering construction [4]. Additionally, stabilizing soils by cement and lime may cause dry reflective cracks [5].

The ionic soil stabilizer (ISS) represents a non-traditional stabilizer for expansive soils supplied in the form of a concentrated liquid, which is also utilized for the treatment of tailings possessing soil-like properties. It is common practice to dilute it with water in a specific proportion and mix it thoroughly with the soil before compacting the soil on site [6,7]. ISS can effectively mitigate the swelling behavior and increase the unconfined compressive strength (UCS) of expansive soil [8,9]. The synergistic stabilization of ISS and calcium-based stabilizers has been suggested as effective measures to improve soil strength and reduce swelling behavior [10]. ISS is a liquid chemical reagent composed mainly of electrolytes and enzymes, which are composed of common cationic groups such as ammonium, calcium, and magnesium ions, as well as anions including phosphate, carboxyl, chloride, and bromide [11,12]. There are differences in the composition of the ISS depending on the manufacturer. However, most of the ISS currently used to solidify soil works through the encapsulation of clay minerals, exchange of interlayer cations, breakdown of clay minerals with expulsion of water from the double layer, and interlayer expulsion with subsequent moisture entrapment [13,14].

The combination of ISS and cement can be used to stabilize soil/tailings, improving its strength and water stability while reducing reflective cracks [15,16]. This method is effective in compensating for the deficiencies of using a single material for solidification [17]. Chen et al. [18] investigated the hydrophobic ionic soil stabilizer, that is, the synergistic effect of the hydrophobic ionic curing agent and cement on the improvement of loess performance, ISS makes more water participate in the hydration reaction. In addition, ISS ions replace the cations on the surface of the loess particles, promoting flocculation and agglomeration between soil particles, which enhances the soil water stability [6,14,19]. Ma et al. [20] used Portland cement and ISS to treat organic-contaminated soil, which produced a 2.3-fold increase in UCS and a significant improvement in the freeze-thaw resistance of the solidified body. This was attributed to the fact that Ca^{2+} in ISS replaced monovalent cations in the soil to reduce the thickness of the double layer of water and increase the attraction between particles.

Existing studies have preliminarily verified the synergistic effect of ISS and calcium-based stabilizers, but the micro-mechanism of their interaction is still lacking multi-scale characterization. In this regard, this paper systematically studied the effects of ISS on cement hydration and surface charge repulsion of tailings particles by means of various characterization methods. It aimed to clarify the mechanism of ISS improving the mechanical properties of cement-stabilized iron tailings soil, so as to better understand the synergic solidification effect of cement on iron tailings soil and find an effective, environmentally friendly, and economic solution for iron tailings soil improvement.

2. Materials and Experimental Methods

2.1. Materials

2.1.1. ISS

ISS used in the test was purchased from Beijing Oriental Yuhong Sand Powder Technology Co., Ltd. (Beijing, China). In this study, ISS was employed to enhance the mechanical properties of cement-stabilized iron tailings soil. The ISS was synthetically formulated within the laboratory environment. The solution contained a notable quantity of Al^{3+}, Ca^{2+} and Na^{+} ions, and maintained a pH range of 6.5 to 7.0. This ISS exhibited favorable solubility in deionized water and possessed a density of 1.13 g/cm^3. ISS is capable of disrupting the electrical double layer structure composed of Na^{+} and K^{+} ions on the surface of iron tailings soil particles, enabling a more compact arrangement under equivalent compaction energy [15,21]. Consequently, this enhances the mechanical properties of the iron tailings soil.

2.1.2. Cement

The cement used in this experiment was purchased from P.O42.5 cement produced by Jinan Shanshui cement factory, Shandong Province. The basic performance indexes of cement are shown in Table 1. Its chemical composition is determined by X-ray fluorescence (XRF) spectroscopy, as shown in Table 2.

Table 1. Physical properties of cement.

Setting Time (min)		Compressive Strength (MPa)		Flexural Strength (MPa)		Fineness
Initial set	Final set	3 d	28 d	3 d	28 d	1.2
156	282	22.1	44.7	5.2	9.6	

Table 2. Chemical composition of cement.

Chemical Composition	CaO	SiO_2	Al_2O_3	Fe_2O_3	MgO	SO_3
Content (%)	58.67	24.16	6.42	4.03	0.83	2.21

2.1.3. Iron Tailings Soil

In the present investigation, the iron tailings soil is taken from Qingdao City, Shandong Province. According to the JTG 3430-2020 [22] normative guidelines, the iron tailings soil exhibited a plastic limit of 14.6% and a liquid limit of 27.8%. The plasticity index, measured as 13.2, falls within the classification of silty clay as defined in the standard GB/T 50547-2022 [23]. The particle size distribution of the dried iron tailings soil was illustrated in Figure 1, and its specific chemical composition is shown in Table 3.

Figure 1. The particle size distribution of iron tailings soil.

Table 3. Chemical composition of iron tailings soil.

Component	Iron Tailings Soil (%)
SiO_2	51.4
Al_2O_3	13.6
Fe_2O_3	10.2
CaO	6.58
MgO	5.07
SO_3	3.84
K_2O	3.06
Na_2O	2.21
P_2O_5	1.03
TiO_2	0.83
MnO	0.17
Loss	2.64

2.2. Mix Proportion and Sample Preparation

2.2.1. Experiment of ISS on Cement Hydration

Table 4 outlines the mixing ratios of ISS and cement, with a fixed water-to-cement ratio of 0.5. ISS dissolved in mixing water, proportionally to cement mass, at ratios (ISS/cement) of 0.17%, 0.33%, and 1.67%, deviating from standard cement admixture practice. Resultant cement pastes are designated as ISS0, ISS017, ISS033, and ISS167. The cement used is P.O42.5 cement produced by Jinan Shanshui cement factory, and the sand used is standard sand, which meets the requirements of Chinese ISO standard sand and contains three levels of composition. Rough sand is 1.0 mm–2.0 mm, medium sand is 0.50 mm–1.0 mm, fine sand is 0.08 mm–0.50 mm, and the content of SiO_2 is >97%.

Table 4. Mix ratio of ISS and cement.

	Cement	Sand	W/C	ISS (%)
ISS0	450	1350	0.5	0
ISS017	450	1350	0.5	$0.17\%m_{cem}$
ISS033	450	1350	0.5	$0.33\%m_{cem}$
ISS167	450	1350	0.5	$1.67\%m_{cem}$

2.2.2. Experiment of ISS on Cement-Stabilized Iron Tailings Soil

The iron tailings soil was solidified with a cement dosage of 6% (by weight of the iron tailings soil). Water for mixing was substituted with varying concentrations of ISS solution ($0.17\%m_{cem}$, $0.33\%m_{cem}$, and $1.67\%m_{cem}$). Subsequently, Proctor compaction tests were conducted to determine the optimum moisture content and maximum dry density of specimens prepared with different ISS solution concentrations.

After determining the optimum moisture content and maximum dry density for specimens at various ISS concentrations, cylindrical compressive strength specimens with both a diameter and height of 50 mm were prepared according to JTG 3441-2024 [24] for testing their unconfined compressive strength. The concrete composition of cement-stabilized iron tailings soil with different ISS content is shown in Table 5.

Table 5. The composition of different cement-stabilized iron tailings soil.

	Maximum Dry Density (g/cm^3)	Optimum Moisture Content (%)	Sample Volume (cm^3)	ISS (%)
ISS0	2.132	13.8	98.125	0
ISS017	2.210	12.7	98.125	$0.17\%m_{cem}$
ISS033	2.271	12.1	98.125	$0.33\%m_{cem}$
ISS167	2.189	13.1	98.125	$1.67\%m_{cem}$

2.2.3. The Effect of ISS on Cement Hydration Under Pressure Compaction

The cement-stabilized iron tailings soil specimens were all prepared via pressure compaction, which constitutes a significant deviation from the conventional casting method employed for ordinary cement paste. To delve deeper into the influence of ISS on cement hydration during pressure compaction, the cement pastes with 1.67% ISS dosage (relative to the binder material) were employed both casting and pressure compaction (at 30 kN) methods for specimen fabrication. A consistent water-to-cement ratio of 0.26 was adopted to ensure the paste's ability to be vibrated and consolidated after pouring, while also facilitating the molding process under pressure [25]. Subsequently, the hydration products of these specimens were analyzed at various curing times through XPS analysis.

2.3. *Methods*

In order to study the mechanism of ISS in the cement system and the cement-stabilized iron tailings soil system, the effects of ISS on the mechanical and microscopic properties of the two systems were tested, respectively.

2.3.1. Test and Characterization Method of ISS in Cement System

Mechanical Properties

Cement mortar specimens with different contents of ISS were prepared using a water-to-cement ratio of 0.5, and the compressive and flexural strengths of different ages (7 d, 28 d, and 60 d) were measured. The specimens were cured under standard conditions according to GB/T 17671-2021 [26] with 20 ± 2 °C and ≥95% RH. According to the standard, the sample is a prismatic sample of 40 × 40 × 160 mm^3. There are three groups of six samples in each group for testing compressive strength. There are three groups of samples used to test the flexural strength, with three samples in each group. The average value of the strength tests under different ages is taken as the final result. The purpose is to study the effect of ISS on strength development and to relate it to the hydration process.

The broken sample was immersed in isopropanol to suspend the hydration process as the recommended samples with the least damage to the hydration products [27].

Isothermal Calorimetry

The hydration heat of cement paste was measured by isothermal calorimeter at 20 ± 0.02 °C. The water-to-cement ratio was 0.5, and the dosages of ISS were 0.17%, 0.33%, and 1.67% by cement weight. The external stirring method was used.

Putting the ampoule into the calorimetric channel, the inevitable heat fluctuation will occur during the period of time when the measurement begins. To minimize the error, the integral of the accumulated heat was started from 0.75 h [28].

X-Ray Diffraction (XRD)

To examine the impact of ISS on phase composition evolution during cement hydration, specimens of varying hydration durations were analyzed using a D8 Advance X-ray diffractometer manufactured by Bruker Analytical Instruments, Karlsruhe, Germany. Utilizing Cu Kα radiation (λ = 0.154 nm). Samples were fractured at designated curing times and immersed in isopropanol, followed by vacuum drying at 40 °C for 48 h and pulverization to a particle size <20 μm. Based on the Rietveld method, different minerals in cement were quantitatively analyzed. The acquired data were processed with V6 version TOPAS software, employing corundum as an external standard, and phase quantification was achieved through Rietveld refinement calculations.

Scanning Electron Microscope (SEM)

The model GeminiSEM360 field emission scanning electron microscope produced by Carl Zeiss AG in Baden-Wurttemberg, Germany was used in this experiment. The sample used to observe the effect of ISS on the morphology of cement hydration products was taken from cement slurry. After reaching the prescribed curing age, the core was broken and cored. The core sample was soaked in isopropyl alcohol for 72 h to terminate hydration, and then dried in a vacuum drying oven at 40 °C. The sample used to observe the effect of ISS on the morphology of hydration products in cement-solidified iron tailings soil was taken from a cylindrical specimen core with a diameter and height of 50 mm. The remaining steps were the same as above. After the sample was dried, the samples were observed on a field emission scanning electron microscope.

Low-Field Nuclear Magnetic Resonance (NMR)

The low field nuclear magnetic resonance analyzer MacroMR12-110H-GS produced by Niumag Company in Suzhou, China was used in this experiment and it assessed porosity and pore distribution in cement-solidified soil, with/without ISS, under 0.3 T magnetic field, 25 mm coil, and 20 MHz frequency. The sample of the cement system is a $2 \times 2 \times 2$ cm^3 cube test block, and the sample of the cement-stabilized iron tailings soil system is a cylindrical test block with a diameter and height of 2 cm. Before the test, isopropyl alcohol was soaked for 72 h to terminate hydration and was dried at 40 °C for 48 h. High-pressure water saturation was then carried out to test the porosity of the sample.

X-Ray Photoelectron Spectroscopy (XPS)

The test samples were taken from the core of the sample in the respective system, and the powder was then tested. XPS was used to analyze the changes of hydration products of cement with ISS in different states. This was conducted utilizing the ESCALAB Xi+ X-Ray Photoelectron Spectroscopy, manufactured by Thermo Fisher Technologies in Massachusetts, USA. Which employed monochromatic Al Kα radiation (operating at 225 W, with a current of 15 mA and a voltage of 15 kV) in conjunction with low-energy electron flooding for effective charge neutralization and compensation. To mitigate the influence of surface charge effects, the characteristic C1s hydrocarbon peak located at a binding energy of 284.80 eV was employed as a reference for calibrating all subsequent binding energy measurements.

2.3.2. Test and Characterization Method of ISS in Cement-Stabilized Iron Tailings System

Proctor Compaction Test

The Proctor compaction test was conducted in accordance with JTG 3441-2024 [24]. Approximately 400–500 g of the mixture was placed as the first layer into the compaction mold and compacted using a compaction apparatus with 27 blows (the compacted sample occupying approximately one-fifth of the total height of the test mold). After compaction, the surface was roughened using a soil scraper, and the second layer of mixture was added and compacted again. This compaction procedure was repeated five times in total. Upon completion of compaction, the top of the mold was leveled off using a straightedge, and the specimen was demolded, weighed, and the corresponding dry density and moisture content were calculated.

Unconfined Compression Strength (UCS)

The automatic road strength tester is used for testing to ensure that the pressure down speed of the indenter is 1.0 mm/min. The specimen is a cylindrical specimen with a diameter and height of 50 mm. On the last day of the specified curing period, the sample was saturated with water for 24 h, and then the surface water of the sample was wiped dry for testing.

Non-Evaporable Water Content

To elucidate the effect of the ISS environment on cement hydration in solidified iron tailings soil, a rigorous investigation of the non-evaporable water content (W_{ne}) in various samples was undertaken. W_{ne} was determined by the differential weight of specimens dried at 105 °C and 950 °C, accounting for contributions from iron tailings soil and cement [29]. The calculation employed the equation:

$$W_{ne} = \frac{W_1 - W_2}{W_2} - L_c - L_i \quad (1)$$

where W_1 was the weight of the specimen after heating 105 °C (g), W_2 was the weight of the specimen after heating 950 °C (g), L_c was the loss of ignition of fly ash (%), and L_i was the loss of ignition iron tailings soil (%).

Zeta Potential

To examine the transformation in surface charge subsequent to the adsorption of ISS onto the surface of iron tailings soil particles, a Zeta-sizer LAB instrument produced by Malvern Instrument Company in the Worcestershire, UK was employed to quantify the Zeta potential of the soil suspended in ISS solutions. The concentrations of the ISS solutions were precisely calibrated at 0%, 0.17%, 0.33%, and 1.67% (based on the mass of ISS utilized). Prior to the addition of the powdered iron tailings soil, the solutions were subjected to magnetic stirring for 5 min to ensure homogeneity. Subsequently, the liquid-solid mixture was agitated at a 50:3 ratio for 120 min at an ambient temperature of 25 °C, allowing for thorough interaction between the ISS and iron tailings soil particles. Following this equilibration period, the supernatant was isolated and subjected to Zeta potential measurements.

3. Results and Analysis

3.1. Mechanical Properties

3.1.1. Effect of ISS on the Hydration Process and Mechanical Properties of Cement

Isothermal Calorimetry and Degree of Cement

The measured thermal power and cumulative heat curves of cement with different ISS contents are shown in Figure 2, and the cumulative hydration heat at different times is recorded in Table 5. The hydration heat release curve shows that ISS delays the hydration process of cement, and the effect is more obvious with the increase of ISS content. In addition, the maximum measured thermal power of the main hydration peak decreased with the increase of ISS content, and the shoulder of the pastes with ISS became more pronounced, indicating an increased second C_3A reaction, which indicated the formation of second ettringite. However, the cement paste containing 1.67% ISS exhibited a broad exothermic peak, merging the peaks of C_3A and C_3S, and the hydration time was delayed by approximately 25 h.

As shown in Table 6, the cumulative heat release of ISS017 and ISS033 increased in the first hour compared with the blank sample. The reason may be that ISS caused the dissolution of the initial phase in the cement, including the gypsum phase, which promoted the formation of the initial ettringite phase. However, when the ISS content was 1.67%, the cumulative hydration heat decreased rapidly at 1 h.

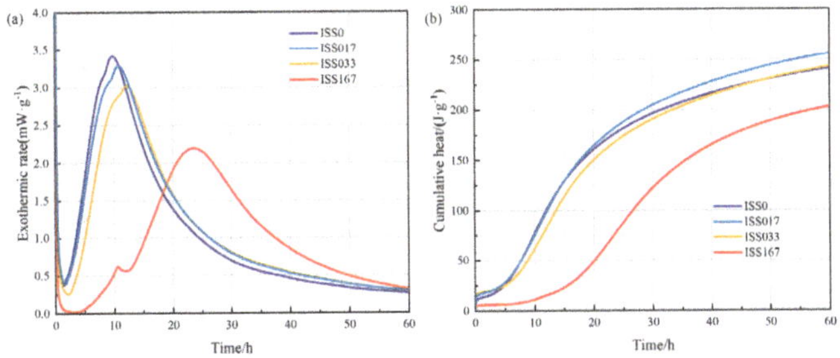

Figure 2. Thermal power and cumulative heat of hydration of cement with 0, 0.17%, 0.33%, and 1.67% ISS (**a**) the relationship between exothermic rate and time and (**b**) the relationship between cumulative heat and time.

Table 6. Cumulative heat of cement pastes with and without ISS (%).

Samples	Cumulative Heat Cement with Different ISS Content (J/g)				
	1 h	10 h	24 h	45 h	60 h
ISS0	13.3	77.4	177.7	224.3	241.3
ISS017	17.3	75.4	183.9	236.7	256.1
ISS033	18.6	62.5	169.8	223.4	243.0
ISS167	5.9	11.6	80.2	178.8	203.1

After 44 h, the cumulative heat measured by 0.17% ISS was slightly higher than that of the reference paste, 0.33% ISS was equivalent to the reference sample, and 1.67% ISS cumulative heat was much lower than the reference.

The Effect of ISS on Mechanical Properties

Figure 3 shows the effect of ISS on the mechanical properties' development of mortar. The compressive and flexural strength of cement mortar was adversely affected by ISS at any curing time. However, the strength of cement mortar increased with the increase of ISS content. The compressive strength of cement mortar at 7 d decreased by 59.8%, 44.1%, and 29.1%, respectively, while the ISS content was 1.67%, 0.33%, and 0.17%, respectively. During the 7 to 28 d curing time, the compressive strength of the samples with ISS exhibited a linear increase, while the strength of ISS0 increased at a slower rate after 28 d. At 7 d, the flexural strength of the specimens with ISS decreased by approximately 17% compared to the blank sample. This suggests that the varying content of ISS had minimal impact on the flexural strength at an early age. However, the flexural strength of the specimens with ISS increased slowly at 28 d.

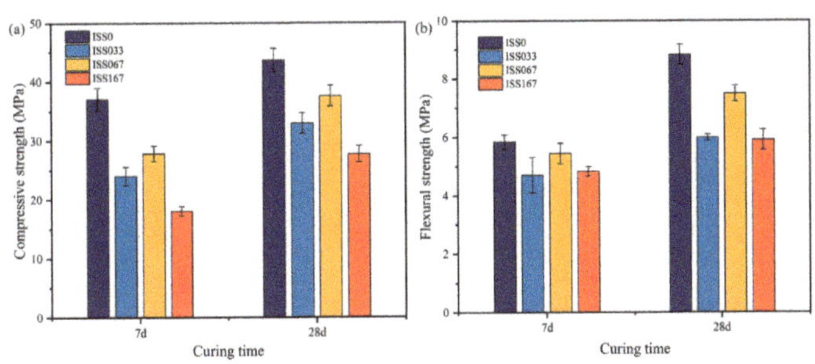

Figure 3. The compressive and flexural strength of ISS cement mortar with different content of ISS cement mortar (**a**) relationship between compressive strength and curing age and (**b**) relationship between flexural strength and curing age.

3.1.2. Effect of ISS on Mechanical Properties of Cement-Stabilized Iron Tailings Soil

Compaction Characteristic

Figure 4 depicts the compaction behavior of cement-stabilized iron tailings soil as a function of varying ISS content. Upon the addition of ISS, a notable increase in the maximum dry density of the samples is observed, accompanied by a corresponding decrease in the optimum moisture content. At an ISS content of 0.33%, the dry density of the specimens attains its maximum value, surpassing that of the control sample by a significant 6.5%. Conversely, when the ISS content is escalated to 1.67%, a discernible downward trend in the dry density is evident, suggesting that an excessive ISS content does not confer any further enhancement in engineering properties. This finding underscores the existence of an optimal ISS dosage within the cement-stabilized iron tailings soil matrix.

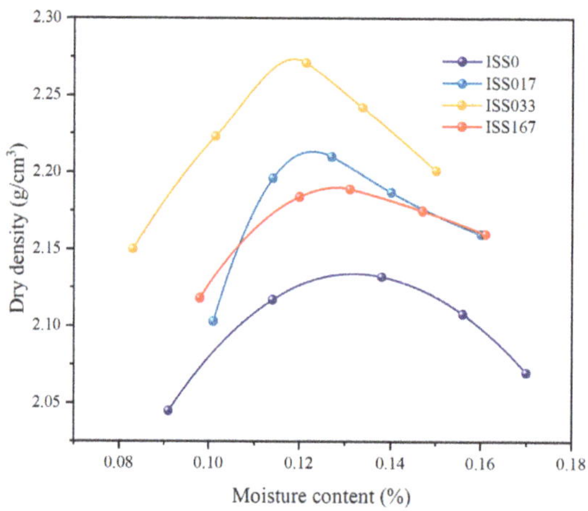

Figure 4. Compaction curve of cement-stabilized iron tailings soil with different ISS content.

UCS

Figure 5 presents the UCS of cement-stabilized iron tailings soil incorporating varying ISS contents. The incorporation of ISS serves to augment the strength characteristics of the cement-stabilized iron tailings soil matrix. When compared to the control specimen, a notable enhancement in UCS is observed, with a 35.3% and 24.1% increase at 7 d and 28 d of curing, respectively, at an ISS content of 0.33%. Furthermore, even at other ISS concentrations, the UCS exhibits a pronounced improvement over the control sample. This phenomenon can be attributed to the ISS's capability to elevate the maximum dry density of the cement-stabilized iron tailings soil. The consequential rise in dry density, indicative of an increase in solid-phase compactness and a corresponding decrease in porosity, thereby underpins the enhancement in mechanical properties.

3.2. *The Pore Distribution of the Hardened Cement Paste*

Figure 6a,b illustrates the pore size distribution and total porosity, respectively, of distinct cement pastes at 7 d. The total porosity of samples incorporated with ISS was consistently higher than that of the control samples. The majority of pores in all specimens were distributed within the 1–100 nm range. With extended curing periods, the pore size distribution of the control samples shifted more towards the 1–20 nm range, whereas the pore size distribution of the ISS-doped samples did not exhibit significant changes. Notably, at an ISS dosage of 0.33%, the total porosity of the specimens was the lowest, and the proportion of detrimental pores was minimized, which correlates with the compressive strength findings. Overall, the incorporation of ISS did not refine the pore structure of the cement paste but rather increased the porosity of the specimens, thereby contributing to the decrease in the strength of the cement paste.

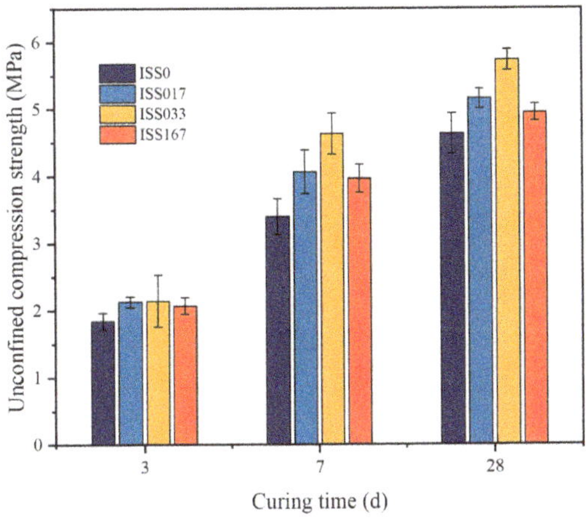

Figure 5. UCS of specimens at different curing times.

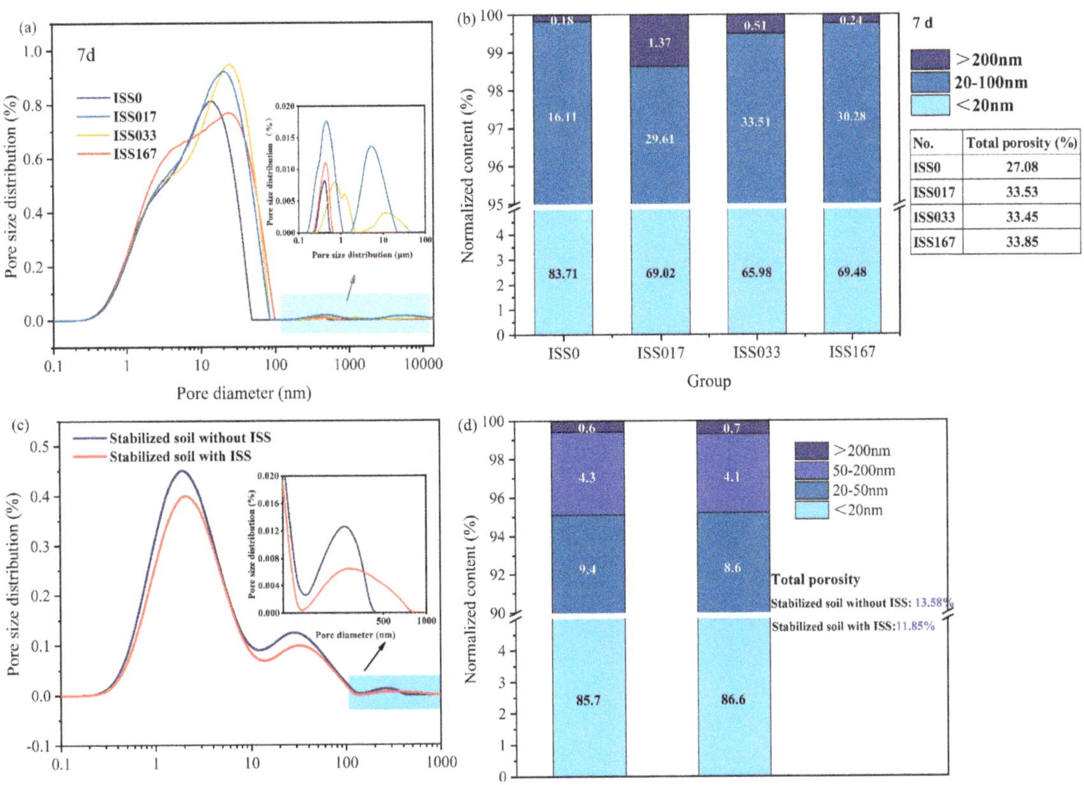

Figure 6. The (**a**) pore distribution and (**b**) total porosity of hardened cement paste; (**c**) pore distribution and (**d**) total porosity of cement-stabilized iron tailings soil.

However, in cement-stabilized iron tailings soil, the samples incorporated with ISS exhibited significantly lower porosity compared to the control samples (Figure 6c,d). Specifically, at an ISS dosage of 0.33%, the porosity was reduced to 11.85%, marking a 13% decrease from the control. This reduced porosity can be attributed to the achievement of higher density and lower water demand in the ISS-modified samples under similar compaction efforts, indicative of a more compact solidified soil structure. Consequently, the incorporation of ISS contributed to the observed lower porosity, which was one of the mechanisms underlying its ability to enhance the performance of cement-stabilized iron tailings soil.

It was evident that ISS exerted an adverse effect on the mechanical properties of cement; paradoxically, it contributed to enhanced strength in cement-stabilized iron tailings soil, partially attributed to the increase in dry density, which was related a reduction in porosity. This raises another question of whether ISS influences cement hydration in the context of cement-stabilized iron tailings systems, and what differences exist in the resulting hydration products. To address this, an analysis of the type and morphology of hydration products is essential.

3.3. *Analysis of Hydration Products*

3.3.1. Mineralogical Analysis

Figure 7 presents the XRD patterns of cement paste incorporating varying concentrations of ISS at distinct curing ages. The incorporation of ISS did not elicit a change in the types of hydration products formed. However, it was observed to delay the emergence of diffraction peaks corresponding to these hydration products. Specifically, at a curing age of 1 day, the sample containing 1.67% ISS exhibited a notably weakened diffraction peak for AFt (Ettringite), with no discernible peak for CH ($Ca(OH)_2$) being detected. Furthermore, the presence of gypsum diffraction peaks suggested that excessive ISS concentrations hindered the dissolution of the gypsum phase. Conversely, at 3 d of curing, samples with ISS concentrations of 0.17% and 0.33% exhibited XRD patterns comparable to those of the control samples, indicating similar phase compositions.

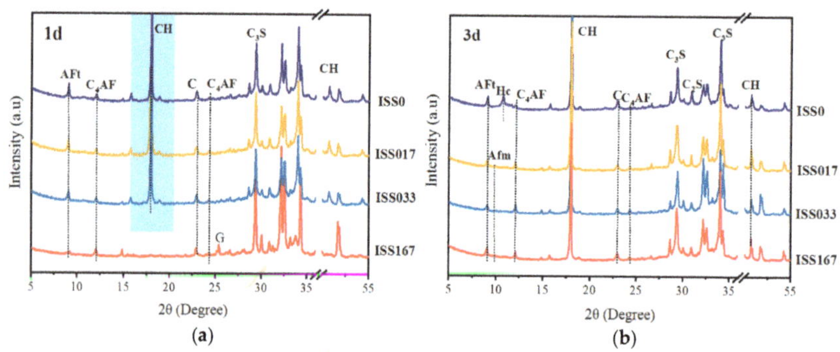

Figure 7. XRD of cement pastes with different ISS content and curing times (**a**) hydration time is 1 d and (**b**) hydration time is 3 d.

However, the XRD patterns of cement-stabilized iron tailings exhibited peaks attributed to ettringite and calcium hydroxide CH at all ages examined (Figure 8), which was notably distinct from the XRD results of cement paste at 1 day. Specifically, under a 1.67% incorporation of ISS, the diffraction peak of CH was barely discernible in the XRD of cement at 1-day age, whereas it was relatively prominent in the XRD of cement-stabilized iron tailings. This observation indicated a mitigated retardation effect of ISS on the early hydration of cement in cement-stabilized soil.

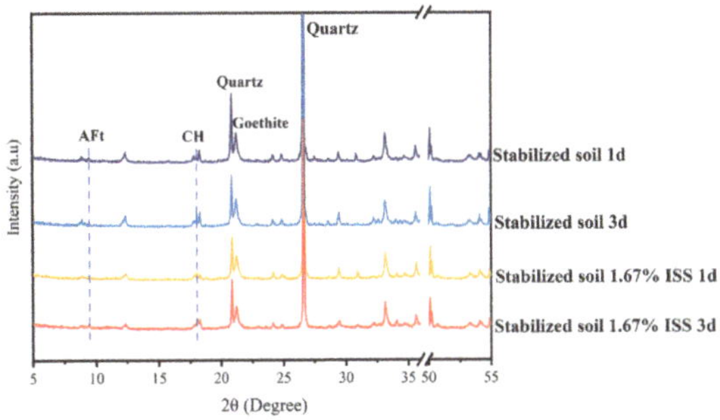

Figure 8. XRD of cement iron stabilized tailings samples with and without ISS.

3.3.2. Morphology Analysis

Figure 9 exhibits 1 d and 3 d SEM micrographs of specimens with varying ISS concentrations. Conspicuous needle-shaped AFt, hexagonal CH platelets, and C-S-H gel are discernible in all samples. At 1 d, the control specimen exhibits extensive AFt formation, enshrouding C-S-H gel and cement particles, whereas AFt precipitation in ISS-incorporated samples is markedly reduced, especially at 1.67% ISS concentration, revealing minimal AFt. Over 3 d, hydration product abundance gradually augments in ISS-containing samples.

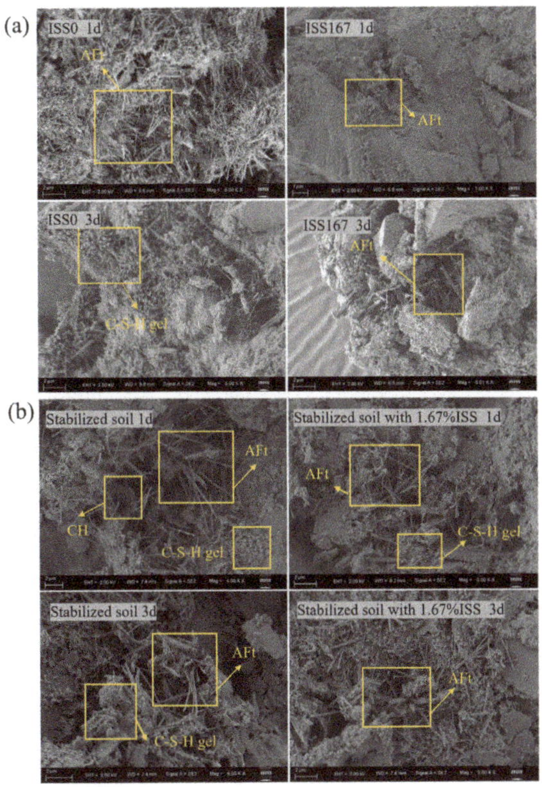

Figure 9. (**a**) SEM images of cement paste with and without ISS; (**b**) SEM images of cement-stabilized iron tailings soil with and without ISS.

However, at both 1 d and 3 d curing times, AFt was observed to intertwine around the iron tailings soil particles in all samples of cement-stabilized iron tailings soil (Figure 9b). In contrast to the cement paste with a dosage of 1.67%, the presence of distinct AFt in the cement-stabilized iron tailings soil at 1 d curing age with an ISS dosage of 1.67% indicated a mitigated early-stage inhibition effect of ISS on hydration products within the iron tailings soil.

ISS exhibits distinct roles in cement paste and cement-stabilized iron tailings soil systems. Consequently, we employed the Rietveld method and non-evaporable water content analysis separately to quantitatively assess the hydration product content in cement paste and cement-stabilized iron tailings soil, respectively.

3.4. *Quantitative Analysis of Hydration Products*

3.4.1. Quantitative Analysis Using the Rietveld Method

Figure 10 presents a quantitative analysis of the degree of mineral hydration and hydration products, as determined through Rietveld refinement, across varying concentrations of ISS. Notably, the consumption rates of C_3S, C_2S, and C_4AF in samples containing ISS were significantly reduced compared to the control sample, indicating that the addition of ISS inhibits the hydration reaction. This inhibitory effect was further corroborated by the observed trends in hydration product formation. Specifically, at 1 day of hydration, the sample containing 1.67% ISS exhibited a negligible quantity of AFt and CH, suggesting a substantial delay in the hydration process. Conversely, the sample with 0.33% ISS demonstrated the highest degree of hydration among all ISS-containing samples, with this trend persisting until the age of 28 d. It is worth noting that in the samples added with ISS, when the ISS content was 0.33%, the hydration products of the specimen were the most, especially the C-S-H gel, which was also reflected in the strength.

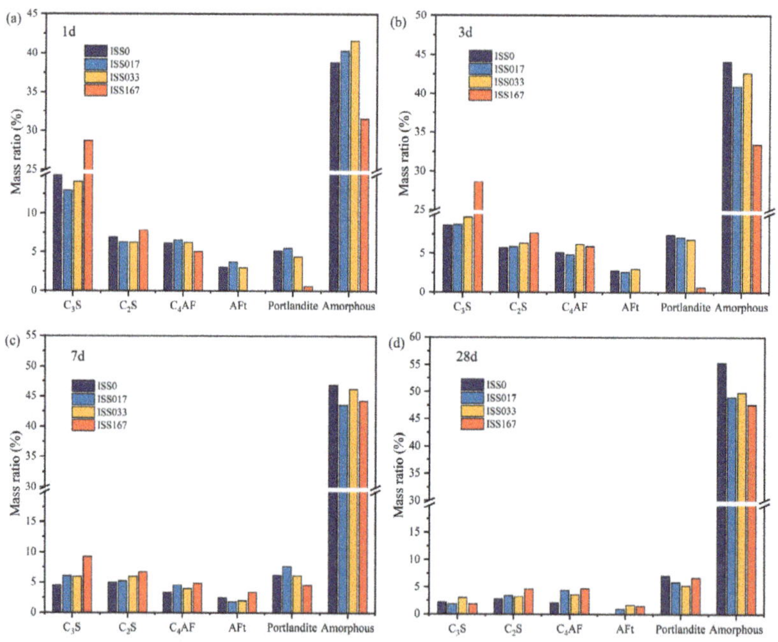

Figure 10. Cement minerals and hydration products in different periods quantified by the Rietveld method (**a**) hydration time is 1 d and (**b**) hydration time is 3 d; (**c**) hydration time is 7 d and (**d**) hydration time is 28 d.

3.4.2. Non-Evaporable Water Content in Cement-Stabilized Iron Tailings Soil

To gain insights into the extent of cement hydration within cement-stabilized iron tailings soil, we conducted an analysis of the non-evaporable water content of these materials. Figure 11 illustrates the variation in non-evaporable water content within cement-stabilized iron tailings soil at 7 d and 28 d curing intervals, contingent upon varying dosages of ISS. A discernible trend emerged, indicating a decline in the non-evaporable water content subsequent to the incorporation of ISS. Specifically, at a 3 d reference point, comparative to the control specimen, the samples with ISS dosages of 0.17%, 0.33%, and 1.67% exhibited respective reductions of 5.1%, 2.5%, and 3.8% in their non-evaporable water content.

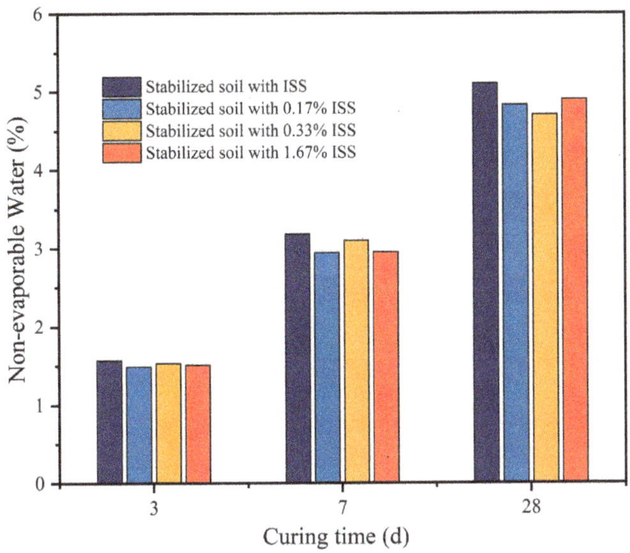

Figure 11. Non-evaporable water content of samples.

The introduction of ISS into cement-stabilized iron tailings soil was observed to attenuate the hydration kinetics of cement, resulting in a weakened hydration process. Notably, even at the highest dosage of 1.67% within the cementitious matrix, ISS failed to contribute appreciably to the overall strength development, highlighting its inability to promote cementitious reactions. Given that the cement-stabilized iron tailings soil was manufactured through a compaction process, characterized by a relatively low water-to-cement ratio, the retarding influence imparted by the admixture on cement hydration became less pronounced.

3.5. The Role of ISS in Cement-Stabilized Iron Tailings Soil

In the cement hydration system, ISS apparently exerts a detrimental effect, leading to a decrease in the mechanical properties of cement and a reduction in the quantity of hydration products. However, in cement-stabilized iron tailings soil, ISS paradoxically enhances its strength. This can be attributed to two aspects: Firstly, ISS increases the dry density of cement-stabilized iron tailings, resulting in improved compactness; secondly, the delaying effect of ISS on cement hydration becomes less pronounced in cement-stabilized iron tailings soil, which may be associated with pressure compaction.

In this section, Zeta potential measurements are utilized to examine the variations in surface charge of iron tailings soil particles, thereby elucidating the influence of ISS on dry density. Moreover, cement specimens were compacted and subjected to XPS analysis to gain a deeper understanding of the attenuation of the delaying effect of ISS on cement hydration under different compaction methods.

3.5.1. Zeta Potential

The implementation of ISS enhances the compaction profile of cement-stabilized iron tailings soil, enabling the specimens to attain a heightened dry density under identical compaction energy inputs. To elucidate the underlying mechanisms responsible for this altered compaction behavior, we conducted a systematic investigation of the Zeta potential on the surface of iron tailings soil particles across varying ISS concentrations. This analysis aimed to characterize the electrostatic repulsion forces operating between the iron tailings soil particles, thereby providing a rationale for the observed modifications in the compaction characteristics [30].

Figure 12 depicts the variation in Zeta potential on the surface of iron tailings soil particles as a function of different ISS dosages. Notably, the application of ISS significantly diminishes the Zeta potential on the surface of iron tailings soil particles. Specifically, at an ISS content of 0.33%, a substantial decrease of 62% in Zeta potential is observed relative to the control sample. However, further increments in ISS content yield diminishing returns in terms of reducing the Zeta potential, suggesting that an optimal adsorption capacity of ISS on the tailings particles had been achieved. This reduction in surface electronegativity of iron tailings soil particles, mediated by ISS, translates into a decrease in repulsive forces between particle surfaces [31]. Consequently, under identical pressure conditions, the particles exhibit a greater propensity for flocculation, ultimately leading to an enhancement in the dry density of the soil.

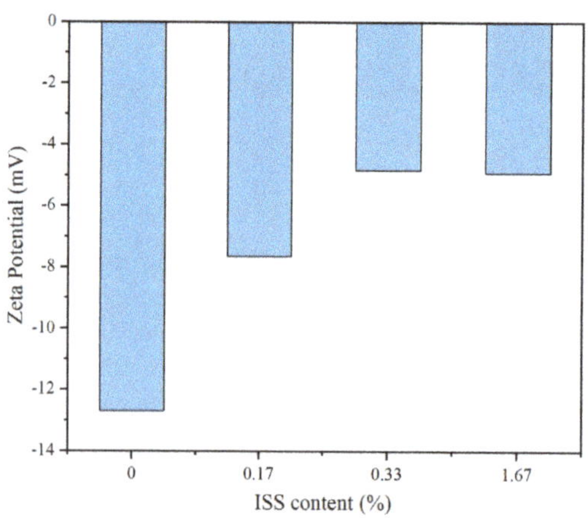

Figure 12. Zeta potential of iron tailings soil particle surface with different ISS content.

3.5.2. Changes of Hydration Products Under Compaction Conditions

Figure 13 presents the XPS spectra of cement samples prepared via casting and pressure-compaction methods. The accuracy of the fitting results was verified by assessing the degree of congruence between the fitted curves and the original spectra. Calcium (Ca) in all samples was identified to exist in three distinct chemical environments, namely, CaO, Ca-OH, and $CaCO_3$, with peak separation and fitting performed for Ca in these varying chemical states [25]. Notably, in samples incorporating industrial starch sludge (ISS), a novel compound, $Ca(COOH)_2$, emerged at 347.4 eV in the spectra, aligning with previous findings [32,33]. This observation reinforces the notion that ISS can form complexes with Ca^{2+} ions in cement, adsorbing onto mineral surfaces and thereby influencing the hydration degree.

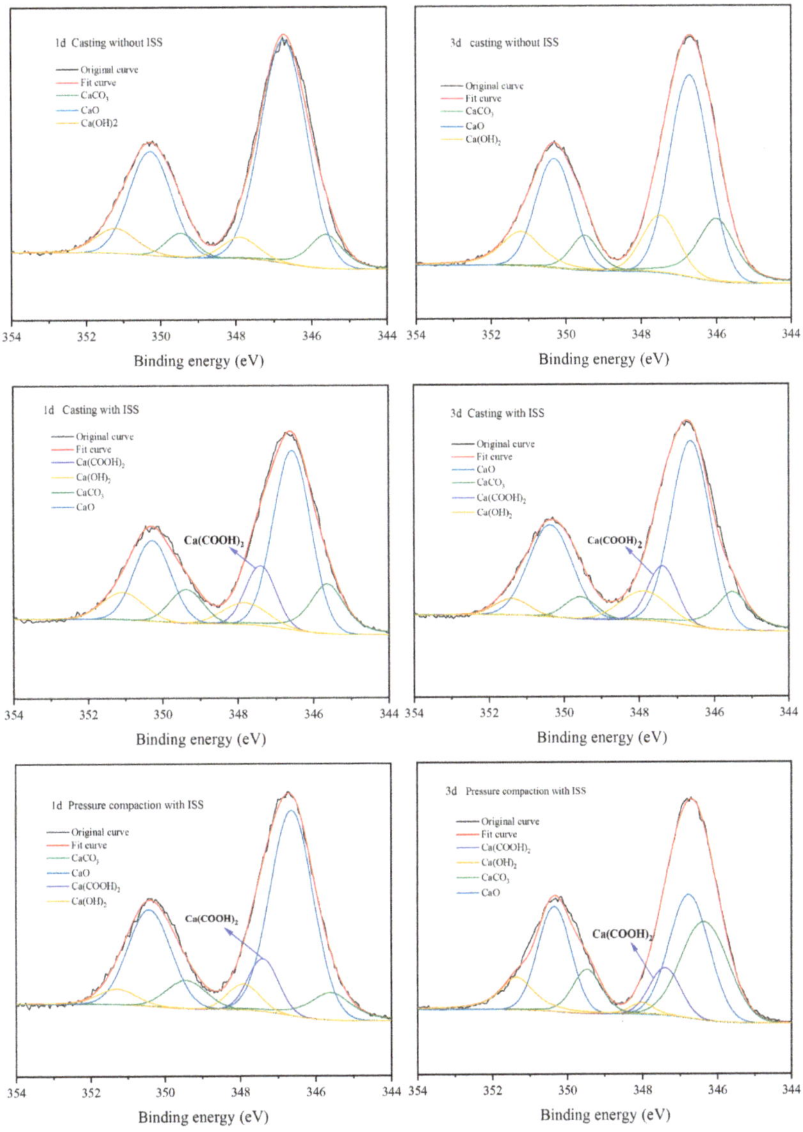

Figure 13. Ca2p peak fitting for different molding methods of cement.

Figure 14 illustrates the proportion of $Ca(COOH)_2$ relative to total calcium under different forming techniques. For cast slurries, the proportions at 1 d and 3 d were 10.6% and 9.6%, respectively, whereas for pressure-compacted samples, these values decreased significantly to 8.5% and 8.2%, respectively. The reduction in $Ca(COOH)_2$ quantity suggests a diminished presence of complexes adsorbed on cement mineral surfaces, subsequently mitigating their impact on cement hydration. Furthermore, within the composite system, cement particles must overcome energy barriers to facilitate hydration reactions. Pressure-compaction enhances interparticle contact area, lowering the energy barrier height and accelerating the reaction rate.

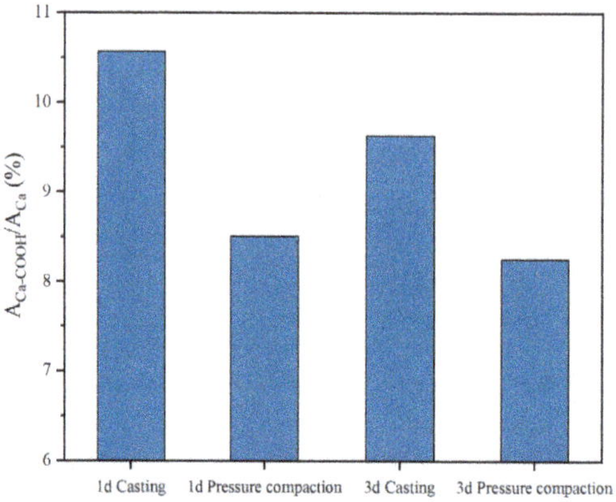

Figure 14. Contents of new phase $Ca(COOH)_2$ in the samples.

3.6. Discussion

3.6.1. The Effect of ISS on the Hydration Properties of Cement

The incorporation of industrial by-product ISS was found to retard the comprehensive progression of cement hydration dynamics, a phenomenon attributed to the presence of carboxyl and hydroxyl functional groups inherent in ISS [34,35]. Specifically, at an ISS concentration of 1.67%, exceeding the conventional range for cement admixtures, the specimens exhibited pronounced retardation in hydration kinetics. Consequently, this elevated dosage led to a notable decrease in the initial heat evolution rate, with the cumulative heat release measured at 1 d hydration time amounting to merely 80.2 J, representing a 54.7% diminution compared to the reference specimen without ISS. Further analysis through XRD revealed the persistence of discernible gypsum phases, suggesting that the hydration process was impeded. This retardation can be rationalized by the excessive adsorption of ISS molecules onto the surface of cement particles, facilitating the formation of an ISS film that obstructs the interfacial contact between water and cement particles, thereby delaying the dissolution of cementitious components. Consequently, at 1 d of hydration, the cement paste containing 1.67% ISS exhibited predominantly weak AFt formation, indicative of a retarded hydration pathway.

At ISS concentrations of 0.17% and 0.33%, the carboxyl group undergoes chemical adsorption onto the active sites of cement particles [36]. This process, facilitated by robust coordination interactions, enables preferential binding of calcium ions present in cement clinker minerals [37]. Consequently, at these ISS concentrations, the cumulative heat release of cement within 24 h surpasses that of the control specimen, corroborated by un-hydrated mineral quantification via Rietveld analysis. Specifically, the consumption of C_3S and C_2S at 24 h is enhanced compared to the control. However, the complexation between ISS and calcium ions, along with potential aluminum ions, forms a covering layer on the cement particle surfaces, which hinders further hydration progression. This phenomenon translates into reduced hydration product formation in cement pastes blended with ISS at 3 d, 7 d, and 28 d, when compared to the control. Furthermore, the porosity of these samples at 7 d and 28 d is notably higher, manifesting macroscopically as a decrease in compressive strength.

The overarching influence of ISS on cement hydration can be concisely encapsulated as follows: In the initial hydration phase, the rapid dissolution of clinker phases, notably gypsum, C_3A, and C_3S, is expedited by the complexation of ISS with calcium and aluminum ions present in the cement matrix. Consequently, there is an elevated consumption of C_3A and C_3S during this initial period. However, the subsequent formation of $Ca(COOH)_2$ complexes on the C_3S surface acts as a hindrance, manifesting as a pronounced inhibitory effect on C_3S hydration within the 3 to 7 d time frame. As cement hydration progresses at a more leisurely pace, the proliferation of hydration products and the consequential expansion of surface area facilitate the desorption of $Ca(COOH)_2$ complexes previously adsorbed around C_3S. Additionally, these hydration products actively consume residual ISS, thereby catalyzing the acceleration of C_3S and C_2S hydration rates.

3.6.2. The Effect of ISS on the Compaction Performance of Cement-Stabilized Soil

Despite its detrimental effects on cement hydration and strength development, ISS exhibits a beneficial role in the context of cement-stabilized iron tailings soil. Specifically, the incorporation of ISS enhances the dry density of the specimen, resulting in an increase in the consolidated bulk density of the stabilized matrix. This, in turn, contributes to a notable improvement in the UCS of the cement-stabilized iron tailings soil.

The surface of iron tailings soil particles is primarily characterized by a double-layer structure dominated by monovalent ions, whereas the introduction of ISS can disrupt this double-layer structure, resulting in a decrease in its thickness [31,38], thereby altering the interparticle interactions. Consequently, the collapse of the electric double-layer structure facilitates the liberation of interlayer water, enabling the achievement of MDD (Maximum dry density) with reduced water content. Compaction experiments confirm a decrease in the OMC (Optimum moisture content) under these conditions.

However, an interesting trend emerges when the ISS content surpasses 0.33%. The dry density of the cement-stabilized iron tailings soil declines, indicating that an excess of ISS on the iron tailings soil surface augments intermolecular repulsive forces. When these repulsive forces surpass the flocculating effect imparted by ISS, the degree of compaction diminishes under a constant compaction effort, as documented in previous works [16,39]. This phenomenon accounts for the observed increase in OMC and the corresponding decrease in MDD beyond 0.33% ISS content. Nevertheless, the overall application of ISS in iron tailings soil base enhancement is advantageous, as it consistently enables the attainment of higher dry densities and lower water contents under equivalent compaction energies. This enhancement is pivotal in optimizing the physical performance indicators of the stabilized soil, thereby underscoring the viability of ISS as a stabilizer in this context.

3.6.3. The Effect of ISS on the Degree of Cement Hydration in Cement-Stabilized Soil

The incorporation of ISS into cement paste has been observed to retard the hydration kinetics of clinker, particularly during the initial stages, resulting in a diminished strength development at an early age (e.g., 3 d) for cement paste containing 1.67% ISS. However, this adverse effect is mitigated in the context of cement-stabilized iron tailings soil. During the assessment of non-evaporable water content, while ISS continues to exert an influence on cement hydration, its impact becomes negligible. This attenuation can be attributed to the unique compaction characteristics of cement-stabilized iron tailings soil, which facilitates the attainment of a dense microstructure that mitigates the disruptive effects of ISS on hydration processes.

The presence of ISS exerts a hindering effect on the hydration of cement, whereby the formation of a complex between ISS and calcium ions encapsulates the surface of C_3S. However, the presence of external pressure alters the amount of complex formation. During the pouring and vibration molding process, the quantity of complexes formed between ISS and Ca^{2+} is significantly higher than that in specimens subjected to compaction molding. This observation implies that in cement-stabilized iron tailings soil, there are fewer complexes covering the surface of cement minerals, thereby reducing their inhibitory effect on cement hydration. Additionally, in the context of compaction molding, the application of external pressure augments the interfacial area between cement particles, thereby reducing the height of the energy barrier and enhancing the reaction rate constant. Consequently, this promotes the hydration of cement, demonstrating that despite ISS's pronounced influence in cement paste, its effects become negligible in cement-stabilized iron tailings soil [25,40].

4. Conclusions

This study investigates the influence of ISS on cement hydration and analyzes its reinforcing effect within cement-stabilized iron tailings soil. The specimens of cement paste and cement-stabilized iron tailings soil were prepared with different contents of ISS. The effects of pores and hydration products on mechanical properties were analyzed, and the mechanism of ISS in cement paste, cement mortar, and cement-stabilized iron tailings soil were also discussed. The results of the subsequent research are summarized as follows:

(1) ISS weakens the mechanical properties of cement. When ISS content is 1.67%, the 7 d compressive strength of cement mortar decreases by 59.8% compared with the reference group. This retardation arises due to carboxyl and hydroxyl moieties in ISS forming complexes with Ca^{2+}, creating a barrier on cement particle surfaces.

(2) ISS reduces the Zeta potential by thinning the double electric layer structure on the surface of iron tailings particles, thereby improving the matrix compactness under the same compaction work and optimizing the mechanical properties of cement-stabilized iron tailings soil. When the ISS content is 0.33%, the 7 d unconfined compressive strength is increased by 35.3% compared with the reference group.

(3) Under the application of pressure, the retardation effect of ISS on cement hydration diminishes. This can be attributed to the reduced reaction between ISS and Ca^{2+} during the pressing and molding process, leading to a decrease in the number of complexes coating cement minerals, which subsequently mitigates the inhibitory effect of ISS on cement hydration.

Author Contributions: Conceptualization, H.L.; Methodology, H.L., J.J., X.L., X.C., J.Z. and L.Z.; Formal analysis, H.L.; Investigation, H.L. and P.G.; Data curation, H.L., P.G. and G.Z.; Writing—original draft preparation, H.L.; Writing—review and editing, X.C.; Supervision, X.C. All authors have read and agreed to the published version of the manuscript.

Funding: This work was supported by the National Natural Science Foundation of China (52341201, 52102020, U22A20126), the Program for Taishan Scholars Program (202211170), Case-by-Case Project for Top Outstanding Talents of Jinan, the 111 Project of International Corporation on Advanced Cement-based Materials (No. D17001), and Young Researchers in Institutes of Higher Education in Shandong.

Institutional Review Board Statement: Not applicable.

Informed Consent Statement: Not applicable.

Data Availability Statement: The original contributions presented in the study are included in the article, further inquiries can be directed to the corresponding author.

Conflicts of Interest: Author Peipei Guo and Gongning Zhai was employed by the company Qingdao Highway Development Center, Qingdao, Shandong. The remaining authors declare that the research was conducted in the absence of any commercial or financial relationships that could be construed as a potential conflict of interest.

References

1. Wang, A.; Zhang, D.; Deng, Y. Lateral response of single piles in cement-improved soil: Numerical and theoretical investigation. *Comput. Geotech.* **2018**, *102*, 164–178. [CrossRef]
2. Qiang, Y.; Chen, Y. Experimental Research on the Mechanical Behavior of Lime-Treated Soil under Different Loading Rates. *Adv. Mater. Sci. Eng.* **2015**, *2015*, 862106. [CrossRef]
3. Jallu, M.; Arulrajah, A.; Saride, S.; Evans, R. Flexural fatigue behavior of fly ash geopolymer stabilized-geogrid reinforced RAP bases. *Constr. Build. Mater.* **2020**, *254*, 119263. [CrossRef]
4. Little, D.N.; Nair, S.; Herbert, B. Addressing Sulfate-Induced Heave in Lime Treated Soils. *J. Geotech. Geoenviron. Eng.* **2010**, *136*, 110–118. [CrossRef]
5. Intaj, F.; Liu, Y.; Wu, Z. Application and Evaluation of Micro-Cracking on Cement-Stabilized Bases at Field Projects in Louisiana. *Transp. Res. Rec.* **2019**, *2673*, 355–364. [CrossRef]
6. Al-Dakheeli, H.; Arefin, S.; Bulut, R.; Little, D. Effectiveness of ionic stabilization in the mitigation of soil volume change behavior. *Transp. Geotech.* **2021**, *29*, 100573. [CrossRef]
7. Katz, L.E.; Rauch, A.F.; Liljestrand, H.M.; Harmon, J.S.; Shaw, K.S.; Albers, H. Mechanisms of Soil Stabilization with Liquid Ionic Stabilizer. *Transp. Res. Rec.* **2001**, *1757*, 50–57. [CrossRef]
8. Arefin, S.; Al-Dakheeli, H.; Bulut, R. Stabilization of expansive soils using ionic stabilizer. *Bull. Eng. Geol. Environ.* **2021**, *80*, 4025–4033. [CrossRef]
9. Gautam, S.; Hoyos, L.R.; He, S.; Prabakar, S.; Yu, X. Chemical Treatment of a Highly Expansive Clay Using a Liquid Ionic Soil Stabilizer. *Geotech. Geol. Eng.* **2020**, *38*, 4981–4993. [CrossRef]
10. Huo, J.; Wang, Z.; Zhang, T.; He, R.; Chen, H. Influences of interaction between cement and ionic paraffin emulsion on cement hydration. *Constr. Build. Mater.* **2021**, *299*, 123951. [CrossRef]
11. Khan, R.A.; Murtaza, M.; Mustafa, A.; Abdulraheem, A.; Mahmoud, M.; Kamal, M.S. Ionic liquids as clay stabilizer additive in fracturing fluid. *Fuel* **2023**, *334*, 126154. [CrossRef]
12. Kaur, G.; Kumar, H.; Singla, M. Diverse applications of ionic liquids: A comprehensive review. *J. Mol. Liq.* **2022**, *351*, 118556. [CrossRef]
13. Petry, T.M.; Das, B. Evaluation of Chemical Modifiers and Stabilizers for Chemically Active Soils—Clays. *Transp. Res. Rec.* **2001**, *1757*, 43–49. [CrossRef]
14. Eren, Ş.; Filiz, M. Comparing the conventional soil stabilization methods to the consolid system used as an alternative admixture matter in Isparta Darıdere material. *Constr. Build. Mater.* **2009**, *23*, 2473–2480. [CrossRef]
15. Soltani, A.; Deng, A.; Taheri, A.; Mirzababaei, M. A sulphonated oil for stabilisation of expansive soils. *Int. J. Pavement Eng.* **2017**, *20*, 1285–1298. [CrossRef]
16. Jia, J.; Lu, X.; Zhu, J.; Wang, J.; Zhang, L.; Cheng, X. Effect of ionic soil stabilizer on the properties of lime and fly ash stabilized iron tailings as pavement base. *Constr. Build. Mater.* **2024**, *450*, 138558. [CrossRef]
17. Zhang, B.; Jiang, W.; Xu, Q.; Yuan, D.; Shan, J.; Lu, R. Experimental feasibility study of ethylene-vinyl acetate copolymer (EVA) as cement-stabilized soil curing agent. *Road Mater. Pavement Des.* **2022**, *23*, 617–638. [CrossRef]
18. Xue, C.; Wang, X.; Lian, B.; Luo, L.; Liu, K. Research on the mechanism of composite improvement of loess based on quantitative analysis of microstructure and mechanical strength. *Constr. Build. Mater.* **2023**, *379*, 131215. [CrossRef]
19. Latifi, N.; Eisazadeh, A.; Marto, A. Strength behavior and microstructural characteristics of tropical laterite soil treated with sodium silicate-based liquid stabilizer. *Environ. Earth Sci.* **2014**, *72*, 91–98. [CrossRef]
20. Ma, F.; Wu, B.; Zhang, Q.; Cui, D.; Liu, Q.; Peng, C.; Li, F.; Gu, Q. An innovative method for the solidification/stabilization of PAHs-contaminated soil using sulfonated oil. *J. Hazard. Mater.* **2018**, *344*, 742–748. [CrossRef]
21. Wu, X.-T.; Qi, Y.; Chen, B. Solidification effect and mechanism of ionic soil stabilizer applied on high-water-content clay. *B. Eng. Geol. Env.* **2021**, *80*, 8583–8595. [CrossRef]
22. Ministry of Transport of the People's Republic of China. *Test Methods of Soils for Highway Engineering*; Ministry of Transport of the People's Republic of China: Beijing, China, 2020.
23. *GB/T 50547-2022*; Technical Standard for Geotechnical Engineering of Tailings Embankmen. Ministry of Housing and Urban-Rural Development, People's Republic of China: Beijing, China, 2022.
24. Ministry of Transport of the People's Republic of China. *Test Methods of Materials Stabilized with Inorganic Binders for Highway Engineering*; Ministry of Transport of the People's Republic of China: Beijing, China, 2024.
25. Wang, B.; Xing, X.; Lin, F.; Wang, S.; Li, F.; Huang, H.; Wei, J.; Yu, Q. Effect of vacuum–vibration–compaction molding on the properties and performances of polymer-modified cementitious material. *Cem. Concr. Compos.* **2024**, *148*, 105450. [CrossRef]
26. *GB/T 17671-2021*; Test Method of Cement Mortar Strength (ISO Method). State Administration for Market Regulation. National Standard of the People's Republic of China: Beijing, China, 2021.
27. Zhang, J.; Scherer, G.W. Comparison of methods for arresting hydration of cement. *Cem. Concr. Res.* **2011**, *41*, 1024–1036. [CrossRef]

28. Yan, Y.; Ouzia, A.; Yu, C.; Liu, J.; Scrivener, K.L. Effect of a novel starch-based temperature rise inhibitor on cement hydration and microstructure development. *Cem. Concr. Res.* **2020**, *129*, 105961. [CrossRef]
29. Ma, Y.; Ye, G. The shrinkage of alkali activated fly ash. *Cem. Concr. Res.* **2015**, *68*, 75–82. [CrossRef]
30. Marchuk, A.; Rengasamy, P.; McNeill, A. Influence of organic matter, clay mineralogy, and pH on the effects of CROSS on soil structure is related to the zeta potential of the dispersed clay. *Soil Res.* **2013**, *51*, 34–40. [CrossRef]
31. Farahani, E.; Emami, H.; Fotovat, A.; Khorassani, R. Effect of different K:Na ratios in soil on dispersive charge, cation exchange and zeta potential. *Eur. J. Soil Sci.* **2019**, *70*, 311–320. [CrossRef]
32. Chen, B.; Wu, K.; Yao, W. Conductivity of carbon fiber reinforced cement-based composites. *Cem. Concr. Compos.* **2004**, *26*, 291–297. [CrossRef]
33. Wang, M.; Wang, R.; Yao, H.; Farhan, S.; Zheng, S.; Wang, Z.; Du, C.; Jiang, H. Research on the mechanism of polymer latex modified cement. *Constr. Build. Mater.* **2016**, *111*, 710–718. [CrossRef]
34. Chaudhari, O.; Biernacki, J.J.; Northrup, S. Effect of carboxylic and hydroxycarboxylic acids on cement hydration: Experimental and molecular modeling study. *J. Mater. Sci.* **2017**, *52*, 13719–13735. [CrossRef]
35. Marchon, D.; Flatt, R.J. 12—Impact of chemical admixtures on cement hydration. In *Science and Technology of Concrete Admixtures*; Aïtcin, P.-C., Flatt, R.J., Eds.; Woodhead Publishing: Sawston, UK, 2016; pp. 279–304.
36. Danner, T.; Justnes, H.; Geiker, M.; Lauten, R.A. Phase changes during the early hydration of Portland cement with Ca-lignosulfonates. *Cem. Concr. Res.* **2015**, *69*, 50–60. [CrossRef]
37. Jolicoeur, C.; Simard, M.-A. Chemical admixture-cement interactions: Phenomenology and physico-chemical concepts. *Cem. Concr. Compos.* **1998**, *20*, 87–101. [CrossRef]
38. Kiventerä, J.; Piekkari, K.; Isteri, V.; Ohenoja, K.; Tanskanen, P.; Illikainen, M. Solidification/stabilization of gold mine tailings using calcium sulfoaluminate-belite cement. *J. Clean. Prod.* **2019**, *239*, 118008.
39. Onyejekwe, S.; Ghataora, G.S. Soil stabilization using proprietary liquid chemical stabilizers: Sulphonated oil and a polymer. *Bull. Eng. Geol. Environ.* **2015**, *74*, 651–665. [CrossRef]
40. Zhu, G.; Luo, Y.; Chen, C.; Shen, K.; Zhang, Y. Effect of metakaolin on the microstructure of cement-based artificial marble slabs incorporating natural marble offcuts under vacuum vibration compaction method. *Constr. Build. Mater.* **2023**, *369*, 130478. [CrossRef]

materials

MDPI

Article

Preparation and Properties of Sulfur-Modified Alite Calcium Sulfoaluminate Cement

Xiaodong Li [1,†], Guodong Kang [1,†], Shang Dou [2], Bing Ma [1], Jin Tang [1], Hao Zhou [1,*], Houhu Zhang [1,*], Jiaqing Wang [3] and Xiaodong Shen [2]

1 Nanjing Institute of Environmental Sciences, Ministry of Ecology and Environment of the People's Republic of China, Nanjing 210042, China; lixiaodong@nies.org (X.L.)
2 State Key Laboratory of Materials-Oriented Chemical Engineering, College of Materials Science and Engineering, Nanjing Tech University, Nanjing 210009, China
3 College of Civil Engineering, Nanjing Forestry University, Nanjing 210037, China
* Correspondence: zhouhao@nies.org (H.Z.); zhh@nies.org (H.Z.)
† These authors contributed equally to this work.

Abstract: Alite calcium sulfoaluminate (ACSA) cement is an innovative and environmentally friendly cement compared to ordinary Portland cement (OPC). The synthesis and hydration of ACSA clinkers doped with gradient sulfur were investigated. The clinker compositions and hydrated pastes were characterized by X-ray diffraction (XRD), isothermal calorimetry, mercury intrusion porosimetry (MIP), and scanning electron microscopy (SEM) to analyze its mineral contents, hydration products, heat release, pore structure, and microstructure. The compressive strength and linear expansion of ACSA mortars were tested for their mechanical properties. Results showed that clinkers doped with 2 wt.% MgO can offset the hurdle that SO_3 caused to the formation of C_3S (tricalcium silicate). Clinkers with varying ratios of C_3S and $C_4A_3\$$ (calcium sulfoaluminate) were obtained, achieving 58–70% C_3S and 2.0–5.6% $C_4A_3\$$ in ACSA through adjusting the KH (lime saturation factor) values and SO_3 dosage. ACSA cement showed better early mechanical properties. The 0.93 KH value with 3% SO_3 dosage in the raw meal, which contains 63.9% C_3S and 2.98% $C_4A_3\$$ in the clinker, reached an optimal compressive strength level at 1d (26.35 MPa) and at 3d (39.41 MPa), marking 30.45% and 18.70% increases compared to PII 52.5. The excellent early strength of ACSA cement may offer promising applications t increasing the incorporation of supplementary cementitious materials, thereby reduce pollution and carbon emissions.

Keywords: $C_4A_3\$$ and C_3S formation; sulfur effects; mineral design; cement hydration

1. Introduction

The cement industry plays a vital role in modern construction [1]. In China, cement production was approximately 20.23 billion tons in 2023, which accounted for over 55% of global cement production [2]. Despite its widespread use, the process of cement manufacturing, particularly the sintering of limestone and clay, is a significant source of carbon dioxide (CO_2) emissions [3], contributing around 26% of total global industrial emissions [2,4]. Thus, the cement industry is a focal point for low-carbon and emission reduction initiatives. There are three main approaches to achieving these goals [5]: (1) improving the durability of concrete during cement application to extend the lifespan of structures [6,7]; (2) enhancing the hydraulic properties of ordinary Portland cement (OPC) clinker to increase the use of supplementary cementitious materials (SCMs) [8,9]; and (3) developing innovative and environmentally friendly cement systems, such as calcined clay limestone cements (LC^3) [10,11], calcium sulfoaluminate cement (CSA) [12,13], and alite–calcium sulfoaluminate cement (ACSA) [14,15].

ACSA cement is an energy-efficient cement that contains an appropriate amount of calcium sulfoaluminate ($C_4A_3\$$) based on conventional Portland cement clinker. $C_4A_3\$$ is a mineral with high early strength [16], which can decrease the shrinkage rate of traditional Portland cement to some extent and improve the physical properties of cement, such as frost resistance, impermeability, and durability [17]. ACSA cements may facilitate an increase in the dosage of supplementary cementitious materials (SMCs) and, thus, achieve the goal of pollution reduction and carbon emissions reduction. SMCs' admixtures can reduce the early performance of cementitious systems [8]; hence, the primary focus of this paper regarding ACSA cement performance is the early strength.

Alite forms abundantly at approximately 1450 °C, while $C_4A_3\$$ decomposes almost completely at temperatures above 1350 °C. A protocol for the coexistence of both components was achieved by introducing a secondary high-temperature treatment after sintering, resulting in the formation of ACSA with good early strength [14]. In general, $C_4A_3\$$ has great early strength performance, while C_3S significantly contributes to the overall mechanical properties in ACSA cement. Therefore, studying the ratio of these two components is crucial for enhancing ACSA cement's performance.

To investigate the effect of calcium silicate phases, previous work [18] has studied the effect of lime saturation factors (KH) on the synthesis and hydration behavior of ACSA clinker. Results showed that clinker with a KH of 0.93 has better mechanical strength. On the other hand, sulfur is an essential component for the formation of $C_4A_3\$$ in the ACSA system. However, the dosage of sulfur in clinker can reduce the content of C_3A and the value of C_3S/C_2S (dicalcium silicate) due to a decrease in the liquid phase's viscosity and surface tension [19], hindering the formation of C_3S and fostering a strong preference for C_2S. Additionally, sulfur can impact the content of C_4AF (brownmillerite). Therefore, MgO was added to compensate for the adverse effects of sulfur on the formation of C_3S [20]. Moreover, the addition of a small amount of MgO and an alkali metal can decrease the sintering temperature, increase the liquid content, and reduce the viscosity of the liquid phase [21]. During the subsequent cooling process, most of the MgO remains in the aluminate and ferrite phases, and small amounts of MgO can disperse as small periclase crystals [22]. Studies have shown that the highest limit for the incorporated amounts of MgO in the clinker minerals is around 2 wt.% [20].

Some research has focused on the effects of KH, MgO, and sulfur on the sintering and mineral phase formation of ordinary Portland cement. However, the impact of MgO addition on the firing of ACSA cement, as well as the synergistic effects of KH and sulfur on the content of C_3S and $C_4A_3\$$ in the ACSA cement and the performance of ACSA cement, have not yet been thoroughly investigated. At the same time, regulating the contents of KH, MgO, and sulfur is of significant importance for the optimization of mineral phases in ACSA clinker and the enhancement of its performance.

In this work, raw meals with and without a 2 wt.% dosage of MgO were annealed along a gradient of temperatures to evaluate MgO effect on clinker sintering. Meanwhile, the clinker composition evolution over KH and sulfur was studied to determine the optimal level of C_3S and $C_4A_3\$$ content. The hydration properties of different contents of sulfur with a KH of 0.93 were studied to compare their strength performance. In summary, this research offers significant understanding into the production and hydration processes of ACSA cement, emphasizing the necessity of refining its composition to optimize its mechanical properties.

2. Materials and Methods

2.1. Sample Preparation

Industrial raw materials, including industrial limestone (from China United Concrete Nanjing Co., Ltd., Nanjing, China), first class fly ash, clay, and silica, were ground to pass through an 80 μm sieve. All chemical reagents, including MgO, K_2O, and $CaSO_4{\cdot}2H_2O$, were purchased from Sigma–Aldrich (St. Louis, MO, USA) and used as received. Chemical compositions of the raw meals are shown in Table 1. The sintering protocol of ACSA clinker has been reported in previous work [18]. A secondary heat treatment at 1270 °C for 1 h, after ordinary sintering at 1470 °C and fast cooling at room temperature, was conducted to achieve the coexistence of alite and ye'elimite [10,18]. The raw meals of M0 and M2 were taken out at 900, 1000, 1100, 1200, 1300, 1400, and 1500 °C, respectively, to evaluate the effect of MgO on clinker sintering. A tungsten carbide vibrating mill was used for two cycles of 10 s each to grind the clinker blocks into fine powders. ACSA cements were prepared by blending the described clinker powders with 4 wt.% of gypsum [18].

Table 1. Chemical compositions of the raw meals (wt.%).

Labels	CaO	SiO_2	Al_2O_3	Fe_2O_3	MgO *	K_2O *	SO_3 *	KH **	SM **	IM **
MKH-Blank	66.27	21.53	5.98	1.88	2.00	0.80	0.00	0.93	2.50	3.10
MKH-1	65.11	20.03	5.62	1.76	2.00	0.80	2.40	0.93	2.50	3.10
MKH-2	63.97	20.11	5.66	1.71	2.00	0.80	2.70	0.93	2.50	3.10
MKH-3	64.22	20.12	5.62	1.70	2.00	0.80	3.00	0.93	2.50	3.10
MKH-4	64.22	20.12	5.62	1.70	2.00	0.80	3.30	0.93	2.50	3.10
LKH-2	64.62	20.42	5.62	1.81	2.00	0.80	2.70	0.90	2.50	3.10
HKH-2	65.72	19.78	5.50	1.77	2.00	0.80	2.70	0.96	2.50	3.10
M0	64.22	20.12	5.62	1.70	0	0.80	3.00	0.93	2.50	3.10
M2	64.22	20.12	5.62	1.70	2.00	0.80	3.00	0.93	2.50	3.10

* The weights of MgO, K_2O, and SO_3 were calculated as exogenous ingredients in the raw meal. ** Lime saturation factor (KH) = (CaO − 1.65 Al_2O_3 − 0.35 Fe_2O_3)/(2.8 SiO_2), silica modulus (SM) = SiO_2/(Al_2O_3 + Fe_2O_3), and iron modulus (IM) = Al_2O_3/Fe_2O_3 [19].

2.2. *Characterization*

2.2.1. X-Ray Diffraction and Rietveld Quantitative Analysis

XRD was performed using a Rigaku Miniflex workstation (Miniflex 600, Rigaku Co., Tokyo, Japan) operating at 40 kV and 15 mA with Cu Kα radiation (λ = 0.1789 nm). Data were collected from 5 °C to 70 °C (2θ) for 13 min with a step size of 0.01 °C. Rietveld [23] quantitative analysis was conducted using GSAS EXPGUI 1.0 version [24] with crystal structures for refinement, including C_3S (alite, ICSD 64759), C_2S (larnite, ICSD 79553), C_4AF (brownmillerite, ICSD 2841), C_3A (ICSD 1841), $C_4A_3\$$ (ye'elimite ICSD 9560), and f-CaO (free lime, ICSD 52783). The background and profile fittings used the shifted Chebyshev polynomial and pseudo-Voigt functions, refining phase fractions, lattice parameters, and profiles (GU, GV, and GW).

2.2.2. Microscopy

Clinker granules were embedded in two-component epoxy, polished successively with sandpaper with a 600, 1200, 3000, and 5000 mesh, and etched with a 1% HNO_3 alcohol solution for 10 s to prepare the surface [25]. Microscopic images were captured using a Leica Microsystems GmbH microscope (Wetzlar, Germany) equipped with a Leica DFC 480 camera.

2.2.3. Particle Size

The clinker powders' particle size distribution (PSD) was measured with Malvern MS2000 laser diffraction granulometry (Malvern Panalytical, Spectris, London, UK) after dispersing in ethanol via ultrasonication for 60 s, with reflection indexes of 1.70 and 1.36 for the clinker and ethanol, respectively.

2.2.4. Compressive Strength Tests

Mortars were prepared in accordance with the Chinese National Standard (GB/T 1346 2011) [26], with a weight ratio of ACSA cement/sand/water = 1/3/0.5. The mortars were placed in a mold (40 × 40 × 160 mm) at 20 ± 1 °C and 95% relative humidity for 24 h. Subsequently, the mortars were removed from the mold and immersed in water at 20 °C until they were ready for strength testing. The compressive strengths of the mortars were tested according to GB/T17671-2021 [27] using an automatic cement strength testing machine (AEC 201, Ruifeng, Shanghai, China) with a maximum load of 200 kN and a constant loading rate of 2.4 kN/s. The average of four tested specimens represented each compressive strength value.

2.2.5. Expansion Tests

Pastes were prepared in accordance with the Chinese National Standard (GB/T 1346 2011) [26]. Samples were added to 4% dihydrate gypsum with a water–cement ratio of 0.35 and set in a 20 × 20 × 80 mm mold. The length of each age was tested, and the expansion is represented by the change rate of the duration relative to the base length. Each sample was formed to 6 pieces and tested averaged.

2.2.6. Calorimetry

Isothermal calorimetry was conducted at 20 °C for 72 h using an eight-channel thermometric air instrument (TAM Air Isothermal Calorimeter, TA Instruments, New Castle, USA) to measure the hydration heat release of pastes. Pastes were prepared with a w/c ratio of 0.5 in plastic ampoules that were externally mixed prior to the test.

2.2.7. Mercury Intrusion Porosimetry (MIP)

Porosity and pore size distribution of hydration pastes were characterized using a Poremaster GT-60 (Quantachrome, Boynton Beach, FL, USA) through mercury intrusion porosimetry (MIP). Regularly shaped, cracked pieces that were not in contact with the mold were chosen for the measurement.

2.2.8. SEM

The microstructures of hydrated pastes were observed using a scanning electron microscope (SEM, Model JSM-5900, JEOL Co., Tokyo, Japan) after fracturing the specimens and attaching them to the specimen holder with carbon tapes. A thin layer of gold was then applied to the specimens to enhance electrical conductivity.

3. Results and Discussion

3.1. Clinker Sintering

3.1.1. Burnability of the Sulfur-Containing Clinker

The evolution of mineral compositions calculated by the Rietveld [23] method from the XRD data is shown in Figure 1. C_3S in sample M2 formed in large quantities after 1300 °C (Figure 1b), which is about 100 °C lower than that of the M0 sample (Figure 1a). Moreover, at 1500 °C, the content of C_3S in the M2 sample is about 65%, which is more than double the C_3S content in M0. It is obvious that the MgO dosage can effectively promote the formation of C_3S in the sulfur-containing clinker. The sulfur in the solid solution of C_2S is four to five times that of C_3S, which makes C_2S in sulfur-containing clinker more stable, thus hindering the formation of C_3S [19]. The addition of MgO can reduce SO_3 content in the solid solution of C_2S, and reduce the obstacle of sulfur to the formation of C_3S [19]. $C_4A_3\$$ formed in large quantities after 1250 °C and vanished after 1400 °C, indicating the necessity of the secondary heat treatment to achieve the coexistence of C_3S and $C_4A_3\$$ [14].

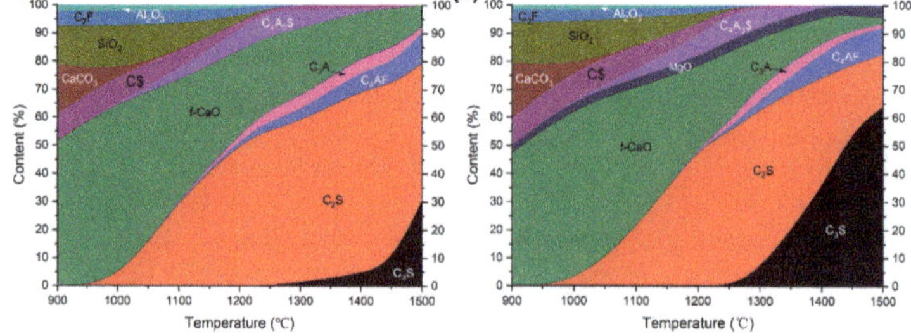

Figure 1. Formation of the clinker during the sintering process: (**a**) M0 and (**b**) M2.

3.1.2. Mineral Compositions of the Clinkers

The XRD spectra, the mineral composition calculated by the Rietveld method, and the contour based on the quantitative results of the clinkers are illustrated in Figure 2, Table 2, and Figure 3, respectively. The peaks of the $C_4A_3\$$ fingerprints at 23.5 °C 2θ on the XRD spectrum gradually enhance with an increase in SO_3 content, while the characteristic peak of C_3A decreases (Figure 2), indicating the formation of $C_4A_3\$$ from C_3A and SO_3.

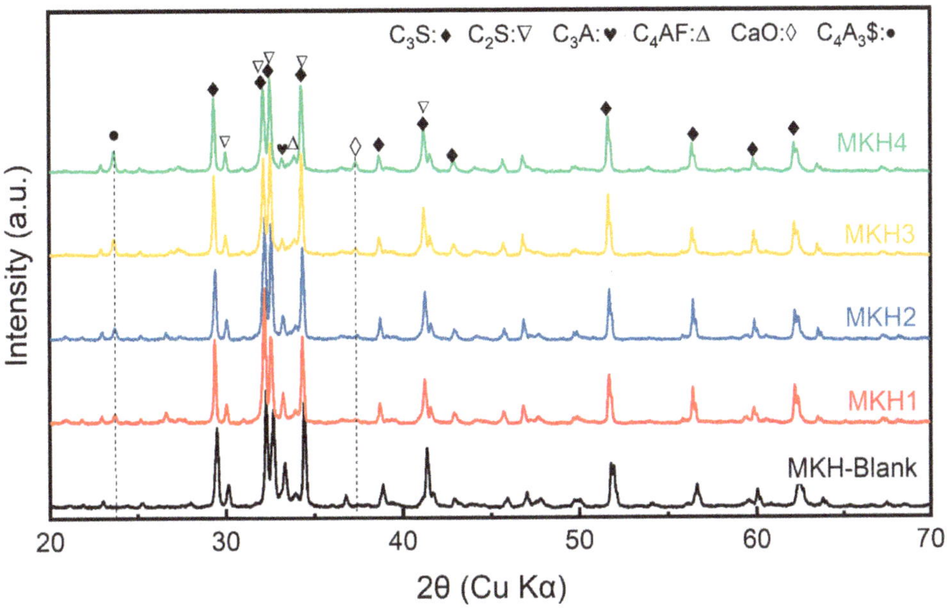

Figure 2. XRD spectra of clinkers with different SO_3 dosage.

Table 2. Mineral compositions of the clinkers calculated using the Rietveld method (%).

Samples	C_3S	C_2S	C_3A	C_4AF	f-CaO	f-MgO	$C_4A_3\$$	R_{wp}
MKH-Blank	68.07	12.94	11.68	4	0.02	3.09	0	8.63
MKH1	63.74	18.17	7.54	4.51	0.51	3.38	2.02	11.87
MKH2	63.90	18.55	5.96	4.57	0.58	3.27	2.98	11.07
MKH3	64.36	18.18	3.32	5.64	0.72	3.27	4.52	10.00
MKH4	64.65	17.62	2.42	4.86	1.38	3.14	5.61	8.51
LKH2	58.23	23.65	6.84	4.61	0	3.58	2.88	8.31
HKH2	70.63	11.69	6.47	4.56	0.50	3.28	3.26	10.81

The fingerprint of f-CaO at 37.5 °C 2θ increases with the SO_3 content [28]. At the same time, the increase in SO_3 content leads to a decrease in the C_3S/C_2S ratio. Belite (C_2S) has a much stronger ability to incorporate sulfur than alite (C_3S), making sulfur-containing belite more stable [19]. This stability inhibits the reaction of C_2S with CaO to form C_3S ($C_2S + CaO \rightarrow C_3S$), thereby hindering the formation of C_3S in the clinker and resulting in more f-CaO in the clinker. In addition, after SO_3 dosage, the fingerprint of C_3S in the 51.2–52.4 °C 2θ region of the two peaks of about the same height changes to one peak with a shoulder, indicating the crystalline transformation of M3 to M1 [29].

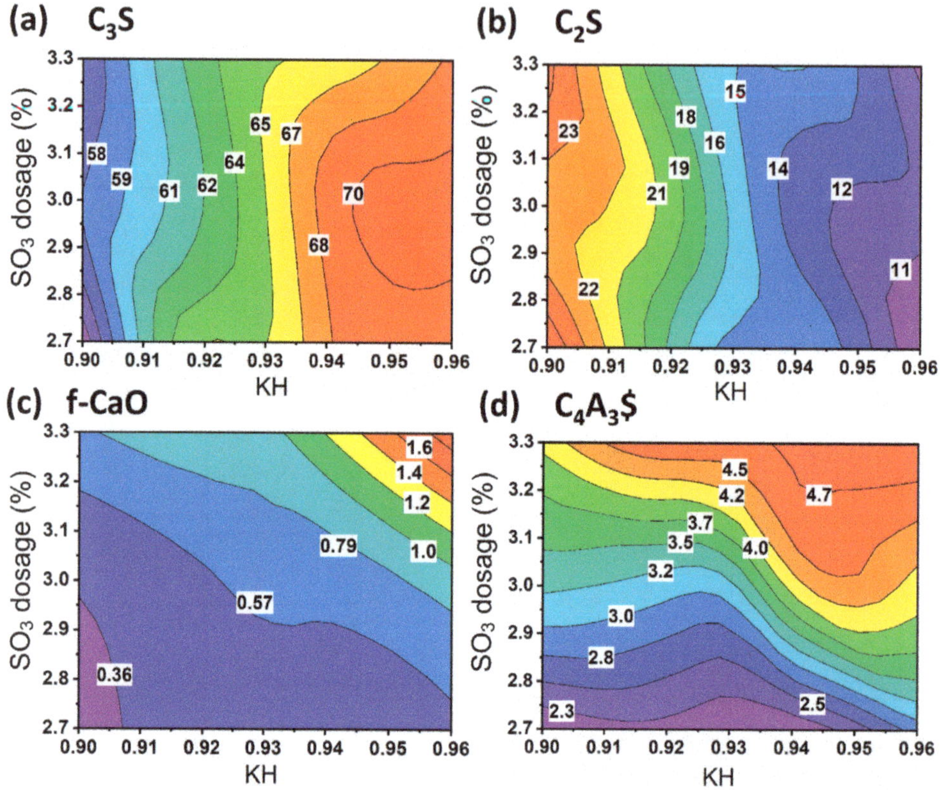

Figure 3. Contour of mineral compositions as a function of KH values and SO_3 content: (**a**) C_3S, (**b**) C_2S, (**c**) f-CaO, and (**d**) $C_4A_3\$$.

Figure 3 shows the contour plots of various minerals as a function of KH values and SO_3 content. Figure 3a demonstrates that the formation of C_3S shifts to the right, indicating that KH values play a decisive role in the formation of C_3S. However, when KH values reach a certain level, SO_3 begins to hinder the formation of C_3S. The trend in C_2S is opposite to that in C_3S, which is consistent with our expectations. For f-CaO and $C_4A_3\$$, Figure 3c,d both show an upward-rightward inclination. It is important to note that the content of f-CaO should not be too high, as it needs to ensure the clinker quality while achieving the formation of $C_4A_3\$$ minerals.

3.1.3. Mineral Microscopy of the Clinkers

The optical microscope images of the samples show that, in the blank sample, most of the C_3S appears as small hexagonal platelets. The addition of SO_3 promotes the growth of large C_3S grains, with the MKH4 sample exhibiting grains exceeding 100 μm in size. Furthermore, with SO_3 doping, most of the C_3S exhibits hexagonal or prismatic shapes with distinct grain boundaries. On the contrary, C_2S consists mostly of small circular grains, contributing relatively less to the overall composition, consistent with quantitative XRD analysis results. Bubble-like free lime is located at the boundaries of C_3S (Figure 4). Meanwhile, the intermediate phases of all clinkers are finely distributed, indicating a moderate liquid phase content and good burning. The brighter and grey phases intermingled in intermediate phases are ferrite and aluminate (F and A in Figure 4). Some samples show slight black speckled stripes, which are speculated from previous results to be $C_4A_3\$$ [18,30].

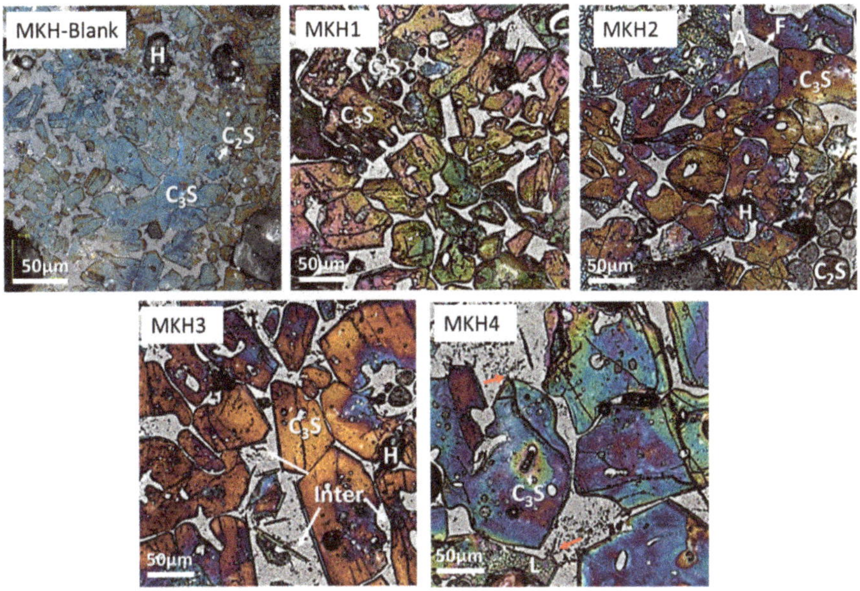

Figure 4. Microscopy images of clinkers. C_3A is alite, C_2S is belite, A is aluminate, F is ferrite, L is free lime, H is hole, and Inter. denotes interstitial phases.

3.2. Mortar Properties

3.2.1. Compressive Strength

Clinkers of varying SO_3 levels were ground using a uniform protocol, yielding particle size distributions with two peaks at 5–10 and 20–30 μm (Figure 5). These powders were then utilized to assess mortar compressive strength and paste hydration characteristics; owing to their similar particle size distribution, their impact on cement performance can be considered negligible.

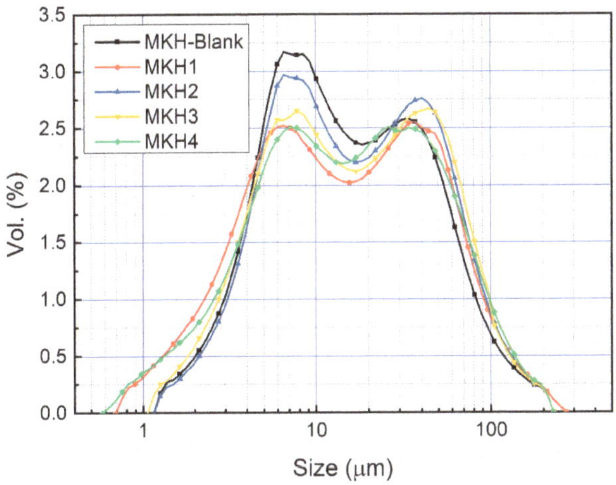

Figure 5. Particle size distribution of ACSA clinkers for mechanical strength measurement.

The compressive strength results for the 1-, 3-, and 28-day curing ages of the samples with the same KH (0.93) but different $C_4A_3\$$ contents are shown in Figure 6. The compressive strength of 1-,3-day curing ages increased significantly due to the existence of $C_4A_3\$$ (MKH1~MKH4). However, an increase in $C_4A_3\$$ content did not necessarily correspond to an increase in clinker strength. Excess $C_4A_3\$$ content can severely affect the later strength of the clinker. For example, MKH4 (with a 5.61% $C_4A_3\$$ content) exhibits a later strength of only 48.26 MPa, a decrease of nearly 7 MPa compared to the blank sample. Furthermore, for both 1-day and 3-day strengths, a higher $C_4A_3\$$ content did not necessarily result in a better performance; instead, an intermediate value tended to yield optimal results. MKH2 exhibited the best overall strength performance: 26.35 MPa at 1 d, 39.41 MPa at 3 d, and 53.15 MPa at 28 d, with 30.45% and 69.02% increases in 1-day strength and 18.70% and 14.46% increases in 3-day strength compared to PII 52.5 and MKH-Blank (Figure 6b). Meanwhile, the compressive strength of the 28-day sample was only decreased by 1.95 MPa compared with MKH-Blank.

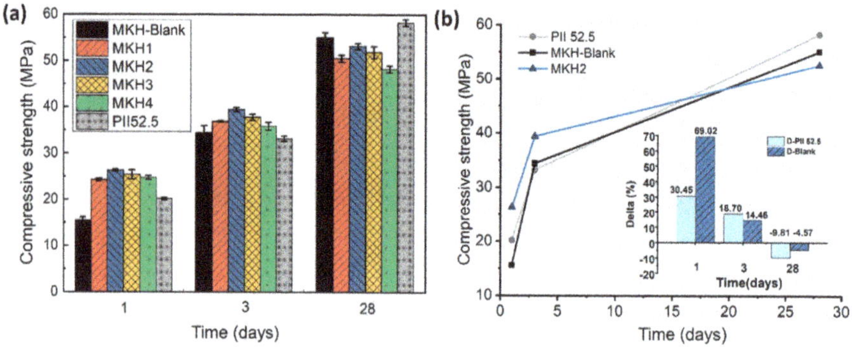

Figure 6. Compressive strength of all the mortars (**a**), compressive strength comparison among PII 52.5, MKH-Blank, and MKH2 (**b**). The inset plot in (**b**) represents the compressive strength increment of MKH2 compared to MKH-Blank and PII 52.5. For D-Blank, Delta = (MKH2 − MKH-Blank)/MKH-Blank; for D-PII 52.5, Delta = (MKH2 − PII 52.5)/PII 52.5.

3.2.2. Expansion Performance

The linear expansion value of the MHK-Blank sample stayed below 0 (Figure 7). The expansion in cement paste increased with the $C_4A_3\$$ content, which corresponds to the formation of ettringite ($C_6A\$_3H_{32}$, AFt) from $C_4A_3\$$ hydration [17]. Apart from MKH4, expansion in the pastes was below 0.1% within 150 days and converged below 0.15% by around 200 days. However, for the MKH4 sample, its expansion was much higher—about two times that of the other samples—and this higher expansion rate may also be the reason for the poorer development of its strength. This expansion property of ACSC cement may be a method of reducing shrinkage and cracking in cement-based materials [31].

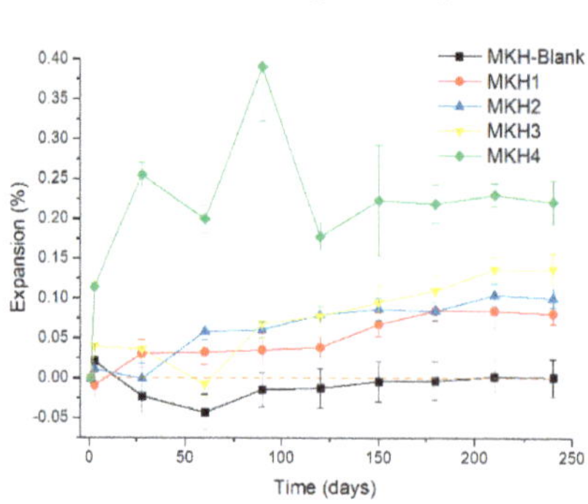

Figure 7. Linear expansion of the mortars.

3.3. *Analysis of Hydration Products*

3.3.1. Heat Release of Hydration

In the Portland cement system, the C_3A and C_3S minerals are involved in early hydration. Whereas, in this matrix, calcium sulfoaluminate reacts with gypsum to form ettringite ($C_6A\$_3H_{32}$, AFt), $C_4A_3\$ + 2C\$H_2 + 34H \rightarrow C_6A\$_3H_{32} + 2AH_3$, releasing a significant amount of heat during early hydration. Figure 8a shows the hydration heat evolution curve of the MKH group clinkers over 72 h, with all samples containing 4% dihydrate gypsum. The main exothermic peak is attributed to the hydration of C_3S. The position of this peak shifts to an earlier time with an increase in $C_4A_3\$$ content, indicating a shorter induction period that is influenced by sulfur content [17]. While the inset depicts the heat release peak when cement samples are in contact with water, primarily from the particle dissolution and the hydration of C_3A and $C_4A_3\$$, the intensity of this peak first increases and then decreases with an increase in $C_4A_3\$$ content. A shoulder peak is obviously observed on MKH2 within 15–20 h of hydration, which corresponds to the transformation of monosulfate ($C_3A\$H_{12}$, AFm) from AFt [32].

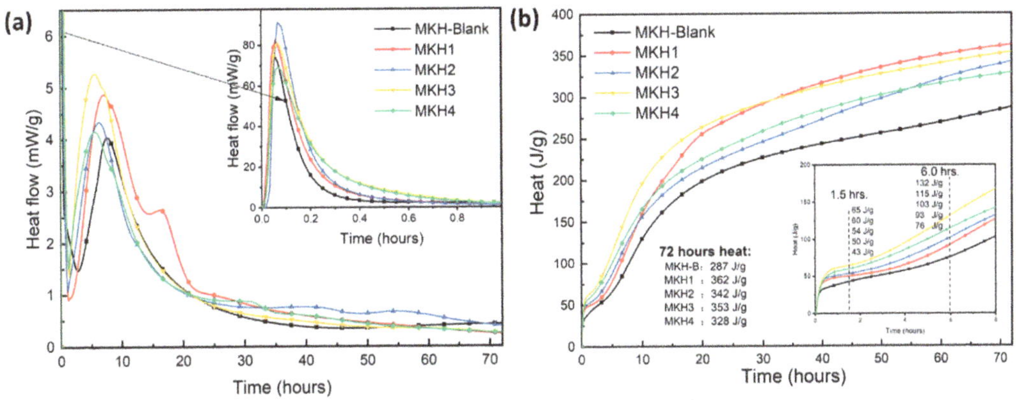

Figure 8. Hydration heat evolution of the clinkers: heat flow (**a**) and accumulative heat (**b**). The inset plot in (**a**) represents the heat release during clinker dissolution within the first hour of hydration.

The cumulative heat release of MKH group cement samples is shown in Figure 8b, with the inset illustrating cumulative heat release within 8 h of hydration. Within this timeframe, except for sample MKH4, the cumulative heat release correlates positively with the content of $C_4A_3\$$; for instance, at 6 h, the heat released by the MKH-Blank, MKH1, MKH2, and MKH3 samples are 76 J/g, 93 J/g, 103 J/g, and 132 J/g, respectively. Regarding the total heat release over 72 h, samples containing $C_4A_3\$$ exhibited significantly higher— at least 50 J/g higher—values compared to the MKH-Blank.

3.3.2. Mineral Composition of the Pastes

XRD analysis was used to characterize the hydration products (Figure 9). The hydration products primarily included calcium hydroxide (CH), AFt, and AFm. The CH mainly originated from the hydration of C_3S, and its presence can enhance the hydration of $C_4A_3\$$ [30]. $C_4A_3\$$ underwent hydration at 1 d to form Aft, and almost all of the $C_4A_3\$$ completed its reaction by 3 d. With the consumption of sulfur, the AFt gradually transformed into AFm. MKH1 shows the highest content of AFm at 1 d (Figure 9a), which is consistent with the heat release results (Figure 8a). In MKH-Blank, AFt completely transformed into AFm by 3 d, while AFt can be observed in the $C_4A_3\$$-containing samples until 28 d. Figure 9d illustrates through the evolution of hydration products over time that CH begins to form significantly between 3 and 6 h after cement hydration starts, while AFt forms in large quantities right at the beginning of hydration. AFm formation occurs later, starting around 3 d of hydration, whereas $C_4A_3\$$ has completely reacted by this period.

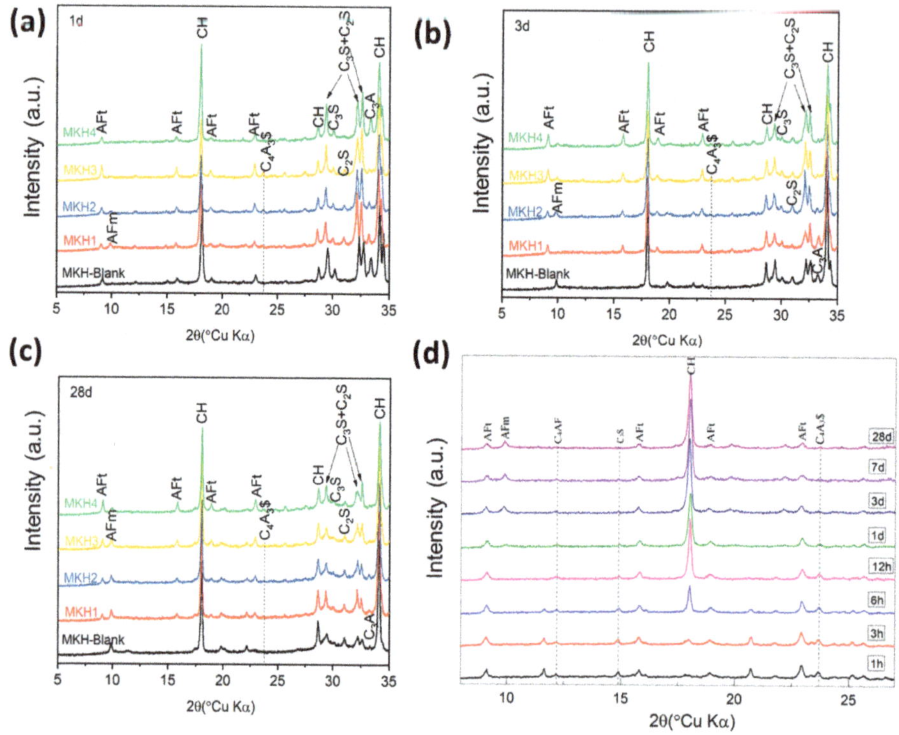

Figure 9. XRD spectra of the hydration pastes at (**a**) 1, (**b**) 3, and (**c**) 28 d. XRD spectra of hydrated MKH2 from 1 h to 28 d (**d**).

3.3.3. Porosity

Due to the presence of the $C_4A_3\$$ mineral, the alite calcium sulfoaluminate cement prepared in this experiment exhibited slight early-stage expansion characteristics and formed a large amount of AFt during early hydration. This compensated for its shrinkage during cement hydration and, to some extent, served as a filler. Figure 10 and Table 3 show the development of the pore structure of samples with and without the $C_4A_3\$$ mineral at various ages of hydration. In the early stages of cement hydration, the total pore volume of MKH2 was significantly smaller than that of MKH-Blank. At 1 day of hydration, the porosity of MKH-Blank was 40.95%, while MKH2's was 34.17%, showing a reduction of 6.78%. This is a major reason for the early development of strength in $C_4A_3\$$-containing cement. The difference in porosity between the two samples gradually decreased over hydration time, making a tendency for their compressive strengths to be consistent (Figure 6). Figure 10 also illustrates the pore size distribution of the cement paste. It shows that, with increasing hydration age, the large pores decrease and the small pores increase in both the MKH-Blank and MKH2 hardened pastes. MKH2 shows a wide pore size distribution at 1 d, a narrow distribution with an obvious peak around 200 nm at 3 d, and two separated peaks at 28 d. While the pattern of pore size distribution of MKH-Blank shows the opposite trend.

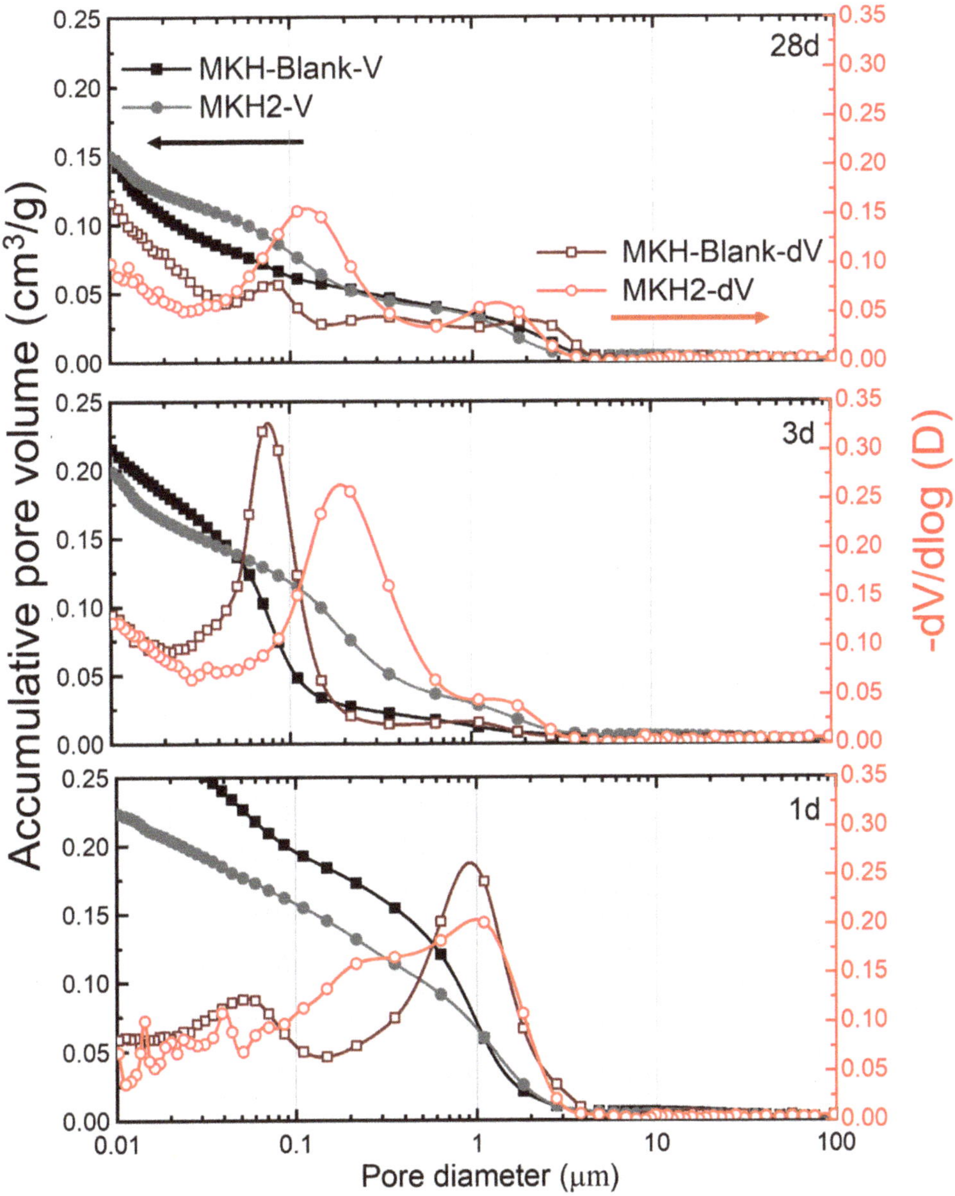

Figure 10. Pore volume and pore size distribution of the hydrated pastes.

Table 3. Total porosity (%) and the pore volume (cm^3/g) of the hydrated pastes.

Samples	Porosity (%)			Pore Volume (cm^3/g)		
	1 Day	3 Day	28 Day	1 Day	3 Day	28 Day
MKH-Blank	40.95	34.00	24.23	0.2968	0.2319	0.1614
MKH2	34.17	31.25	24.42	0.2217	0.2134	0.1586

3.3.4. Microstructure

Typical needle-shaped AFt [33] in the $C_4A_3\$$-containing sample can be observed in the SEM images, with fine and elongated crystals at 1 day (Figure 11b). Due to the early hydration product of C_3S being a C-S-H gel, this enhances the bonding between AFt and other hydration products [30]. The needle-shaped AFt crystals also serve as a framework of support in the C-S-H gel, explaining why the compressive strength of the MKH2 samples at 1 day is significantly higher than that of MKH-Blank. Figure 11c also confirms the presence of petal-like AFm [34], which is in accordance with the mineral composition results (Figure 9c).

Figure 11. SEM of (**a**) MKH0-1d, (**b**) MKH2-1d, (**c**) MKH0-28d, and (**d**) MKH2-28d.

4. Conclusions

A great amount of C_3S and $C_4A_3\$$ could coexist in a burning system with a sintering temperature of 1470 °C and a secondary heat treatment temperature of 1270 °C, while the content of f-CaO is within a desirable range. Clinkers doped with 2 wt.% MgO can offset the hurdle that SO_3 may cause to the formation of C_3S. Through adjusting the KH values (0.90–0.96) and SO_3 dosage (2.4–3.3 wt.%) in clinker preparation, clinkers with varying ratios of C_3S and $C_4A_3\$$ were obtained, achieving 58–70% C_3S and 2.0–5.6% $C_4A_3\$$ in ACSA clinkers. Based on the above experimental results and discussion, the following conclusions can be drawn:

Due to the early hydration formation of abundant AFt by calcium sulfoaluminate, ACSA cement shows a higher hydration heat release, with a 72-hour heat release that is 50–70 J/g higher than the blank sample.

$C_4A_3\$$ is almost completely reacted within 3 days of hydration, accompanied by gradual AFm formation. Meanwhile, ACSA cement shows better early mechanical properties.

The 0.93 KH value with 3% SO_3 dosage in the raw meal (sample MKH2), which contains 63.9% C_3S and 2.98% $C_4A_3\$$ in the clinker, reaches an optimal compressive strength level: at 1 d (26.35 MPa) and at 3 d (39.41 MPa), marking 30.45% and 18.70% increases compared to PII 52.5.

Because of the higher early age strength of ACSA cement, this cement has the potential to increase the utilization of supplementary cementitious materials, thereby reducing pollution and carbon emissions. Furthermore, the controlled expansion characteristics of ACSA cement may decrease shrinkage and cracking phenomena in cementitious materials, consequently enhancing the service life and durability of concrete structures.

Although it is possible to achieve the coexistence of the difficult-to-coexist C_3S and $C_4A_3\$$ at a firing temperature of 1470 °C and a secondary heat treatment temperature of 1270 °C, this condition is not easy to realize in actual cement production. Future research should focus on exploring conditions at a temperature between 1270 °C and 1470 °C to achieve the coexistence of C_3S and $C_4A_3\$$ through a single firing process. This study has significantly improved the early strength of ACSA cement by regulating KH values and the dosage of SO_3, and it has explained the reasons behind this enhancement in early strength. However, the factors affecting the strength of ACSA are not limited to the contents of C_3S and $C_4A_3\$$, as well as the regulation of KH values and SO_3 dosage. Future research should focus on other factors, such as the synergistic optimization of mineral contents and the optimization of ACSA cement particle size, to achieve superior performance in the 28 d strength of ACSA cement.

Author Contributions: X.L.: Writing—original draft, conceptualization; G.K.: writing—original draft, formal analysis; S.D.: formal analysis, data curation; B.M.: writing—review and editing, data curation; J.T.: writing—review and editing, data curation; H.Z. (Hao Zhou): supervision, writing—review & editing; H.Z. (Houhu Zhang): resources, conceptualization; J.W.: writing—review and editing, formal analysis; X.S.: resources, conceptualization. All authors have read and agreed to the published version of the manuscript.

Funding: This research was funded by the Special Fund of the Chinese Central Government for Basic Scientific Research Operations in the Commonwealth Research Institute (GYZX220301) and the Special Fund of the Carbon Peak and Carbon Neutrality Research Institute that is supported by the Nanjing Institute of Environmental Sciences, Ministry of Ecology and Environment (ZX2023SZY060). As well, funding provided by the National Key R&D Program of China (2022YFC3901405), the Priority Academic Program Development of the Jiangsu Higher Education Institutions (PAPD), and the Jiangsu National Synergetic Innovation Centre for Advanced Materials (SICAM) is also appreciated.

Institutional Review Board Statement: Not applicable.

Informed Consent Statement: Not applicable.

Data Availability Statement: The original contributions presented in this study are included in the article; further inquiries can be directed to the corresponding authors.

Conflicts of Interest: The authors declare no conflicts of interest.

Nomenclature

C = CaO, S = SiO_2, A = Al_2O_3, F = Fe_2O_3, \$ = SO_3, nomenclature is not suitable if the compounds are described in their full representation [32].

References

1. Kirgiz, F.P.D. Reduction in ecology, environment, economy and energy in concrete industry using waste materials. *J. Adv. Compos. Mater. Constr. Environ. Nano Technol.* **2020**, 1–7.
2. Guo, Y.; Luo, L.; Liu, T.; Hao, L.; Li, Y.; Liu, P.; Zhu, T. A review of low-carbon technologies and projects for the global cement industry. *J. Environ. Sci.* **2024**, *136*, 682–697. [CrossRef] [PubMed]

3. Xi, F.; Davis, S.J.; Ciais, P.; Crawford-Brown, D.; Guan, D.; Pade, C.; Shi, T.; Syddall, M.; Lv, J.; Ji, L.; et al. Substantial global carbon uptake by cement carbonation. *Nat. Geosci.* **2016**, *9*, 880–883. [CrossRef]
4. International Energy Agency (IEA). *CO2 Emissions in 2023*; IEA: Paris, France, 2024.
5. Gartner, E. Industrially interesting approaches to "low-CO_2" cements. *Cem. Concr. Res.* **2004**, *34*, 1489–1498. [CrossRef]
6. Li, C.; Li, J.; Ren, Q.; Zheng, Q.; Jiang, Z. Durability of concrete coupled with life cycle assessment: Review and perspective. *Cem. Concr. Compos.* **2023**, *139*, 105041. [CrossRef]
7. Tayeh, B.A.; Hamada, H.M.; Almeshal, I.; Bakar, B.H.A. Durability and mechanical properties of cement concrete comprising pozzolanic materials with alkali-activated binder: A comprehensive review. *Case Stud. Constr. Mater.* **2022**, *17*, e01429. [CrossRef]
8. Juenger, M.C.G.; Siddique, R. Recent advances in understanding the role of supplementary cementitious materials in concrete. *Cem. Concr. Res.* **2015**, *78*, 71–80. [CrossRef]
9. Skibsted, J.; Snellings, R. Reactivity of supplementary cementitious materials (SCMs) in cement blends. *Cem. Concr. Res.* **2019**, *124*, 105799. [CrossRef]
10. Scrivener, K.; Martirena, F.; Bishnoi, S.; Maity, S. Calcined clay limestone cements (LC3). *Cem. Concr. Res.* **2018**, *114*, 49–56. [CrossRef]
11. Barbhuiya, S.; Nepal, J.; Das, B.B. Properties, compatibility, environmental benefits and future directions of limestone calcined clay cement (LC3) concrete: A review. *J. Build. Eng.* **2023**, *79*, 107794. [CrossRef]
12. Chaunsali, P.; Mondal, P.; Bullard, J. Influence of Calcium Sulfoaluminate (CSA) Cement Content on Expansion and Hydration Behavior of Various Ordinary Portland Cement-CSA Blends. *J. Am. Ceram. Soc.* **2015**, *98*, 2617–2624. [CrossRef]
13. Winnefeld, F.; Martin, L.H.J.; Müller, C.J.; Lothenbach, B. Using gypsum to control hydration kinetics of CSA cements. *Constr. Build. Mater.* **2017**, *155*, 154–163. [CrossRef]
14. Zea-Garcia, J.D.; De la Torre, A.G.; Aranda, M.A.G.; Santacruz, I. Processing and characterisation of standard and doped alite-belite-ye'elimite ecocement pastes and mortars. *Cem. Concr. Res.* **2020**, *127*, 105911. [CrossRef]
15. Zea-Garcia, J.D.; Santacruz, I.; Aranda, M.A.G.; De la Torre, A.G. Alite-belite-ye'elimite cements: Effect of dopants on the clinker phase composition and properties. *Cem. Concr. Res.* **2019**, *115*, 192–202. [CrossRef]
16. Ben Haha, M.; Winnefeld, F.; Pisch, A. Advances in understanding ye'elimite-rich cements. *Cem. Concr. Res.* **2019**, *123*, 105778. [CrossRef]
17. Hargis, C.W.; Telesca, A.; Monteiro, P.J.M. Calcium sulfoaluminate (Ye'elimite) hydration in the presence of gypsum, calcite, and vaterite. *Cem. Concr. Res.* **2014**, *65*, 15–20. [CrossRef]
18. Li, X.; Ma, B.; Ji, W.; Dou, S.; Zhou, H.; Zhang, H.; Wang, J.; Hu, Y.; Shen, X. Impact of Lime Saturation Factor on Alite-Ye'Elimite Cement Synthesis and Hydration. *Materials* **2024**, *17*, 3035. [CrossRef]
19. Li, X.; Xu, W.; Wang, S.; Tang, M.; Shen, X. Effect of SO3 and MgO on Portland cement clinker: Formation of clinker phases and alite polymorphism. *Constr. Build. Mater.* **2014**, *58*, 182–192. [CrossRef]
20. Staněk, T.; Sulovský, P. The influence of the alite polymorphism on the strength of the Portland cement. *Cem. Concr. Res.* **2002**, *32*, 1169–1175. [CrossRef]
21. Segata, M.; Marinoni, N.; Galimberti, M.; Marchi, M.; Cantaluppi, M.; Pavese, A.; De la Torre, Á.G. The effects of MgO, Na2O and SO3 on industrial clinkering process: Phase composition, polymorphism, microstructure and hydration, using a multidisciplinary approach. *Mater. Charact.* **2019**, *155*, 109809. [CrossRef]
22. Long, G.R. Microstructure and chemistry of unhydrated cements. Philosophical Transactions of the Royal Society of London. *Ser. A Math. Phys. Sci.* **1983**, *310*, 43–51.
23. Scrivener, K.L.; Fullmann, A.; Gallucci, E.; Walenta, G.; Bermejo, E. Quantitative study of Portland cement hydration by X-ray diffraction/Rietveld analysis and independent methods. *Cem. Concr. Res.* **2004**, *34*, 1541–1547. [CrossRef]
24. Toby, B.H. EXPGUI, a graphical user interface for GSAS. *J. Appl. Crystallogr.* **2001**, *34*, 210–213. [CrossRef]
25. Campbell, H.D. *Microscopical Examination and Interpretation of Portland Cement and Clinker*; PCA: Washington, DC, USA, 1999.
26. *GB/T1346-2011*; Test Methods for Water Requirement of Normal Consistency, Setting Time and Soundness of the Portland Cement. China Standard Press: Beijing, China, 2011.
27. *GB/T17671-2021*; Test Method of Cement Mortar Strength (ISO Method). China Standard Press: Beijing, China, 2021.
28. Kothari, C.; Garg, N. Quantitative phase analysis of anhydrous Portland cement via combined X-ray diffraction and Raman imaging: Synergy and impact of analysis parameters. *Cem. Concr. Res.* **2024**, *186*, 107662. [CrossRef]
29. Dvořák, K.; Všianský, D.; Ravaszová, S.; Jančíků, A. Synthesis of M1 and M3 alite polymorphs and accuracy of their quantification. *Cem. Concr. Res.* **2023**, *163*, 107016. [CrossRef]
30. Li, X.; Shen, X.; Xu, J.; Li, X.; Ma, S. Hydration properties of the alite-ye'elimite cement clinker synthesized by reformation. *Constr. Build. Mater.* **2015**, *99*, 254–259. [CrossRef]
31. Sant, G.; Lothenbach, B.; Juilland, P.; Le Saout, G.; Weiss, J.; Scrivener, K. The origin of early age expansions induced in cementitious materials containing shrinkage reducing admixtures. *Cem. Concr. Res.* **2011**, *41*, 218–229. [CrossRef]
32. Taylor, H.F.W. *Cement Chemistry*; Thomas Telford: London, UK, 1997.

33. Colombo, A.; Geiker, M.; Justnes, H.; Lauten, R.A.; De Weerdt, K. The effect of calcium lignosulfonate on ettringite formation in cement paste. *Cem. Concr. Res.* **2018**, *107*, 188–205. [CrossRef]
34. Xu, F.; Li, Z.; Li, T.; Wang, S. The Mechanical Properties and Microstructure of Tailing Recycled Aggregate Concrete. *Materials* **2024**, *17*, 1058. [CrossRef]

materials

MDPI

Article

Optimizing Cement Content in Controlled Low-Strength Soils: Effects of Water Content and Hydration Time

Yilian Luo [1], Liangwei Jiang [1,2], Libing Qin [1,3], Qiang Luo [1,2], David P. Connolly [4] and Tengfei Wang [1,2,*]

1 School of Civil Engineering, Southwest Jiaotong University, Chengdu 610031, China
2 Key Laboratory of High-Speed Railway Engineering, Ministry of Education, Southwest Jiaotong University, Chengdu 610031, China
3 Mingyang Yuheng Engineering Management Group Co., Ltd., Chengdu 610017, China
4 School of Civil Engineering, University of Leeds, Leeds LS2 9JT, UK
* Correspondence: w@swjtu.edu.cn

Abstract: The Ethylene Diamine Tetra-acetic Acid (EDTA) titration test is widely used for determining cement content, but its reliability is influenced by the hydration process of cement, which is affected by factors such as water content and hydration time. Despite their importance, these factors have received limited attention in existing research. This study explores the relationships between the volume of titrant required for stabilization, cement content, water content, and hydration time. Using a regression orthogonal test, the primary and secondary relationships, as well as the interdependencies among these factors, are analyzed. Results reveal a negative linear relationship between the titrant volume and both water content and hydration time. Cement content, water content, and hydration time are identified as the most significant factors, with minimal interdependencies observed. Within the test parameters, calculated values exhibit an error margin below 2.4%. Deviations of 2.9% in water content and 86 min in hydration time correspond to an approximate 0.5% change in cement content. These findings offer valuable insights for optimizing cement content detection in Controlled Low-Strength Material (CLSM) mixes, promoting more sustainable construction practices.

Keywords: Controlled Low-Strength Material (CLSM); cement content; EDTA titration; sustainability; circular economy; material recycling and repurposing

1. Introduction

The expansion of underground construction has led to a significant increase in urban excavation activities, generating substantial amounts of waste soil [1–4]. This situation presents dual challenges in managing waste soil efficiently and mitigating its environmental impact [5–8]. Embracing the principles of the circular economy, Controlled Low-Strength Material (CLSM) has the potential to be a sustainable solution. As defined by the American Concrete Institute (ACI) Committee, CLSM is a self-compacting, cementitious material that repurposes engineering waste soil, thereby addressing both waste reduction and the utilization of recycled materials in construction, particularly for backfilling in restricted spaces [6,9–11].

CLSM not only aids waste soil management but also exemplifies sustainable construction practices. Applications of CLSM in areas such as ditch backfills, structural fills, pavement bases, and bridge reclamations [12–15] demonstrate its versatility and contribution to the circular economy by promoting material reuse and reducing the environmental footprint. Similarly to other cement-based materials, CLSM's strength mainly comes from the gel material formed during the hydration reaction between cement and soil particles, leading to a stable structure [5,9,16,17]. Cement content is the basis for ensuring the quality of construction works and is a key factor in determining the hardening strength of CLSM to avoid construction accidents caused by quality problems. This underscores the criticality of precise cement content to achieve the desired material properties while maintaining sustainability goals. Therefore, the precise measurement of cement is crucial [18–20].

Cement content can be measured using various methods, including directly via calcium meter, the EDTA titration method, the neutralization heat titration method, and the acid-base neutralization titration method. Among them, the range of detecting cement content directly through a calcium meter is strictly limited, generally in the range of 3% to 10%, with certain limitations and low popularity of the instrument. Neutralization thermal titration and acid-base neutralization titration use complicated chemical reagents, their testing steps are more cumbersome, and their testing time is longer. The EDTA titration method, in particular, is attractive due to its low testing cost, speed, accuracy, and broad applicability, making it a common choice for cement-based soil modification projects [21,22].

Research into the factors affecting EDTA titration test data have been extensive. For example, Shen et al. [23] analyzed cement-stabilized granular materials used in pavement bases and found particle size significantly affects data dispersion. Their work suggested that sieving particles below 2.36 mm could enhance the reliability of titration data. Wang et al. [24] observed a nonlinear relationship between cement content and the volume of titrant used at high cement content, with the volume of titrant being inversely proportional to particle size. As particle size increases, the variability among titration samples also grows. Zhang et al. [25] focused on cement content in cement mixing piles, noting a rapid decrease in the volume of titrant used during the early maintenance stages and a reduced increase in titrant volume relative to cement content at a higher content. They also found that NH_4Cl's efficiency in extracting Ca^{2+} diminishes with increased cement content. Furthermore, Zhang et al. [26] identified an age effect in the measurement of content in cement-modified expansive soil, indicating the measured values are lower than the actual cement content for ages over 2 h and content above 4%, thus requiring adjustments to the calibration curve.

These studies underscore the need for a comprehensive analysis of factors like particle size, age, and cement content on titration data. Thus, this research aims to refine this analysis by examining the impact of cement content, water content, and hydration time on the volume of titrant used. By employing a regression orthogonal experimental design, the primary and secondary relationships among these factors are identified, and the significance of their interactions are studied. The resulting regression analysis, which considers cement content, water content, and hydration time, undergoes significance and misfit testing. Its accuracy is verified against measured data, offering a reliable basis for cement content detection in CLSM and for calibrating correction curves.

2. Materials and Methods

This section outlines the methodology used in this investigation from the procurement and characterization of materials to the experimental procedures. This includes an overview of the raw materials used, the core principles of the EDTA titration test, and the experimental processes. Further, it uses both one-factor-at-a-time (OFAT) and orthogonal experimental designs.

2.1. Materials

The red-bed mudstone studied was sourced from a construction project in Sichuan Province. The mudstone, notable for its brownish-red hue, was pulverized and sifted through a 5 mm geotechnical sieve. P·O 42.5 ordinary Portland cement served as the curing agent along with tap water. The particle size distributions of the red-bed mudstone and cement are presented in Figures S1 and S2, respectively (Supplementary Information). Table 1 presents the basic properties of the materials.

Table 1. Comparative index properties of red-bed mudstone and cement.

Material	Property	Value	Unit
Red-Bed Mudstone	Specific Gravity	2.69	-
	Max. Dry Density	1.95	g/cm^3
	Optimum Moisture Content	10.65	%
	Liquid Limit (LL)	31.5	%
	Plastic Limit	17.1	%
Cement	Initial Setting Time	182	min
	Final Setting Time	249	min
	Standard Consistency	28.4	%
	3-Day Flexural Strength	5.4	MPa
	3-Day Compressive Strength	25.2	MPa

2.2. Principles of EDTA Titration Test

Ethylenediaminetetraacetic acid (EDTA) titration is a prevalent analytical method in chemistry for measuring metal ion concentrations in solutions. This technique is based on chelation, where EDTA, as a ligand, binds with metal ions to allow their quantity to be measured. The versatility and accuracy of the EDTA titration test make it commonly used in various disciplines, including biochemistry, environmental science, and materials engineering. This section focuses on the application of EDTA titration in evaluating the calcium content in cement clinker, covering the chemical reactions involved, the significance of pH in the process, and the indicators used to identify the titration's endpoint.

Cement clinker primarily comprises tri-calcium silicate (C_3S), di-calcium silicate (C_2S), tri-calcium aluminate (C_3A), and tetra-calcium ferro aluminate (C_4AF), as outlined in Equations (1)–(4). Following the hydration reaction, calcium manifests in several forms, including hydration products like calcium silicate gel and calcium hydroxide, alongside free calcium ions (Ca^{2+}) and unreacted calcium [27–29].

$$3CaO{\cdot}SiO_2 + nH_2O \rightarrow xCaO{\cdot}SiO_2{\cdot}yH_2O + (3-x)Ca(OH)_2 \quad (1)$$

$$2CaO{\cdot}SiO_2 + mH_2O \rightarrow xCaO{\cdot}SiO_2{\cdot}yH_2O + (2-x)Ca(OH)_2 \quad (2)$$

$$2(3CaO{\cdot}Al_2O_3) + 27H_2O \rightarrow 4CaO{\cdot}Al_2O_3{\cdot}19H_2O + 2CaO{\cdot}Al_2O_3{\cdot}8H_2O \quad (3)$$

$$4CaO{\cdot}Al_2O_3{\cdot}Fe_2O_3 + 4Ca(OH)_2 + 22H_2O \rightarrow 2[4CaO{\cdot}(Al_2O_3{\cdot}Fe_2O_3){\cdot}13H_2O] \quad (4)$$

In an environment of weakly acidic NH_4Cl solution, calcium in the form of $Ca(OH)_2$ is extracted as free Ca^{2+} ions, predominantly originating from C_3S and C_2S, as these are the clinker components capable of producing $Ca(OH)_2$ post-hydration. The process for Ca^{2+} extraction is described by Equation (5).

$$Ca(OH)_2 + 2NH_4Cl \rightarrow 2NH_3{\cdot}H_2O + CaCl_2 \quad (5)$$

Subsequently, a 1.8% sodium hydroxide solution, which includes tri-ethanol amine, is incrementally introduced to the Ca^{2+} solution for testing. The solution's pH is adjusted to fall within the strongly alkaline range of 12.5 to 13.0. The core of the EDTA titration method is the chelation reaction where EDTA-2Na captures Ca^{2+} ions within an alkaline setting. This environment also ensures that other ions potentially present in the cement, such as Fe^{3+} and Al^{3+}, which could otherwise interfere with the accuracy of the test, do not react with EDTA-2Na, thereby minimizing testing errors [28].

The main purpose of adding tri-ethanol amine to sodium hydroxide solutions is to assist in adjusting the pH of the solution. Since the accuracy and reproducibility of the reaction depends on maintaining constant alkaline conditions, tri-ethanol amine is used as a buffer to help stabilize the pH (i.e., the rate of change in pH decreases when a certain amount of acid or alkali is added to the solution containing the buffer, avoiding drastic changes in pH that could affect the test results).

The addition of tri-ethanol amine to the sodium hydroxide solution serves primarily to modulate the solution's pH. The consistency and repeatability of the reaction are contingent upon the maintenance of a stable alkaline environment, hence the use of tri-ethanol amine as a buffer. This buffer mitigates fluctuations in pH when acids or bases are introduced, ensuring the titration test remains unaffected by abrupt pH changes. EDTA-2Na exhibits a higher affinity for Ca^{2+} compared to calcium-carboxylate sodium. Throughout the titration, EDTA-2Na binds with Ca^{2+} in the solution under examination, liberating calcium-carboxylate sodium ions and culminating in a pure blue solution, indicating the titration endpoint, as described in Equation (6).

$$Ca^{2+} + \text{EDTA-2Na} \rightarrow \text{EDTA-Ca} + 2Na^{+} \quad (6)$$

The chemical principle underlying this titration process hinges on the pH range being in the 12.5 to 13.0 range, within which EDTA-2Na readily forms a colorless complex with Ca^{2+}. In contrast, calcium-carboxylate sodium creates a less stable rose-red complex with Ca^{2+}. As the titration progresses, and the endpoint is approached, the standard solution incrementally captures the Ca^{2+} from the red complex. This gradual process changes the solution's color to pure blue, marking the completion of the titration.

2.3. EDTA Titration Test Procedure

The goal of the EDTA titration test is to benchmark against the designed cement content of the actual project, creating samples with ±2% and ±4% variations in the designed dosage. Each sample group is titrated to determine the volume of titrant used, which is then plotted on a calibration curve. Samples from the site are then tested using standard titration to determine the volume used and locate the corresponding cement dosage on the calibration curve.

The EDTA titration test is useful for evaluating the composition and quality of CLSM mixes. Thus, this section details the methodology, starting with the initial preparation of the CLSM mixes and culminating in the creation of a calibration curve that relates cement dosage to the volume of titrant used. This procedure is important for the accurate measurement and analysis needed to understand the material characteristics.

(1) The process begins with the preparation of CLSM mixes: the necessary quantities of each material are weighed according to specified ratios to guarantee mix uniformity. Initially, cement and water are mixed to react and form a slurry, stirring 25 times along the container circumference. Subsequently, soil is incorporated, followed by another 25 rounds of stirring. The CLSM mix is now complete.

(2) For sample preparation and calcium ion extraction, two samples of CLSM titration are prepared in parallel, each weighing 100 g, which are dissolved in 200 mL of a 10% weakly acidic NH_4Cl solution. The mix is stirred for 3 min at a rate of 110–120 RPM to ensure thorough mixing. After reaction and mixing, the solution is allowed to settle until the upper layer is clear and transparent, and it is then transferred to a beaker for further stirring.

(3) To prepare the solution for testing, 10.0 mL of the clear liquid is pipetted from 1 to 2 cm below the surface and transferred into a 200 mL conical flask. A 50 mL measure of 1.8% sodium hydroxide solution containing tri-ethanolamine is gradually added to the flask. The solution's pH is adjusted to move it into the strongly alkaline range of 12.5–13.0, followed by the addition of a small dosage of calcium-carboxylate sodium (approximately 0.05 g) until the solution turns rose-red.

(4) During the titration test, disodium EDTA standard solution is added from a burette, with the initial volume V_1 recorded. The solution is shaken well during titration, and the color change is carefully observed. As the solution shifts to purple, the titration is slowed, and when it turns pure blue, indicating the endpoint, the final volume V_2 is recorded. The difference V_2–V_1 represents the volume of titrant which indicates stabilization is complete.

(5) To construct the calibration curve, specimens are titrated using the same method across different volumes of cement, and the volume of titrant used is recorded. The average volume of titrant used for CLSM samples of the same proportion is plotted against the cement dosage (%).

Figure 1 shows the test procedure and details of the EDTA titration method.

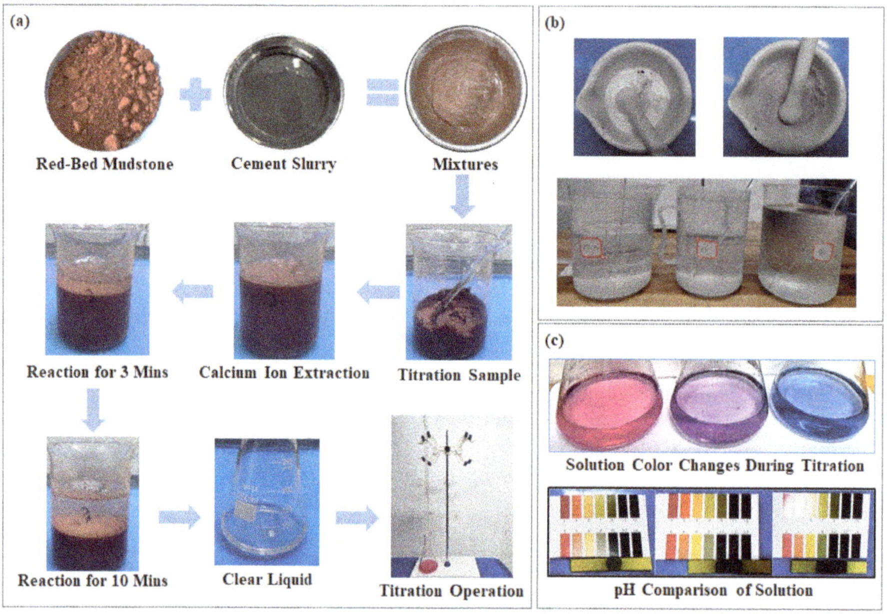

Figure 1. EDTA titration test: (**a**) Procedure; (**b**) Reagent preparation; (**c**) Phenomena observed during testing.

2.4. Test Program

The test program is divided into two methodologies, the "one-factor-at-a-time" (OFAT) experimental design and the orthogonal design. The OFAT approach involves varying a single factor while keeping others constant to isolate and understand its individual impact on the outcome. This method helps in identifying key variables and their direct effects. In contrast, the orthogonal design adopts a more complex strategy, allowing for the simultaneous variation in multiple factors. This method evaluates both the individual and combined effects of each factor with fewer trials, offering an additional understanding of the interactions at play. It should be noted that a selection of OFAT test groups is used in orthogonal experiments to enhance efficiency.

2.4.1. One-Factor Testing Methodology

CLSMs are characterized by their high water content compared to typical cement-stabilized materials, with the preparation, casting, and transportation phases influencing the volume of titrant used (V). This research studies the influence of three specific variables, cement content, water content and hydration time, and their effect on the volume of titrant needed for stabilization. Cement content (C) is defined as the mass ratio of cement to dry soil, water content (W) represents the mass ratio of water to the total solid content (dry soil plus cement), and hydration time (T) starts from the moment the CLSM mix is completed. The experimental protocol was as follows:

(1) Maintaining a constant water content at 40%, the cement content was adjusted from 6% to 18% in increments of 2%, resulting in the preparation of seven CLSM sample groups. These were then subject to titration tests to explore the correlation between the volume of titrant used and cement, ignoring the effect of hydration time.
(2) With the cement content set at 10%, the water content was varied from 38% to 50% in 2% increments, forming seven fluid curing soil samples. This part of the study concentrated on the relationship between the volume of titrant used and water content, again excluding the influence of time.
(3) The cement content and moisture content were fixed at 10% and 40%, respectively. Titration samples were prepared and tested at intervals ranging from 0 to 300 min post-preparation. The objective was to study fluctuations in titrant volume as a function of hydration time. Table 2 summarizes the OFAT experimental design.

Table 2. OFAT experimental design overview.

Scenario ID	Cement Content (*C*)/%	Water Content (*W*)/%	Hydration Time (*T*)/min
1	6, 8, ... 18	40	0
2	10	38, 40, ... 50	0
3	10	40	0, 30, 60, 120, 180, 240, 300

2.4.2. Orthogonal Testing Methodology

The orthogonal experimental design integrates the strengths of orthogonal experimentation and regression analysis to evaluate the interactions, primary and secondary effects, and significance of multiple factors [30–32]. Selecting suitable experiments across various factor levels facilitates the development of a regression equation with precise and robust statistical attributes using a minimal number of trials [33–35].

The study considers three factors, cement content (x_1), water content (x_2), hydration time (x_3), and their interaction effects (x_1x_2, x_1x_3, x_2x_3). Based on the characteristics of the raw soil and the project requirements, the ranges of cement content (x_1), water content (x_2), and hydration time (x_3) are set between 7% and 13%, 40% and 48%, and 90 and 210 min, respectively. The orthogonal design regression method determines the relationship between the volume of titrant and each factor, which is then predicted and evaluated via regression.

For a given factor's value range [x_{j1}, x_{j2}], where x_{j1} and x_{j2} represent the lower and upper levels, respectively, the mean x_{j0} is computed and designated as the factor's zero level. The difference between the upper level x_{j2} and zero level x_{j0}, or between zero level x_{j0} and lower level x_{j1}, defines the factor's increment Δ_j. Factor levels are linearized and their codes calculated by substituting the levels x_{j0}, x_{j1}, and x_{j2} into the designated equations, as detailed in Table S1.

Upon acquiring the factor level coding table, the next steps involved selecting an orthogonal array table for the experimental design and outlining the experimental program. This process took into account the three primary factors, cement content, water content and hydration time, along with the effect of their interaction on the volume of titrant needed to indicate stabilization. These factors and interactions collectively occupied six columns in the orthogonal table, necessitating the use of an L_8 (2^7) orthogonal table to organize the experiment. Subsequent to the code conversion, a regression orthogonal table was formulated. In adherence to orthogonal table design principles, the level codes z_1, z_2, and z_3 were placed in columns 1, 2, and 4. The corresponding interaction effects, z_1z_2, z_1z_3, and z_2z_3, were then assigned to columns 3, 5, and 6. To determine the total number of experiments, three zero level tests were performed, bringing the count to eleven. Table 3 illustrates the regression orthogonal array design for the experiment, omitting any empty columns for clarity. Figure 2 provides an overview of the procedure for the orthogonal experimental design as part of the full experimental program.

$$x_{j0} = \frac{x_{j1} + x_{j2}}{2} \tag{7}$$

$$\Delta_j = x_{j2} - x_{j0} = x_{j0} - x_{j1} \tag{8}$$

$$z_j = \frac{x_j - x_{j0}}{\Delta_j} \tag{9}$$

Table 3. Regression orthogonal array in experimental design.

ID	z_1	z_2	z_1z_2	z_3	z_1z_3	z_2z_3	x_1/%	x_2/%	x_3/min
1	1	1	1	1	1	1	13	48	210
2	1	1	1	−1	−1	−1	13	48	90
3	1	−1	−1	1	1	−1	13	40	210
4	1	−1	−1	−1	−1	1	13	40	90
5	−1	1	−1	1	−1	1	7	48	210
6	−1	1	−1	−1	1	−1	7	48	90
7	−1	−1	1	1	−1	−1	7	40	210
8	−1	−1	1	−1	1	1	7	40	90
9	0	0	0	0	0	0	10	44	150
10	0	0	0	0	0	0	10	44	150
11	0	0	0	0	0	0	10	44	150

Note: z_1, Factor 1; z_2, Factor 2; z_1z_2, z_1z_3, z_2z_3, interaction; z_3, Factor 3; x_1, cement content; x_2, water content; x_3, hydration time.

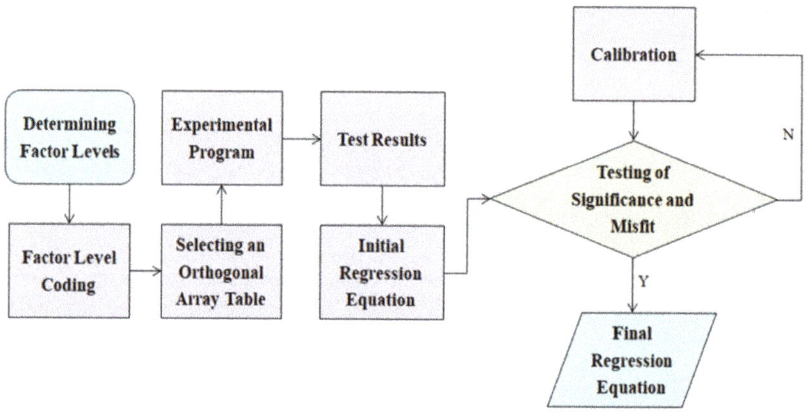

Figure 2. Orthogonal experimental design.

3. Results and Analysis

This section details the results of the experiments and their analysis. It begins by exploring the effects of the cement content, water content, and hydration time on the volume of titrant needed for stabilization using the OFAT method. It then progresses to examine the findings from the regression orthogonal tests. This includes the development and adjustment of a regression equation, assessment of its significance, and error analysis. It offers insights into the various factors that affect the detection of the cement content in CLSM.

3.1. One-Factor Test Results

3.1.1. Cement Content

Figure 3 illustrates the relationship between the volume of titrant needed for stabilization and cement content, showing a positive linear correlation. As the cement increases, so do the quantities of tri-calcium silicate and di-calcium silicate, leading to a higher production of $Ca(OH)_2$ through the hydration reaction. This increase in $Ca(OH)_2$ consequently elevates the level of soluble Ca^{2+}, resulting in an additional volume of titrant required.

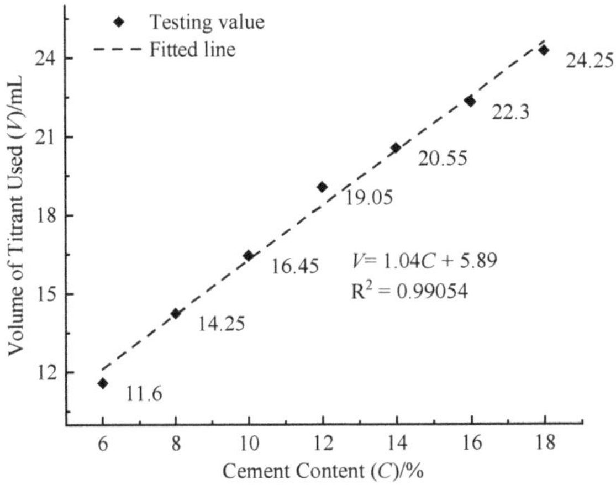

Figure 3. Relationship between volume of titrant used (V) and cement content (C).

3.1.2. Water Content

Figure 4 presents the relationship between the volume of titrant used and water content, showing a negative linear correlation. The hydration reaction of cement clinkers, influenced by factors such as the content and ratio of the clinkers, ion concentration, temperature, and mixing conditions, is inherently complex. CLSM usually obtains good flow performance with high water content, which is convenient for pouring construction. Under the condition of ultra-high water/cement ratio, there is more water around the cement particles, and the hydration products are diluted in the early formation. This dilution effect reduces the deposition of cement hydration products on the surface of the particles and hinders the further hydration reaction of cement. An increase in water content dilutes the ion concentration within the cement hydration reaction, thereby decelerating the reaction rate. This reduction in reaction velocity leads to a decrease in $Ca(OH)_2$ production, subsequently lowering the amount of soluble Ca^{2+} and reducing the volume of titrant needed.

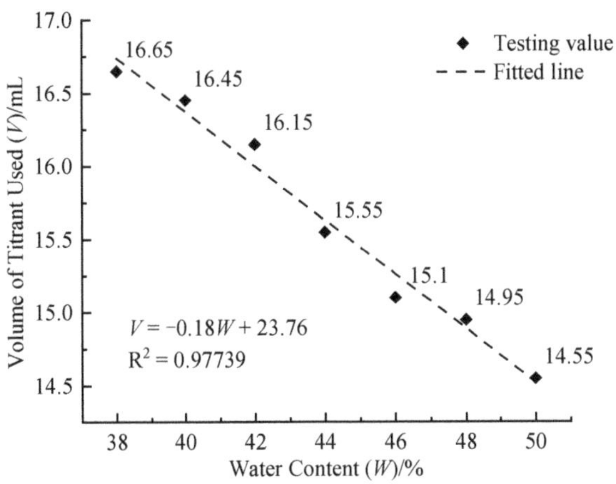

Figure 4. Relationship between volume of titrant used (V) and water content (W).

3.1.3. Hydration Time

Figure 5 shows the correlation between the volume of titrant used and hydration time, which exhibits a negative linear relationship. As the hydration time of the cement progresses, some of the Ca^{2+} ions react with soil minerals to form new compounds, altering the original structure. This transformation contributes to the gradual formation of a stable monolithic structure, reducing the NH_4Cl solution's extraction rate of Ca^{2+}. The resultant decrease in extracted Ca^{2+} lowers the required volume of titrant needed to detect stabilization.

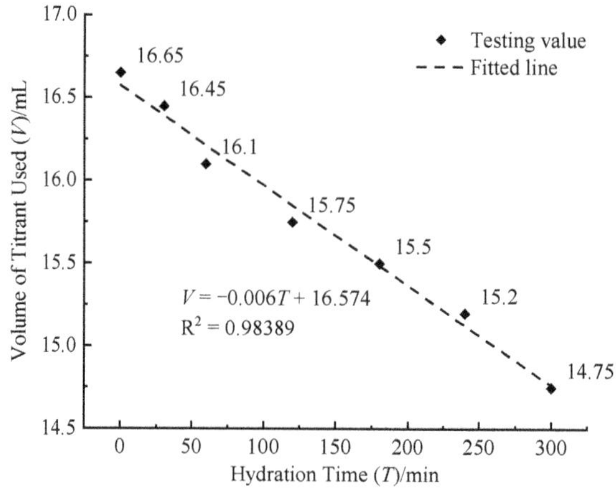

Figure 5. Relationship between volume of titrant used (V) and hydration time (T).

3.2. Orthogonal Test Results

CLSM samples were prepared in accordance with the ratios outlined in Table 4, requiring two samples for each specified ratio. The titration data encompass 11 groups of parallel tests, with the error margin for each group less than 5%. This level of precision indicates the titration results are reliable and comply with the allowable error margins for repeatability tests as stipulated in the specifications.

Table 4. Orthogonal testing data.

ID	Initial Volume V_1/mL	Final Volume V_2/mL	Volume of Titrant Used/mL	Repeat Initial Volume V_1/mL	Repeat Final Volume V_2/mL	Repeat Volume of Titrant Used/mL	Average Volume of Titrant Used/mL	Error/%
1	17.8	36.3	18.5	3.2	21.4	18.4	18.45	0.54
2	12.8	31.7	18.9	3.8	23.0	19.2	19.05	1.57
3	21.8	41.7	19.9	4.0	23.8	19.8	19.85	0.50
4	23.3	44.0	20.7	3.0	24.0	21.0	20.85	1.44
5	24.0	36.2	12.2	36.3	48.4	12.1	12.15	0.82
6	24.4	37.2	12.8	2.2	14.8	12.6	12.70	1.57
7	15.3	28.5	13.2	29.8	43.3	13.5	13.35	2.25
8	0.5	14.8	14.3	14.3	28.4	14.1	14.20	1.41
9	27.7	43.8	16.1	2.2	18.2	16.0	16.05	0.62
10	18.6	34.4	15.8	4.6	20.5	15.9	15.85	0.63
11	12.6	28.5	15.9	3.8	19.9	16.1	16.00	1.25

3.2.1. Regression Analysis

Out of the total 11 trials conducted, 8 were bi-level trials set at the upper and lower levels ($m_c = 8$), and 3 were zero level trials ($m_0 = 3$). The regression equation is formulated as shown in Equation (10), where V represents the predicted volume of titrant used, a, b_j, and b_{kj} denote the regression coefficients, and x_k and x_j are the influencing factors.

$$V = a + \sum_{j=1}^{3} b_j x_j + \sum_{k<j} b_{kj} x_k x_j,\ j \neq k \tag{10}$$

The volume of titrant used in each trial is represented as V_i (i = 1, 2, . . . 11), with the regression coefficients a, b_j, and b_{kj} derived from Equation (10). Within these equations, z_{ji} represents the z_j column and $(z_k z_j)_i$ represents the $z_k z_j$ column, as detailed in Table 4.

$$a = \frac{1}{11}\sum_{i=1}^{11} V_i \tag{11}$$

$$b_j = \frac{\sum_{i=1}^{11} z_{ji} V_i}{m_c},\ j = 1,2,3 \tag{12}$$

$$b_{kj} = \frac{\sum_{i=1}^{11} (z_k z_j)_i V_i}{m_c},\ k < j,\ k = 1,\ 2 \tag{13}$$

Using the average volume of titrant used (V_i) from the trials listed in Table 5 and integrating this with the regression orthogonal array design from Table 4, a calculation table for the regression orthogonal experimental design is shown in Table S2. The regression coefficients were determined according to Equations (11)–(13), resulting in the coded regression equation presented in Equation (14).

$$V = 16.2273 + 3.2250z_1 - 0.7375z_2 - 0.3750z_3 - 0.0625z_1z_2 - 0.0250z_1z_3 + 0.0875z_2z_3 \tag{14}$$

Table 5. Comparative analysis of test values and calculated volumes of titrant used.

ID	C/%	W/%	T/min	Test Value/mL	Calculated Value/mL	Absolute Error/mL	Relative Error/%
1	13	48	210	18.45	18.34	0.11	0.60
2	13	48	90	19.05	19.09	0.04	0.21
3	13	40	210	19.85	19.81	0.04	0.20
4	13	40	90	20.85	20.56	0.29	1.39
5	7	48	210	12.15	11.89	0.26	2.14
6	7	48	90	12.70	12.64	0.06	0.47
7	7	40	210	13.35	13.36	0.01	0.07
8	7	40	90	14.20	14.11	0.09	0.63
9	10	44	150	16.05	16.23	0.18	1.12
10	10	44	150	15.85	16.23	0.38	2.40
11	10	44	150	16.00	16.23	0.23	1.44
12	8	46	120	14.00	13.90	0.10	0.71
13	10	42	180	16.15	16.41	0.26	1.61
14	12	44	90	18.95	18.75	0.20	1.06
15	12	50	0	19.20	18.21	0.99	5.16
16	12	42	0	18.65	19.68	1.03	5.52
17	16	38	240	22.15	23.22	1.07	4.83

Through this methodology, the relative impact of each factor and their interactions were determined based on the absolute values of the partial regression coefficients. The order of influence was found to be as follows: cement content > water content > hydration time > interaction. Notably, the volume of titrant used exhibited a positive correlation with cement content and a negative correlation with both water content and hydration time.

3.2.2. Significance and Misfit Testing

The total sum of squared deviations for the volume of titrant needed to indicate stabilization (V_i) in the experimental data were calculated using Equation (15), with a degree of freedom (f_T) of $n - 1 = 10$. The squared deviations for the regression of the primary term were determined using Equation (16), each with a degree of freedom (f_j) of 1. Similarly, the squared deviations for the regression of the interaction term were computed using Equation (17), with each interaction term's degree of freedom (f_{kj}) also being 1.0.

$$S_T^2 = \sum_{i=1}^{11} (V_i - \overline{V})^2 = \sum_{i=1}^{11} V_i^2 - \frac{1}{11}\left(\sum_{i=1}^{11} V_i\right)^2 \tag{15}$$

$$S_j^2 = m_c b_j^2,\ j = 1,2,3 \tag{16}$$

$$S_{kj}^2 = m_c b_{kj}^2,\ k < j,\ k = 1,2 \tag{17}$$

The regression sum of squared deviations is obtained by summing the squared deviations of the primary and interaction terms as per Equation (18). The degree of freedom for the regression sum of squared deviations (f_R) equals the total degrees of freedom for the primary and interaction terms, calculated as $f_R = m = 3f_j + 3f_{kj} = 6$. The residual sum of squares is derived from Equation (19), with degrees of freedom (f_e) calculated as $n - m - 1 = 4$.

$$S_R^2 = \sum S_j^2 + \sum S_{kj}^2 \tag{18}$$

$$S_e^2 = S_T^2 - S_R^2 \tag{19}$$

The mean square (MS), or the average sum of squared deviations, is calculated by dividing the sum of squared deviations (S^2) for each source of variation by its corresponding degree of freedom (f), as outlined in Equation (20). The F value for the significance test of each factor and the regression equation is determined using Equation (21), with MS_e representing the mean square of the residuals. The results of the ANOVA (Analysis of Variance) are presented in Table S3 (See the Supplementary Materials).

$$MS = \frac{S^2}{f} \tag{20}$$

$$F = \frac{MS}{MS_e} \tag{21}$$

From Table S3, it is evident that at the significance level $\alpha = 0.01$, only the factors z_1, z_2, and z_3 are significant, whereas none of the interactions (z_{12}, z_{13}, z_{23}) showed significant effects. The coded regression Equation (14) passed the significance test, indicating that these factors significantly influence the trial outcomes relative to the residual sum of squares and that the regression equation adequately fits the test data at the trial points. However, a misfit test is also necessary to validate the fit across the entire range of factor level trials.

With three zero level trials ($m_0 = 3$), the repeatability error was calculated using Equation (22), and the trial error's degree of freedom (f_{e1}) was $m_0 - 1 = 2$. The sum of misfit squares, calculated using Equation (23), corresponds to a degree of freedom (f_{Lf}) of $f_e - f_{e1} = 2$. The significance test value (F_{Lf}) for the sum of misfit squares, calculated using Equation (24), adheres to the F distribution's degrees of freedom for the significance test (f_{Lf}, f_{e1}).

$$S_{e1}^2 = \sum_{i=1}^{3} (V_{0i} - \overline{V}_0)^2 = 0.02167 \tag{22}$$

$$S_{Lf}^2 = S_e^2 - S_{e1}^2 = 0.28138 \tag{23}$$

$$F_{Lf} = \frac{S_{Lf}^2/f_{Lf}}{S_{e1}^2/f_{e1}} = \frac{0.28138/2}{0.02167/2} = 12.98 \tag{24}$$

At a significance level of $\alpha = 0.1$, given $F_{0.1}(2,2) = 9.0 < F_{Lf} = 12.98$, the regression equation significantly fails the misfit test. This indicates that the existing regression model does not conform to the actual situation of the relationship between the test index and the influencing factors. The interaction between the factors should not be considered, and the previous coded regression model Equation (14) needs to be revised.

3.2.3. Calibration and Re-Testing the Regression Equation

When the regression equation fails the misfit test, a correction is necessary. Factors with a mean square (MS) smaller than the mean square of the error (MS_e) are reclassified as errors, forming a new error term. Subsequently, the F-value for each factor is recalculated and a significance analysis is conducted. As indicated in Table S3, the mean squares of the three interactions (z_{12}, z_{13}, and z_{23}) were less than the error mean squares and were thus reclassified as errors. The corrected ANOVA results are displayed in Table S4.

A F_{Lf} value of 6.97 was calculated and at a significance level of $\alpha = 0.1$, where $F_{0.1}(5, 2) = 9.29$ exceeds F_{Lf}, the misfit test is deemed not significant, suggesting the regression equation is plausible. The three interactions, now classified as errors, are removed from the regression equation without impacting the remaining regression coefficients. The updated coded regression equation is presented in Equation (25). Equation (26) is derived from Equation (9) and integrated into Equation (25), yielding the final regression equation shown in Equation (27).

$$V = 3.2250z_1 - 0.7375z_2 - 0.3750z_3 + 16.27 \tag{25}$$

$$\begin{cases} z_1 = \frac{x_1 - 10}{3} \\ z_2 = \frac{x_2 - 44}{4} \\ z_3 = \frac{x_3 - 150}{60} \end{cases} \tag{26}$$

$$\begin{aligned} V &= 1.075x_1 - 0.184375x_2 - 0.00625x_3 + 14.5273 \\ &= 1.075C - 0.184375W - 0.00625T + 14.5273 \end{aligned} \tag{27}$$

3.2.4. Error Analysis

To evaluate the accuracy of the regression equation, six sample groups with factor levels distinct from those in the regression orthogonal test were subjected to titration tests. The results, juxtaposed with the calculated values, are detailed in Table 5. The initial groups, numbered 1 through 11, belong to the regression orthogonal test category, where the absolute error varies between 0.01 mL and 0.38 mL, with relative errors spanning from 0.07% to 2.40%. These figures fall below the 5% repeatability error threshold set for parallel testing. Conversely, groups 12 to 17, designated as error test groups, demonstrate varying degrees of relative error. Specifically, for groups 12 to 14, which remained within the factor level range of the regression orthogonal test, the relative error peaked at a modest 1.61%. However, for groups 15 to 17, where factor levels surpassed the test group's range, the relative error was higher, with the lowest recorded being 4.83%. On the one hand, this is because when the level of factors taken is outside the range of the regressive orthogonal test group, more variables and uncertainties will be introduced into the test, and the random error may increase, resulting in the increased instability of the prediction results. On the other hand, when the factor range exceeds the limit, we do not grasp the change law of the test index with the influencing factors, and we do not consider the design of the regression orthogonal experiment scheme on this basis. This difference emphasizes the robustness of the regression equation within the predefined factor level range of the test group, while emphasizing the significant increase in the prediction error beyond this range.

Using Equation (27), the impact of changes in water content (Δw) and the extension of hydration time (Δt) on the cement content detection error (Δc) was analyzed, as depicted in Figure 6. An increase in water content from 2.9% to 5.8% or an extension in hydration time from 86 to 172 min reduced the cement content detection by approximately 0.5% to 1%, consistent with the trends observed in the volume of titrant. The regression equation thus allows for a comprehensive consideration of the effects of water content fluctuation and construction time on cement dosage detection in CLSM preparation, offering support for maintaining construction quality.

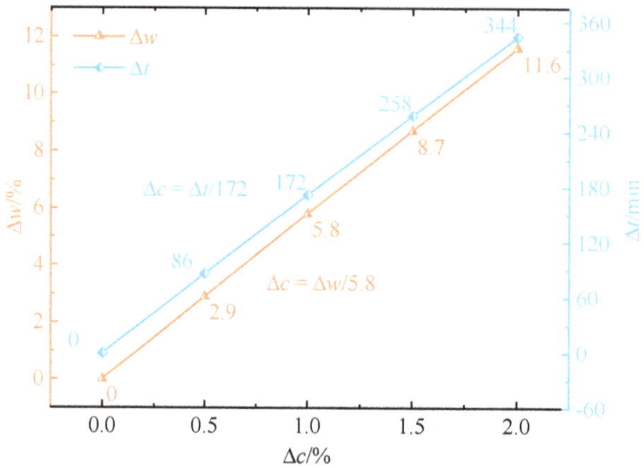

Figure 6. Variation in cement content detection error (Δc) with changes in water content (Δw) and hydration time (Δt).

4. Conclusions

In this study, red-bed mudstone was repurposed to develop sustainable Controlled Low-Strength Material (CLSM) mixes, contributing to circular economy principles by promoting the sustainable use of resources. The research examined the impacts of cement content, water content, and hydration time on the volume of titrant used for cement content detection, employing a regression orthogonal experimental design to evaluate these factors comprehensively. The findings underscore the critical role of precise cement content determination in achieving material efficiency and sustainability in construction practices.

Firstly, the volume of titrant required for detection stabilization was positively correlated with the cement content, with a correlation coefficient of 1.075. It was inversely correlated with water content and hydration time, with correlation coefficients of 0.184375 and 0.00625, respectively. Among these factors, cement content emerged as the most significant, followed by water content, and then hydration time. The interactions between these variables were found to be negligible, indicating that each factor independently influences the detection process.

Secondly, a regression equation was developed to characterize the combined effects of cement content, water content, and hydration time. An error analysis revealed that the error was within the range of 0.07% to 2.40% for factor levels included in the regression orthogonal test group. However, for data points outside the tested range, error rates increased between 4.83% and 5.52%, highlighting the model's limitations when applied beyond the experimental scope.

Thirdly, fluctuations in water content and hydration time during CLSM preparation were shown to affect the accuracy of cement content detection. Specifically, the results indicated that variations in water content from 2.9% to 5.8% or hydration time from 86 to 172 min could lead to discrepancies ranging from 0.5% to 1% in detected cement content when compared to the standard curve employed in the test protocol. This underscores the importance of maintaining strict control over these parameters during preparation to ensure consistent and accurate test results.

Lastly, the EDTA titration method was validated as an effective tool for detecting the cement content of CLSM. The study identified the variation rules of the test results with influencing factors and introduced reasonable corrections, confirming the method's feasibility. However, it was noted that additional factors, beyond the primary ones considered here, could influence the detection results. Different raw materials require separate and rigorous testing. This study provides an example for researchers and highlights the need for the comprehensive consideration of influencing factors in subsequent experiments, supported by comparative validation.

Supplementary Materials: The following supporting information can be downloaded at: https://www.mdpi.com/article/10.3390/ma17235915/s1, Figure S1: Particle size distributions of red-bed mudstone; Figure S2: Particle size distributions of cement; Table S1: Factor level coding for variables; Table S2: Data analysis for the regression orthogonal experimental design; Table S3: Variance analysis of factors and interactions; Table S4: Variance analysis after factor reassessment.

Author Contributions: Conceptualization, L.J. and Q.L.; Formal analysis, Y.L. and L.Q.; Investigation, Y.L.; Resources, L.Q.; Writing—original draft, Y.L.; Writing—review & editing, D.P.C. and T.W.; Supervision, L.J. and D.P.C.; Project administration, Q.L.; Funding acquisition, T.W. All authors have read and agreed to the published version of the manuscript.

Funding: This research was funded by the Major Program of Natural Science Foundation of Sichuan Province (Grant No. 2024NSFSC0003), the National Natural Science Foundation of China (Grant No. 52078435), the Natural Science Foundation of Sichuan Province (Grant No. 2023NSFSC0391), and the Overseas Expertise Introduction Project for Discipline Innovation ("111 Project", Grant No. B21011).

Institutional Review Board Statement: Not applicable.

Informed Consent Statement: Not applicable.

Data Availability Statement: The data presented in this study are available on request from the corresponding author. The data are not publicly available due to confidentiality agreements with our research partners.

Conflicts of Interest: Author Libing Qin was employed by the company Mingyang Yuheng Engineering Management Group Co., Ltd. The remaining authors declare that the research was conducted in the absence of any commercial or financial relationships that could be construed as a potential conflict of interest.

References

1. Chittoori, B.; Puppala, A.J.; Pedarla, A.; Vanga, D.P.R. Durability Studies on Native Soil-Based Controlled Low Strength Materials. In Proceedings of the Ground Improvement and Geosynthetics, Shanghai, China, 26–28 May 2014; American Society of Civil Engineers: Shanghai, China, 2014; pp. 249–257.
2. Wang, C. A comprehensive review on mechanical properties of green controlled low strength materials. *Constr. Build. Mater.* **2023**, *363*, 129611. [CrossRef]
3. Chittoori, B.; Puppala, A.J.; Raavi, A. Strength and Stiffness Characterization of Controlled Low-Strength Material Using Native High-Plasticity Clay. *J. Mater. Civ. Eng.* **2014**, *26*, 04014007. [CrossRef]
4. Chen, W.; Qu, S.; Lin, L.; Luo, Q.; Wang, T. Ensemble Learning Methods for Shear Strength Prediction of Fly Ash-Amended Soils with Lignin Reinforcement. *J. Mater. Civ. Eng.* **2023**, *35*, 04023022. [CrossRef]
5. Chompoorat, T. Engineering properties of controlled low-strength material (CLSM) as an alternative fill material. *Constr. Build. Mater.* **2018**, *259*, 277–285.
6. Do, T.M.; Kim, Y. Engineering properties of controlled low strength material (CLSM) incorporating red mud. *Geo-Engineering* **2016**, *7*, 7. [CrossRef]
7. Feng, J.; Yin, G.; Tuo, H.; Niu, Z. Parameter optimization and regression analysis for multi-index of hybrid fiber-reinforced recycled coarse aggregate concrete using orthogonal experimental design. *Constr. Build. Mater.* **2021**, *267*, 121013. [CrossRef]
8. Dalal, P.H.; Patil, M.; Dave, T.N.; Iyer, K.K.R. An experimental study on controlled low strength material (CLSM) for utilization as sustainable backfill. *Mater. Today Proc.* **2022**, *65*, 1178–1185. [CrossRef]
9. Lee, K.-J.; Kim, S.-K.; Lee, K.-H. Flowable Backfill Materials from Bottom Ash for Underground Pipeline. *Materials* **2014**, *7*, 3337–3352. [CrossRef]
10. Liu, H.; Luo, Q.; El Naggar, M.H.; Zhang, L.; Wang, T. Centrifuge Modeling of Stability of Embankment on Soft Soil Improved by Rigid Columns. *J. Geotech. Geoenviron. Eng.* **2023**, *149*, 04023069. [CrossRef]
11. Kim, Y.; Do, T.M.; Kim, H.; Kang, G. Utilization of excavated soil in coal ash-based controlled low strength material (CLSM). *Constr. Build. Mater.* **2016**, *124*, 598–605. [CrossRef]
12. Khadka, S.D.; Okuyucu, O.; Jayawickrama, P.W.; Senadheera, S. Controlled low strength materials (CLSM) activated with alkaline solution: Flowability, setting time and microstructural characteristics. *Case Stud. Constr. Mater.* **2023**, *18*, e01892. [CrossRef]

13. Qian, J.; Hu, Y.; Zhang, J.; Xiao, W.; Ling, J. Evaluation the performance of controlled low strength material made of excess excavated soil. *J. Clean. Prod.* **2019**, *214*, 79–88. [CrossRef]
14. Ling, T.-C.; Kaliyavaradhan, S.K.; Poon, C.S. Use of Native Soil in Place of Aggregate in Controlled Low Strength Material (CLSM). *Constr. Build. Mater.* **2018**, *158*, 535–548. [CrossRef]
15. Decký, M.; Remišová, E.; Mečár, M.; Bartuška, L.; Lizbetin, J.; Drevený, I. In situ Determination of Load Bearing Capacity of Soils on the Airfields. *Procedia Earth Planet. Sci.* **2015**, *15*, 11–18. [CrossRef]
16. Zhang, S.; Jiao, N.; Ding, J.; Guo, C.; Gao, P.; Wei, X. Utilization of waste marine dredged clay in preparing controlled low strength materials with polycarboxylate superplasticizer and ground granulated blast furnace slag. *J. Build. Eng.* **2023**, *76*, 107351. [CrossRef]
17. Feng, G.; Luo, Q.; Wang, T.; Connolly, D.P.; Liu, K. Frequency Spectra Analysis of Vertical Stress in Ballasted Track Foundations: Influence of Train Configuration and Subgrade Depth. *Transp. Geotech.* **2024**, *44*, 101167. [CrossRef]
18. Mahamat, A.A.; Dayyabu, A.; Sanusi, A.; Ado, M.; Obianyo, I.I.; Stanislas, T.T.; Bih, N.L. Dimensionnal stability and strength appraisal of termite hill soil stabilisation using hybrid bio-waste and cement for eco-friendly housing. *Heliyon* **2022**, *8*, e09406. [CrossRef]
19. Teerawattanasuk, C.; Voottipruex, P.; Horpibulsuk, S. Improved heavy metal immobilization of compacted clay by cement treatment. *Heliyon* **2021**, *7*, e06917. [CrossRef]
20. Zhang, L.; Hu, Q.; Luo, Q.; Connolly, D.P.; Wang, T. Assessing Workability of Ready-Mixed Soils Derived from Excess Spoil. *J. Mater. Civ. Eng.* **2024**, *36*, 04024331. [CrossRef]
21. Halliday, C.G.; Leonard, M.A. Selective and precise photometric titration of metal ions with EDTA using masking agents. Part III. Determination of calcium, lead and zinc in mixtures using simple inorganic ions as masks. *Analyst* **1987**, *112*, 1731. [CrossRef]
22. Hayashi, H. Determination of phosphorus and tungsten in heteropoly acids by EDTA-titration. *Talanta* **1982**, *29*, 943–945. [CrossRef] [PubMed]
23. Shen, W.; Zhou, M.; Yang, Z.; Yu, X. A Study on EDTA Titration Testing Method of Cement Dosage of Cement Stabilized Granular Soil. *Highway* **2007**, *52*, 46–50.
24. Wang, B.; Zhang, Y.; Song, W.; Chen, Y. Experimental Study on the Factor that EDTA Titration Detects the Cement Content of Cement Mixing Pile. *Sci. Technol. Eng.* **2015**, *15*, 270–274.
25. Zhang, F.; Zhang, L.; Yang, L.; Feng, T.; Chen, L.; Zhong, X. The Cement Content Measurement of Cement Mixing Piles with EDTA Titration Method. *KSCE J. Civ. Eng.* **2018**, *22*, 4306–4315. [CrossRef]
26. Zhang, C.; Liu, S.; Zhang, X. Age Effects on EDTA Titration in Detecting Cement Content for Treatment of Expansive Soil. *South-to-North Water Transf. Water Sci. Technol.* **2012**, *10*, 76–79.
27. Hošková, Š.; Tichá, P.; Demo, P. Determination of Ca^{2+} Ions at early stage of hydrating cement paste. *Ceramics* **2009**, *53*, 76–80.
28. Kim, J.; Vipulanandan, C. Effect of pH, sulfate and sodium on the EDTA titration of calcium. *Cem. Concr. Res.* **2003**, *33*, 621–627. [CrossRef]
29. Mounanga, P.; Khelidj, A.; Loukili, A.; Baroghel-Bouny, V. Predicting $Ca(OH)_2$ content and chemical shrinkage of hydrating cement pastes using analytical approach. *Cem. Concr. Res.* **2004**, *34*, 255–265. [CrossRef]
30. Pallavi; Joshi, S.; Singh, D.; Kaur, M.; Lee, H.-N. Comprehensive Review of Orthogonal Regression and Its Applications in Different Domains. *Arch. Comput. Methods Eng.* **2022**, *29*, 4027–4047. [CrossRef]
31. Han, X.; Guan, J.; Zhu, Q.; Han, Z. Application of quadratic regression orthogonal design to optimization surface heating for control wet steam condensation flow in nozzle. *Case Stud. Therm. Eng.* **2022**, *34*, 101987. [CrossRef]
32. Yang, Z.; Mou, X.; Ji, H.; Mao, Y.; Song, H. Orthogonal experimental study on the compressibility characteristics of bank sandy silt based on freeze-thaw effects. *Heliyon* **2024**, *10*, e31545. [CrossRef] [PubMed]
33. Wang, X.; Feng, J.; Wang, H.; Hong, S.; Zheng, S. Stress Regression Analysis of Asphalt Concrete Deck Pavement Based on Orthogonal Experimental Design and Interlayer Contact. *IOP Conf. Ser. Mater. Sci. Eng.* **2018**, *322*, 042026. [CrossRef]
34. Wu, M.; Liu, F.; Lin, Y.; Wang, M.; Zhou, S.; Zhang, C.; Mu, Y.; Han, G.; Hao, L. Parameter Optimization via Orthogonal Experiment to Improve Accuracy of Metakaolin Ceramics Fabricated by Direct Ink Writing. *Chin. J. Mech. Eng. Addit. Manuf. Front.* **2023**, *2*, 100098. [CrossRef]
35. Wu, M.; Jiang, S. Parameter Optimization of Piston Rod Rough Turning Based on Quadratic Orthogonal Regression. *IOP Conf. Ser. Mater. Sci. Eng.* **2018**, *382*, 042053. [CrossRef]

materials

MDPI

Article

The Synergistic Effect of Calcained Coal-Series Kaolinite and Limestone on the Hydration of Portland Cement

Jin Tang [1], Yue Yu [1], Yuanqing Bu [1], Bing Ma [1], Hao Zhou [1,*], Rong Zhou [1], Jiaqing Wang [2] and Houhu Zhang [1,*]

[1] Nanjing Institute of Environmental Sciences, Ministry of Ecology and Environment of the Peopole's Republic of China, Nanjing 210042, China; tangjin@nies.org (J.T.); yuyue@nies.org (Y.Y.); byq@nies.org (Y.B.); myyanb@aliyun.com (B.M.); zhourong@nies.org (R.Z.)

[2] College of Civil Engineering, Nanjing Forestry University, Nanjing 210037, China; jiaqingw@njfu.edu.cn

* Correspondence: zhouhao@nies.org (H.Z.); zhh@nies.org (H.Z.)

Abstract: Limestone calcined clay cement (LC^3) presents a promising alternative material due to its reduced CO_2 emissions and superior mechanical properties compared to traditional Portland cement (PC). This study investigates the synergistic effect of calcined coal-series kaolinite (CCK) and limestone (LS) on the hydration behavior of cement, specifically focusing on varying mass ratios. The combination of CCK and LS promotes the formation of strätlingite and carboaluminates, which enhances early-age strength development. Additionally, the inclusion of CCK facilitates the formation of carboaluminates during later stages of hydration. After 56 days of hydration, the content of carboaluminates is over 10%wt. This stimulation of secondary hydration products significantly refines the evolution of pore structure, with the harmful large pores gradually transformed into harmless medium pores and gel pores, leading to marked improvements in compressive strength from 7 to 28 days. Replacing 45% PC with CCK and LS at mass ratio of 7 to 2, the compressive strength of blends reaches 47.2 MPa at 28 days. Overall, the synergistic interaction between CCK and LS presents unique opportunities to minimize the CO_2 footprint of the cement industry without compromising early and long-term performance.

Keywords: calcined coal-series kaolinite; limestone; synergistic effect; carboaluminates

1. Introduction

The most effective strategy for mitigating the carbon footprint of the cement industry involves the partial substitution of clinker with supplementary cementitious materials (SCMs) [1,2]. In recent decades, the use of blended cements has increased to address the need for clinker reduction [3,4]. However, the application of traditional SCMs, such as fine limestone (LS), fly ash, and slag, is limited in their replacement levels, which restricts the potential for achieving a significant reduction in the clinker factor beyond 15% [5,6].

The synergistic effect of limestone and calcined clay contributes to the formation of additional hydration products and refines the pore structure of the cement during hydration [7]. Kaolinite-rich clays have gained increased attention for their potential to reduce the clinker factor in cement production. Compared to reference Portland cement (PC), limestone calcined clay cement (LC^3) exhibits similar or superior mechanical properties, enhanced pore structure evolution, and improved resistance to alkali–silica reactions and chloride penetration starting from 7 days of curing, even when employing a high replacement level of 50% [8,9]. Furthermore, LC^3 can reduce CO_2 emissions by 30% to 40% at a consistent mass level compared to PC [10,11].

Previous studies have indicated that the kaolinite content of clay is a significant factor influencing the mechanical, hydration, and durability performance of LC^3 [12–14]. However, kaolinite-rich clay also has high-value applications in industries such as ceramics, paper-making, and rubber. In contrast, clay with low kaolinite content is widely available globally [15]. Recent studies have focused on the effects of low-grade calcined clay and limestone on the hydration and mechanical properties of PC [14,16]. Key parameters investigated include the kaolinite content in the clay, the proportion of each raw material, and the cement substitution level [17,18]. Given that clay is a valuable resource in China, developing alternative clay materials as SCMs has positive implications for sustainable construction practices [19].

Coal-series kaolinite, a byproduct of coal mining, constitutes approximately 10% to 15% of annual coal production emissions, with reserves in China exceeding 4 billion tons to date. This type of kaolinite is rich in kaolinite, while impurities include quartz, carbon, and calcium carbonate, among others [20,21]. Similar to kaolinite clay, coal-series kaolinite undergoes thermal activation at temperatures between 500 °C and 900 °C, converting into metakaolin, which can serve as a sustainable pozzolan for cementitious phases [22]. During cement hydration, coal-series kaolinite reacts with portlandite to produce additional hydration products, such as calcium silicate hydrate (C-S-H), calcium aluminate hydrate (C-A-H), and calcium aluminosilicate hydrate (C-A-S-H), as represented in Equations (1) and (2).

$$AS_2 + 6CH + 9H \rightarrow 2C - S - H + C_4AH_{13} \quad (1)$$

$$AS_2 + 3CH + 6H \rightarrow C - S - H + C_2ASH_8 \quad (2)$$

where AS_2 means metakaolin, CH means portlandite, and H indicates water.

It is well established that LS reacts with tricalcium aluminate (C_3A) to form hemicarboaluminate (HC) and monocarboaluminate (MC), which improves the pore structure of the system during hydration [23,24]. This reaction promotes the stabilization of ettringite rather than its conversion into monosulfoaluminate. Lothenbach et al. reported that aluminum uptake in calcium silicate hydrate (C-S-H) plays a significant role in the LS–Portland cement system, indicating that the reactivity of the LS depends on the aluminum content (from C_3A and tetracalcium aluminoferrite (C_4AF)) and the sulfate content of the PC [25,26]. The pozzolanic reaction of CCK introduces additional aluminum into the gel during cement hydration, suggesting that a greater substitution of clinker can be achieved in the ternary system [27,28]. The synergistic effect of CCK and LS enhances the mechanical properties of the system, allowing for a reduction in clinker content by up to 50%, as illustrated in Equation (3) [29].

$$A + C\overline{C} + 3CH + H \rightarrow C_3A{\cdot}C\overline{C}{\cdot}H_{12} \quad (3)$$

where A indicates Al in hydrated gel, $C\overline{C}$ means calcite, $C_3A{\cdot}C\overline{C}{\cdot}H_{12}$ is MC.

Several studies have examined the effects of cement replacement levels, the quality of calcined clay, the replacement ratios of calcined clay, and other influencing factors on the strength development of LC^3 [30,31]. The synergistic effect of calcined clay and limestone optimized the pore structure evolution and hydration products; the workability and durability of the LC^3 improved [32,33].

Our previous research indicated that the incorporation of metakaolin consistently promotes the formation of HC and MC, refining the pore structure of blends during cement hydration [34]. When CCK and LS are mixed in a 2:1 ratio to substitute 30% of the cement, the mechanical properties of the sample exceed those of reference PC after 7 days of hydration [35]. Cardinard et al. reported that low-grade calcined clay exhibits a comparable degree of reactivity to that of high-grade calcined clay during long-term hydration [36]. Kaminskas et al. found that the pozzolanic reaction of calcined clay leads to the formation of strätlingite (C_2ASH_8), contributing to strength development at an early age [37]. The formation of HC and MC further enhances strength development at later ages. Additionally, Sun et al. discovered that the addition of calcium hydroxide (CH) effectively enhances the reaction of calcined clay, significantly improving the formation of carboaluminates [8].

Previous research has indicated that the formation of carboaluminates significantly influences pore structure and strength development [38,39]. Investigating the reaction between LS and its relationship with pore structure evolution and strength development is essential for understanding the hydration mechanisms of LC^3. In this study, we manipulate and control the mass ratio of CCK and LS to develop novel LC^3 with enhanced properties.

2. Materials and Methods

2.1. Raw Materials and Mixture Design

In this study, clinker, limestone, coal-series kaolinite, and gypsum were utilized to prepare cementitious materials. Clinker and gypsum were procured from Nanjing China United Cement Co., Ltd., while the LS was obtained from Suzhou Conch Cement Co., Ltd., and the coal-series kaolinite was sourced from a coal mine in Yunnan Province. The chemical composition of the raw materials, along with the mineral composition of the clinker, is presented in Table 1. The phase compositions of the raw materials were analyzed using X-ray diffraction (XRD), as illustrated in Figure 1. Prior to the experiments, the raw materials were ground, and Figure 2 depicts the particle size distribution curve of the raw materials. The D50 values for the clinker, calcined coal-series kaolinite, and limestone were 17.59 μm, 18.84 μm, and 6.17 μm, respectively.

Table 1. Chemical composition of raw materials and the XRD–Rietveld refinement of clinker.

		SiO_2	Al_2O_3	Fe_2O_3	TiO_2	CaO	MgO	SO_3	K_2O	Na_2O	TOL
Chemical Composition (%)	Clinker	21.29	5.66	3.78	0.29	65.99	1.16	0.37	0.50	0.32	0
	Gypsum	6.13	1.64	0.44	0.09	31.98	1.21	36.35	0.35	0.12	21.07
	Limestone	0.85	0.32	0.31	0.027	54.19	0.89	0.035	0.12	0.011	43.06
	CCK-raw	44.79	23.32	7.80	1.96	2.13	1.07	0.32	1.14	0.16	18.78
	CCK-750	50.22	28.83	9.03	2.52	2.26	1.24	0.39	1.33	0	3.33
Phase composition (%)		C_3S	C_2S	C_3A	C_4AF	CaO	$CaSO_4$		Blaine specific surface		
	Clinker	66.03	9.26	12.11	8.62	1.95	0.06		314 m^2/kg		

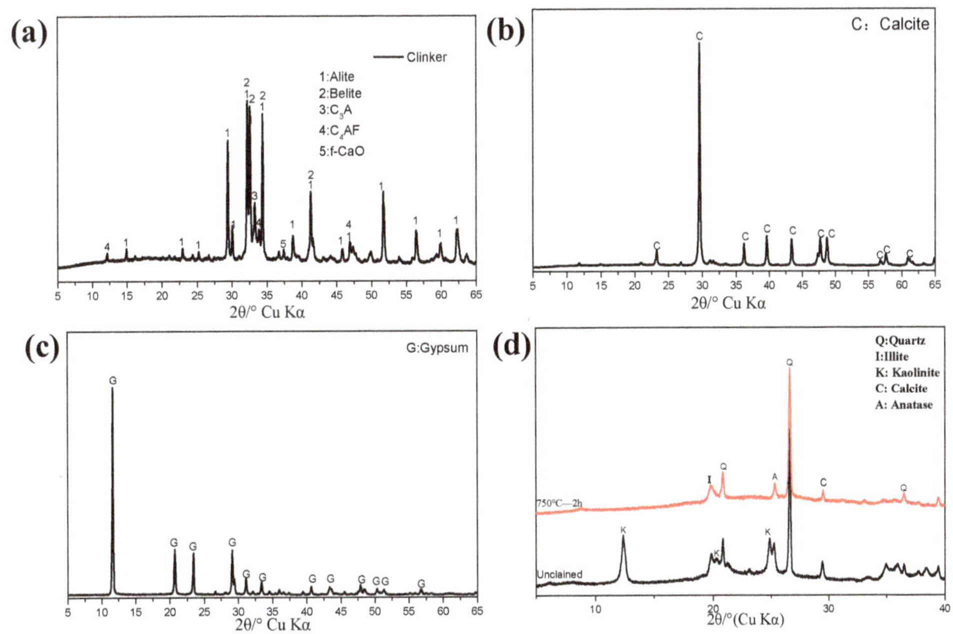

Figure 1. XRD patterns of (**a**) clinker, (**b**) limestone, (**c**) gypsum, and (**d**) coal-series kaolinite before and after 750 °C 2 h calcination.

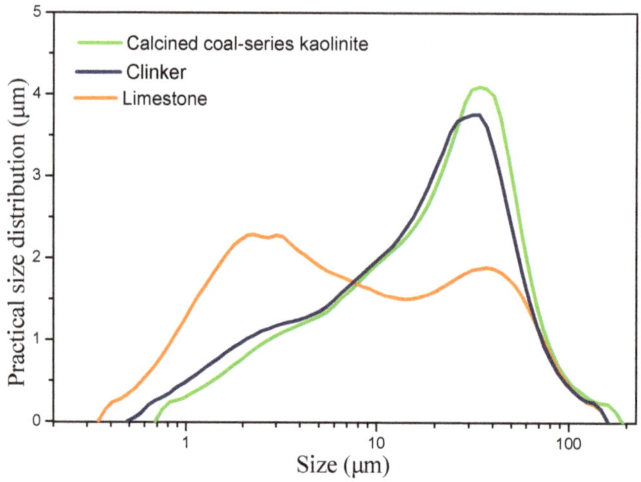

Figure 2. Particle size distribution curve of raw materials.

The coal-series kaolinite was calcined at 750 °C for 2 h in a muffle furnace to remove carbon and transform the kaolinite into metakaolin. The chemical compositions and X-ray diffraction (XRD) patterns of the coal-series kaolinite before and after calcination are presented in Table 1 and Figure 1d. Utilizing Equation (4), the kaolinite content in the coal-series kaolinite was calculated to be 47.75% by evaluating the dehydroxylation process of kaolinite during thermogravimetric (TG) analysis, as outlined in Equation (5).

$$AS_2H_2 \rightarrow AS_2 + 2H \tag{4}$$

$$Wt.\%_{kaolinite} = \frac{W_{350} - W_{575}}{W_{350}} \times \frac{258}{36} \times 100\% \tag{5}$$

where AS_2H_2 indicates kaolinite. $Wt.\%_{kaolinite}$ means the kaolinite content of coal-series kaolinite, W_{350} and W_{575} mean the weight of sample at 350 °C and 575 °C during TG experiment.

The ternary cementitious system was formulated using clinker, gypsum, LS, and CCK, maintaining a constant substitution rate of 45% for the latter two components. The detailed composition of the blends is provided in Table 2. All raw materials were accurately weighed according to the mix design and were subsequently combined in a mixing tank for 24 h to produce homogeneous cementitious materials.

Table 2. Mixture design.

	Clinker (%)	Gypsum (%)	CCK (%)	LS (%)
PC	95	5	-	-
PLC	52.25	2.75	-	45
PGC	52.25	2.75	45	-
PGLC1	52.25	2.75	15	30
PGLC2	52.25	2.75	22.5	22.5
PGLC3	52.25	2.75	30	15
PGLC4	52.25	2.75	35	10

2.2. Sample Preparations

Cement pastes were prepared using a laboratory mixer. In this experiment, the water-to-cement ratio was controlled as 0.5. The weighed blends and water were added to the mixer and homogenized at 500 rpm for 2 min. The prepared pastes were then placed into plastic tubes and cured in a curing box maintained at 98% humidity and 20 °C until the time of testing. At specified time intervals, the cured samples were precisely sectioned and immersed in alcohol for 2 days to halt the hydration process. Subsequently, the slices were vacuum-dried at 40 °C for 2 days.

Standard mortars were cast in accordance with the Chinese Standard GB/T 17671-1999 based on the mixture design to evaluate strength evolution during hydration. A total of 450 g of blends, 225 g of water, and 1350 g standard sand were used as raw materials. The mortars were poured into steel molds and mechanically vibrated for 1 min. Afterward, the molds were stored at 20 °C for 1 day. Strength testing was conducted at 3, 7, and 28 days post-casting, with the mortar strength determined by calculating the mean strength of three samples. A consistent water-to-blend ratio of 0.5 was applied across all mixtures in this study.

2.3. Characterization Methods

The phase composition for raw materials and hydration productions were measured by X-ray diffraction (XRD) measurement. XRD (Rigaku SmartLab 3000A diffractometer) was performed at a scanning rate of 5°/min over scanning range 2θ of 5°–70°; the step size was controlled as 0.01°.

The content of hydration products and the hydration degree of clinker during the hydration process was characterized by Q-XRD approach. α-Al_2O_3 (NISR standard SRM-676a) was selected as the standard material [40]. The degree of hydration (DoH) was determined by comparing the contents of clinker's primary minerals (C_3S, C_2S, C_3A, and C_4AF) before and after the hydration process, as shown in Equation (6).

$$\mathrm{DoH} = 1 - \frac{W_{C_3S}(t) + W_{C_2S}(t) + W_{C_3A}(t) + W_{C_4AF}(t)}{W_{C_3S}(t_0) + W_{C_2S}(t_0) + W_{C_3A}(t_0) + W_{C_4AF}(t_0)} \tag{6}$$

where $W_{C_3S}(t_0)$, $W_{C_2S}(t_0)$, $W_{C_3A}(t_0)$, and $W_{C_4AF}(t_0)$ represent the initial content of mineral phases in anhydrous blends before hydration, and $W_{C_3S}(t)$, $W_{C_2S}(t)$, $W_{C_3A}(t)$ and $W_{C_4AF}(t)$ represent the content of minerals in sample hydrated at t.

The heat release during early hydration at 20 °C was measured in an isothermal calorimeter (TAM Air). The cement pastes with water-to-cement ratio of 0.5 were mixed with a laboratory mixer at 500 rpm for 2 min. Then, 6 g of paste was placed in a plastic ampoule, sealed, and put in the calorimetry for 3 d.

Thermogravimetric analysis (TGA) was employed to determine the amount of bound water and CO_2 released while heating. This measurement condition of TG (Mettler Toledo TGA/DSC1) was as follows: The heating rate, the test temperature rate, and the atmosphere are 10 °C/min, 30 °C to 1000 °C, and nitrogen gas at 30 mL/min. The formation hydration products of the cement pastes were analyzed by monitoring the mass change of sample during heating process. The bound water content of the blends was obtained by calculating mass changes between 50 °C to 400 °C, which can be used to characterize the amount of hydration products at different hydration age. Typically, the content of CH in hydrated sample was evaluated based on the weight loss range from 400 °C to 470 °C, with the anhydrous sample mass at 550 °C as reference, according to Equation (7).

$$\mathrm{CH_{dry}} = \frac{W_{400} - W_{470}}{W_{550}} \times \frac{74}{18} \tag{7}$$

where W_{400}, W_{470}, and W_{550} indicate the mass of sample at 400 °C, 470 °C, and 550 °C during thermal analysis.

The pore structure evolution of the samples during the hydration was characterized using Mercury intrusion porosimeter (MIP) method (Quanta chrome PoreMaster 60 GT), which can generate pressure in the range of 1.5 KPa to 350 KPa. Approximately 1 g of prepared sample (3 to 4 pieces) was used in this test.

3. Results and Discussion

3.1. Effect of CCK and LS on the Mechanical Properties

Figure 3 illustrates the mechanical properties of cement mortars hydrated for 3, 7, and 28 days. The strength development of the PLC sample was significantly impaired by the dilution effect of LS, resulting in a reduction of compressive strength by 62.7%, 62.4%, and 54.2% at 3, 7, and 28 days, respectively, compared to the reference PC. Previous studies have indicated that LS can enhance the hydration of cement and generate additional hydration products; however, in this case, the dilution effect of LS was the predominant factor influencing the results. The compressive strength of PGC exceeded that of PLC at all hydration ages. According to Equations (1) and (2), the formation of abundant calcium–alumino–silicate hydrate (C-(A)-S-H) gel contributed to the enhanced strength development observed in PGC [41].

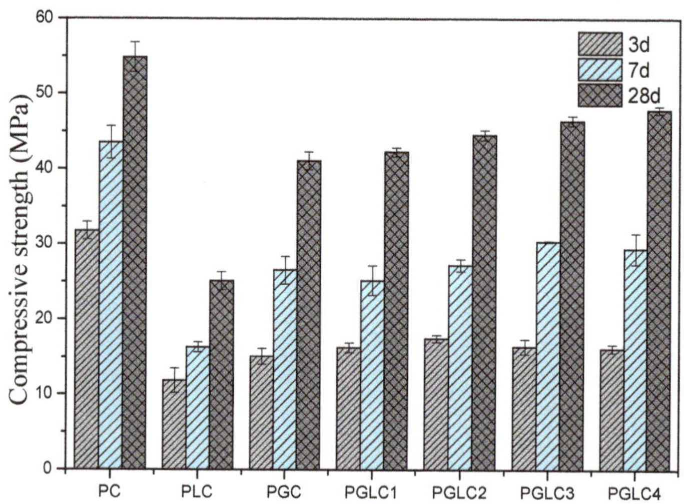

Figure 3. Effect of CCK and LS on the compressive strength of cement.

At early stages of hydration, the nucleation and synergistic effects of CCK and LS enhance the mechanical properties, resulting in the strength of PGLC being approximately 5 MPa higher than that of PLC at 3 days. The mass ratio of CCK to LS shows a minimal influence on strength development during the early hydration period, as all PGLC samples exhibited similar compressive strengths at 3 days. With the exception of PGLC1, the strengths of PGLC2, PGLC3, and PGLC4 were slightly higher than that of PGC, with this trend becoming more pronounced after 28 days of hydration. This outcome indicates that the synergistic effects of CCK and LS are more advantageous to the overall strength development of the system. The incorporation of CCK and LS as SCMs is an effective way to reduce the CO_2 footprint of cement industry. Conversely, at later hydration ages, the mass ratio of CCK to LS becomes the dominant factor influencing strength. As the mass ratio of CCK to LS increased from 1:2 to 7:2, the compressive strength of PGLC rose from 42.6 MPa to 47.2 MPa after 28 days of hydration. Although the CCK content in PGLC4 is significantly higher than in PGLC2, the strength of the blend only increased by 4.2 MPa (9.6%), indicating that there is an optimal dosage range for CCK to maximize strength development.

3.2. Effect of CCK and LS on the Cement Hydration

The isothermal calorimetry results for the various samples are displayed in Figure 4. Both CCK and LS accelerated the hydration of cement, particularly the hydration of tricalcium silicate (C_3S). Notably, the peak of the acceleration period decreased with increasing

CCK content, suggesting that the incorporation of CCK requires more free water than the cement itself during the mixing process.

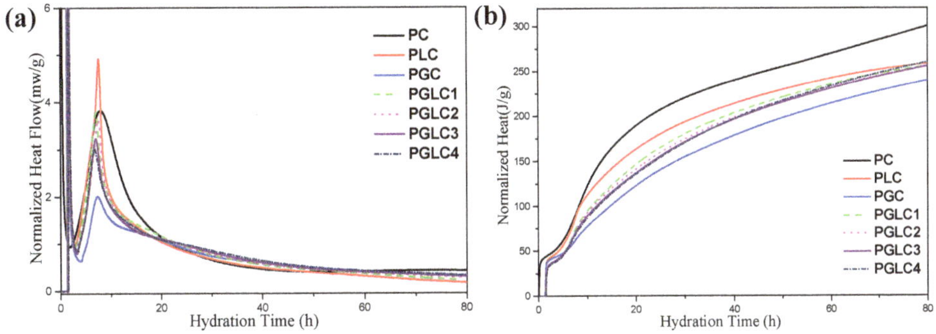

Figure 4. The isothermal calorimetry curves of per gram of blends (**a**) rate of hydration and (**b**) cumulative heat of hydration.

After 1 day of hydration, the heat released by the PGLC blends exceeded that of the reference PC and PLC (see Figure 4a). This enhanced heat release can be attributed to the pozzolanic reactions of CCK and the formation of carboaluminates.

The additional hydration products contributed to the higher strength results observed at 3 days (refer to Figure 3). During the first 3 days of hydration, PLC exhibited a higher total heat release compared to the PGLC blends, indicating that the hydration of C_3S played a dominant role in the early hydration phase.

However, after 80 h of hydration, the total heat release from PLC and PGLC blends became similar, with PGLC demonstrating a stronger heat release capacity in the later hydration stages. After 80 h of hydration, PGLC 4 blends represented 86.2% of the heat that reference PC released, it can be illustrated by the nucleation effect of SCMs promotes the hydration of C_3S in early hydration age. This enhanced heat release is likely responsible for the further strength development of PGLC at later hydration ages.

In contrast, PGC exhibited the lowest heat release at 80 h, approximately 12 J less than the PGLC blends, after 80 h of hydration. The incorporation of CCK significantly suppressed the early hydration of C_3S, while the dilution effect of LS optimized the early hydration process within the PC–CCK system.

Figure 5 presents the X-ray diffraction (XRD) patterns of the blends after different hydration ages. The hydration of PC generates ettringite, which subsequently transforms into monosulfoaluminate. The presence of portlandite (CH) is maintained throughout the hydration process. As shown in Figure 5a, hydrates (HC) were observed in all blends containing LS after 1 day of hydration; the incorporation of CCK only slightly influenced the intensity of the HC diffraction peaks. This limited reaction of CCK suggests that the aluminum (Al) in the system at this stage primarily comes from the hydration of tricalcium aluminate (C_3A).

Similar to the reference PC, the transformation from ettringite to monosulfoaluminate was observed in PGC due to the lack of LS after 3 days of hydration. The conversion of HC to MC can also be detected, with the incorporation of CCK stimulating the formation of carboaluminates; the intensity of the diffraction peaks for both HC and MC was higher than that of PLC. This trend continued at 7 days of hydration, where a more pronounced MC peak was observed alongside a decrease in the HC peak. HC persisted in PGC from 3 days onward, likely due to the carbonization of CH and LS within the CCK (refer to Figure 1d). At 28 days, the abundance of Al in CCK promoted the formation of stratlingite (C_2ASH_8) in PGC, which remained present even after 56 days of hydration.

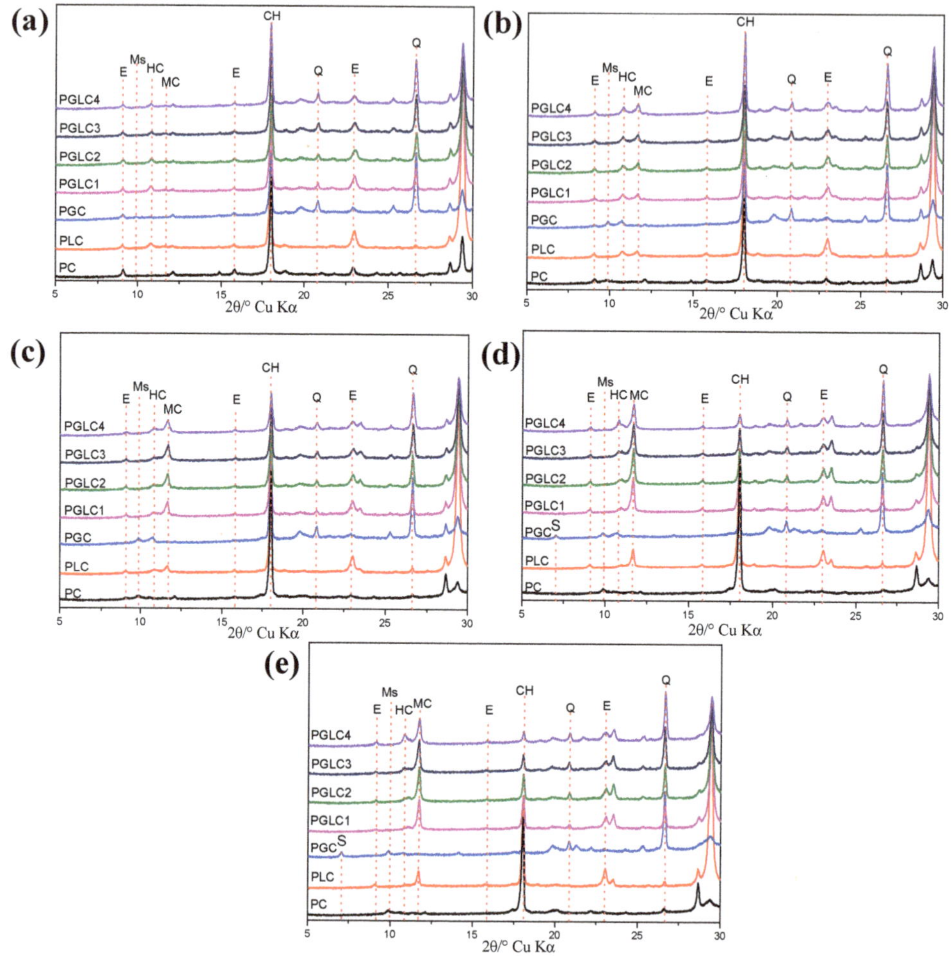

Figure 5. XRD patterns of blends after (**a**) 1 d, (**b**) 3 d, (**c**) 7 d, (**d**) 28 d, and (**e**) 56 d of hydration. The main peak of ettringite (E), monosulfoaluminate (Ms), hemicarboaluminate (HC), monocarboaluminate (MC), ferrite (F), portlandite (CH), strätlingite (S), and quartz (Q) were observed.

However, the incorporation of LS appeared to inhibit the formation of stratlingite, as no corresponding diffraction peaks (d = 12.55, 7.05°) were found in the LS-containing mixtures. Increased Al content also stabilized the existence of HC. As the content of CCK increased, the intensity of the CH diffraction peak decreased, attributable to the pozzolanic reaction of CCK during the later stages of hydration.

The content of carboaluminates throughout hydration (at 1, 3, 7, 28, and 56 days) was determined using X-ray diffraction (XRD) refinement, as illustrated in Figure 6. After 1 day of hydration, the content of hydrates (HC) decreased with increasing CCK content, which can be attributed to the aluminum (Al) available at that time being primarily derived from the hydration of tricalcium aluminate (C_3A). Thus, the incorporation of CCK had a minor effect on the early formation of HC.

Excluding the PGLC4 blend, the HC content reached its maximum at 3 days of hydration before subsequently declining. As the CCK content increased, further formation of HC was observed at 7 days, indicating that the additional CCK enhanced the reactivity of LS. The transformation of HC into monosulfoaluminate (MS) was notably evident throughout the hydration process, as the content of HC gradually decreased while the content of MC increased during the later stages of hydration.

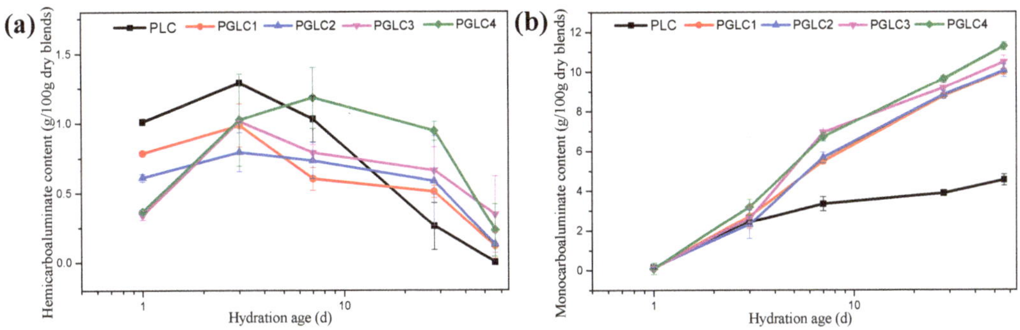

Figure 6. The formation of (**a**) HC and (**b**) MC during cement hydration, calculated by XRD refinement measurement.

The continued formation of MC observed in the later hydration periods suggests that the synergistic effects of CCK and LS exert a long-term influence on cement hydration. An interesting phenomenon can be observed from Figure 6b: The content of MC of all PGLC blends was over 10% after 56 days of hydration, indicating that the formation of MC is critical for the late strength development of the system. The results indicated that the CCK and LS are an excellent combination of SCMs. However, increasing the CCK content from 15% to 35%, the content of MC only increased by about 2%; increasing the content of CCK has a limited effect on the content of MC at a late hydration age.

In addition to HC and MC, the content of ettringite and the degree of hydration of the clinker can be calculated through X-ray diffraction refinement, as presented in Figure 7. The incorporation of LS indirectly stabilized the presence of ettringite, with all samples exhibiting similar patterns of ettringite evolution (Figure 7a). The presence of CCK appeared to have a limited impact on the existence of ettringite, as its content was primarily influenced by the gypsum content.

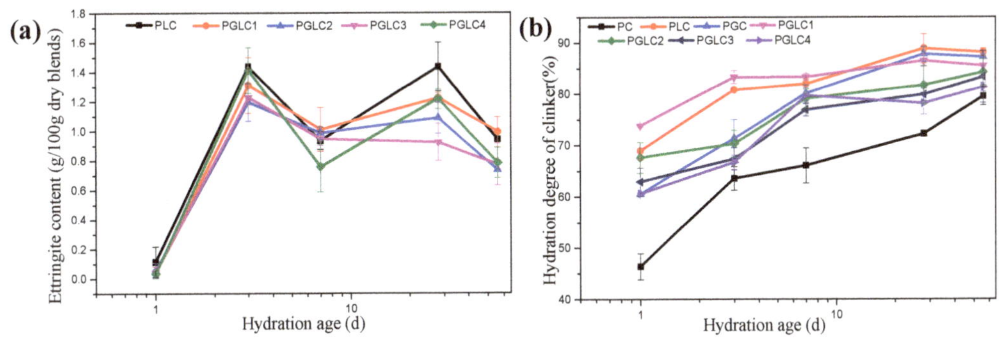

Figure 7. The (**a**) formation of ettringite and (**b**) DoH of clinker during cement hydration, calculated by XRD refinement measurement.

Both CCK and LS enhanced the hydration of the clinker at all hydration stages, which can be attributed to the nucleation effects of SCMs. In comparison to CCK, the addition of more LS significantly improved the degree of hydration results. In correspondence with the compressive strength and carboaluminates results, the synergistic effects of CCK and LS play a dominant role in further strength development during the late stages of hydration.

Figure 8 illustrates the thermogravimetric (TG) curves and the differential thermogravimetric (DTG) curves after hydration for 3 days and 28 days. Analysis of the mass changes during TG allows for the determination of the portlandite and bound water content in the blends at each hydration age, as shown in Figure 9. The incorporation of CCK stimulated the formation of carboaluminates after 3 days, with this trend further enhanced at 28 days. Additionally, as indicated in Figure 8b, the synergistic effect of CCK and LS contributes to the generation of additional C-(A)-S-H gel by consuming portlandite. This increase in hydration products leads to improved mechanical properties.

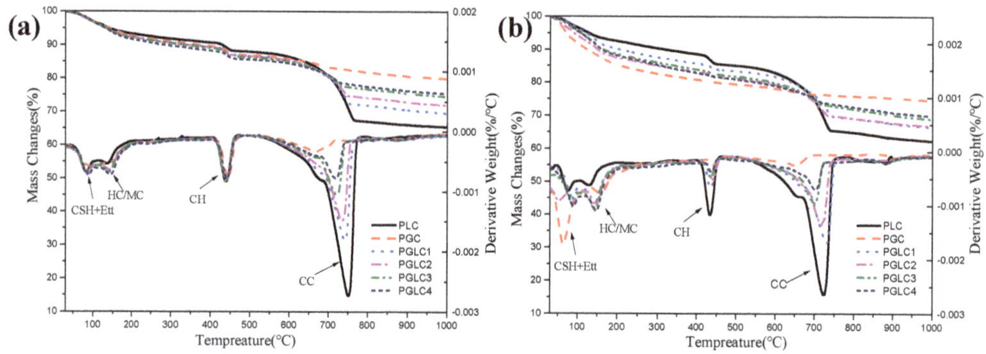

Figure 8. The thermogravimetric and different thermogravimetric resulted after hydrated (**a**) 3 d and (**b**) 28 d.

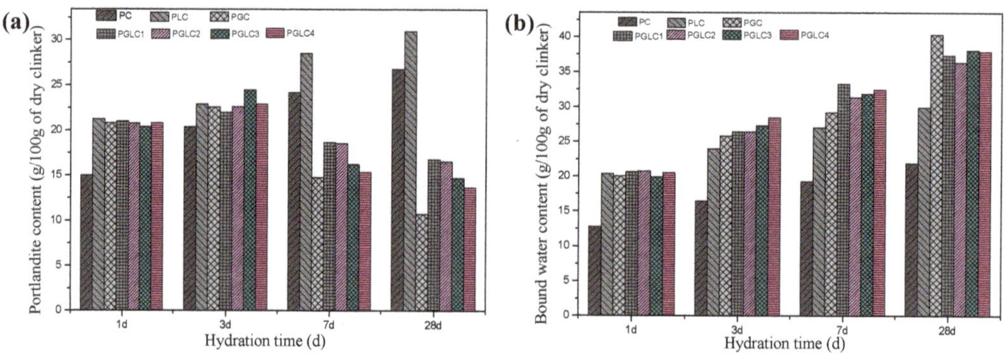

Figure 9. The (**a**) portlandite content and (**b**) bound water content after 1, 3, 7, and 28 d of hydration.

The content of portlandite and bound water further illustrates the role of the combined use of CCK and LS in hydration. After 28 days of hydration, the PGC blend displayed the highest content of C-(A)-S-H gel, as evidenced by its elevated bound water content and reduced portlandite content. This result indicates that the depletion of portlandite restricts further pozzolanic reactions within the PGC system. In contrast, the PGLC blend exhibited superior mechanical properties compared to PGC, attributable to the filling effect of LS and the refinement of pore structure by carboaluminates. For the PGLC samples, enhanced pozzolanic reactions were observed with increasing incorporation of CCK. Although PGLC samples maintained a similar bound water content after 28 days of hydration, the formation of carboaluminates appears to significantly contribute to the improvement in mechanical properties.

3.3. *Effect of CCK and LS on the Cement Pore Structure*

Based on the relationship between pore size and the overall performance of the cement system, the pore structure in the hydrated samples can be categorized into gel pores (<0.01 μm), medium pores (0.01–0.5 μm), and large pores (>0.5 μm). Figure 10 illustrates the pore size distribution Figure 10a,b and porosity distribution Figure 10c,d of the samples after 7 days and 28 days of hydration. After 7 days of hydration, the incorporation of CCK and LS slightly refined the pore distribution of the system. The samples exhibited similar critical pore sizes, which corresponded to the strength and X-ray diffraction results. The addition of CCK and LS converted harmful large pores into less harmful medium pores during the hydration period. Notably, the differential volume curves for the ternary system shifted clearly to the left; the tendency of the shift is related to the mass ratio of CCK to LS. Furthermore, the critical pore size decreased at 28 days, indicating that this refinement of the pore structure contributes significantly to the improvement of system strength.

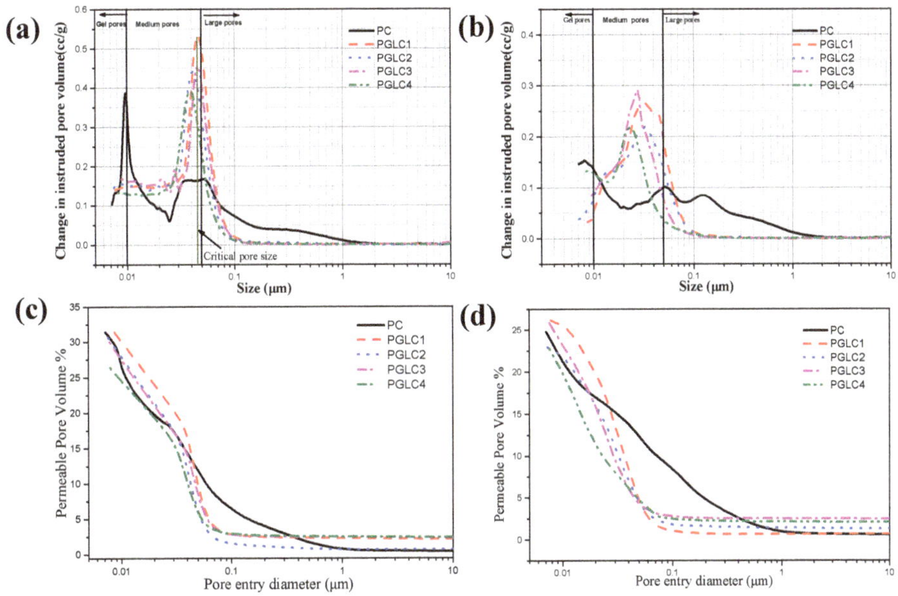

Figure 10. Pore size distribution of blends after (**a**) 7 d and (**b**) 28 d of hydration; porosity distribution of blends after (**c**) 7 d and (**d**) 28 d of hydration.

At 28 days, the DoH of the clinker is comparable to the ternary system. When considered alongside hydration and strength results, the presence of additional hydration products is identified as a key factor influencing pore refinement and strength development at later stages of hydration. As the mass ratio of CCK to LS increases, the generation of additional hydration products facilitates improved pore structure evolution. This suggests that carboaluminate plays a significant role in refining the pore structure of cement during the later hydration period. Compared to PGLC3, PGLC4 demonstrates a slight improvement in both pore structure and mechanical properties. To mitigate the CO_2 footprint of the cement industry, a replacement ratio of 2:1 for CCK and LS is recommended as an effective option within this system.

4. Conclusions

This paper investigates the effects of varying mass ratios of CCK and LS with a 45% cement substitution. Based on the results regarding hydration, mechanical properties, and pore structure, the following conclusions can be drawn:

(1) The dilution effect of CCK and LS dominate the strength development at early hydration age, as all ternary blends represent a similar compressive strength of cement at 3 days; however, the synergistic effect of CCK and LS significantly promotes further strength development at a late hydration age. The highest compressive strength reached 47.2 MPa at 28 days. An optimal mass ratio of CCK to LS exists, wherein a ratio of 2:1 enables the blends to achieve comparable strength development.
(2) CCK and LS notably influence the early hydration of cement within the first day, as evidenced by the inverse relationship between heat release and CCK content. Additionally, both CCK and LS promote a more pronounced exothermic process during hydration; all PGLC blends show similar total heat release results, and PGLC 4 represents 86.2% of total heat release as compared to reference PC after 80 h of hydration.
(3) Both CCK and LS enhance the DoH of clinker throughout the hydration process, stimulating the formation of additional C-(A)-S-H and carboaluminates by consuming portlandite. The content of MC in the system is higher than 10 wt% after 56 days of hydration, and the content of CCK in raw materials only has slight impact on the content of MC at late hydration age. The additional hydration products optimized the pore structure evolution at a late hydration age; the reduction of harmful pores favors the further strength development of the system at late hydration.
(4) This work establishes that the combination of CCK and LS is an ideal design for SCMs. However, it is important to consider that the incorporation of CCK increases the CO_2 footprint of these blends, necessitating an optimization of the mass ratio of CCK to LS. In terms of balancing CO_2 emissions and blend performance, a CCK–to–LS ratio of 2:1 exhibits comparable performance under the conditions explored in this study. This work provides a new insight into the application of LC^3 systems and reduces the CO_2 footprint in the cement industry.

Author Contributions: J.T.: writing—original draft, conceptualization. Y.Y.: data curation, formal analysis. H.Z. (Houhu Zhang): resources, conceptualization. B.M.: formal analysis, data curation. H.Z. (Hao Zhou): supervision, writing—review and editing. R.Z.: writing—review and editing, formal analysis. J.W.: writing—review and editing, data curation. Y.B.: resources, conceptualization. All authors have read and agreed to the published version of the manuscript.

Funding: The authors are grateful for the financial support of the National Key Research and Development Program of China (No. 2022YFC2105405), Jiangsu Funding Program for Excellent Postdoctoral Talent (2024ZB239). The special fund project for the Special Fund of Chinese Central Government for Basic Scientific Research Operations in commonweal Research Institute (GYZX220301) and the Special Fund of Carbon Peak and Carbon Neutrality Research Institute supported by Nanjing Institute of Environmental Sciences, Ministry of Ecology and Environment (ZX2023SZY060) are also appreciated.

Institutional Review Board Statement: Not applicable.

Informed Consent Statement: Not applicable.

Data Availability Statement: The original contributions presented in the study are included in the article; further inquiries can be directed to the corresponding authors.

Conflicts of Interest: The authors declare no conflict of interest.

References

1. Shah, V.; Parashar, A.; Mishra, G.; Medepalli, S.; Krishnan, S.; Bishnoi, S. Influence of cement replacement by limestone calcined clay pozzolan on the engineering properties of mortar and concrete. *Adv. Cem. Res.* **2020**, *32*, 101–111. [CrossRef]
2. Wang, Y.; Lei, L.; Shi, C. Effect of diethanolisopropanolamine and ethyldiisopropylamine on hydration and strength development of cement-fly ash-limestone ternary blend. *Cem. Concr. Compos.* **2024**, *145*, 105354. [CrossRef]
3. Gong, Y.; Shao, C.; Wang, Z.; Liu, R.; Zhang, Q.; Ren, Z.; Zhu, Z.; Zhao, D.; Bai, J.; Tian, J.; et al. Feasibility study on the preparation of ternary cement with calcined slag powder and seashell powder as supplementary cementitious materials. *Constr. Build. Mater.* **2023**, *409*, 134223. [CrossRef]
4. Lin, R.S.; Liao, Y.; Han, Y.; Oh, S.; Park, K.B.; Yang, H.M.; Wang, X.Y.; Yang, B.; Meng, L.Y. Low-CO_2 Optimization Design of Quaternary Binder Containing Calcined Clay, Slag, and Limestone. *Materials* **2023**, *16*, 6385. [CrossRef]

5. Zhang, Y.; Ling, T.-C. Reactivity activation of waste coal gangue and its impact on the properties of cement-based materials—A review. *Constr. Build. Mater.* **2020**, *234*, 117424. [CrossRef]
6. Marchetti, G.; Rahhal, V.; Pavlík, Z.; Pavlíková, M.; Irassar, E.F. Assessment of packing, flowability, hydration kinetics, and strength of blended cements with illitic calcined shale. *Constr. Build. Mater.* **2020**, *254*, 119042. [CrossRef]
7. Varhen, C.; Dilonardo, I.; de Oliveira Romano, R.C.; Pileggi, R.G.; de Figueiredo, A.D. Effect of the substitution of cement by limestone filler on the rheological behaviour and shrinkage of microconcretes. *Constr. Build. Mater.* **2016**, *125*, 375–386. [CrossRef]
8. Sun, J.; Zunino, F.; Scrivener, K. Hydration and phase assemblage of limestone calcined clay cements (LC^3) with clinker content below 50%. *Cem. Concr. Res.* **2024**, *177*, 107417. [CrossRef]
9. Zunino, F.; Scrivener, K. Insights on the role of alumina content and the filler effect on the sulfate requirement of PC and blended cements. *Cem. Concr. Res.* **2022**, *160*, 106929. [CrossRef]
10. Barbhuiya, S.; Nepal, J.; Das, B.B. Properties, compatibility, environmental benefits and future directions of limestone calcined clay cement (LC^3) concrete: A review. *J. Build. Eng.* **2023**, *79*, 107794. [CrossRef]
11. Bae, S.-C.; Yang, K.-H.; Jang, S.Y.; Kwon, S.-J.; Wang, X.-Y. Sustainable design of low-CO_2 hybrid concrete incorporating calcined clay and limestone powder. *J. Build. Eng.* **2024**, *92*, 109798. [CrossRef]
12. Qian, X.; Ruan, Y.; Jamil, T.; Hu, C.; Wang, F.; Hu, S.; Liu, Y. Sustainable cementitious material with ultra-high content partially calcined limestone-calcined clay. *Constr. Build. Mater.* **2023**, *373*, 130891. [CrossRef]
13. Golaszewska, M.; Giergiczny, Z. Study of the Properties of Blended Cements Containing Various Types of Slag Cements and Limestone Powder. *Materials* **2021**, *14*, 6072. [CrossRef] [PubMed]
14. Tamma, V.R.; Pancharathi, R.K.; Bibekananda, M.; Pallapothu, S.N.R.G. Strength and microstructure characteristics of low-grade (LG) limestone-based cements for a sustainable concrete. *Environ. Dev. Sustain.* **2023**, *26*, 22907–22927. [CrossRef]
15. Blouch, N.; Rashid, K.; Ju, M. Exploring low-grade clay minerals diving into limestone calcined clay cement (LC^3): Characterization−Hydration−Performance. *J. Clean. Prod.* **2023**, *426*, 139065. [CrossRef]
16. Scrivener, K.; Avet, F.; Maraghechi, H.; Zunino, F.; Ston, J.; Hanpongpun, W.; Favier, A. Impacting factors and properties of limestone calcined clay cements (LC^3). *Green Mater.* **2019**, *7*, 3–14. [CrossRef]
17. Wang, Y.-S.; Wang, X.-Y. Multi-characterizations of the hydration, microstructure, and mechanical properties of a biochar–limestone calcined clay cement (LC^3) mixture. *J. Mater. Res. Technol.* **2023**, *24*, 3691–3703. [CrossRef]
18. Hu, Y.; Xiong, L.; Yan, Y.; Geng, G. Performance of limestone calcined clay cement (LC^3) incorporating low-grade marine clay. *Case Stud. Constr. Mater.* **2024**, *20*, e03283. [CrossRef]
19. Li, J.; Wang, J. Comprehensive utilization and environmental risks of coal gangue: A review. *J. Clean. Prod.* **2019**, *239*, 117946. [CrossRef]
20. Alujas, A.; Fernández, R.; Quintana, R.; Scrivener, K.L.; Martirena, F. Pozzolanic reactivity of low grade kaolinitic clays: Influence of calcination temperature and impact of calcination products on OPC hydration. *Appl. Clay Sci.* **2015**, *108*, 94–101. [CrossRef]
21. Taylor-Lange, S.C.; Lamon, E.L.; Riding, K.A.; Juenger, M.C.G. Calcined kaolinite–bentonite clay blends as supplementary cementitious materials. *Appl. Clay Sci.* **2015**, *108*, 84–93. [CrossRef]
22. Zunino, F.; Scrivener, K. Microstructural developments of limestone calcined clay cement (LC^3) pastes after long-term (3 years) hydration. *Cem. Concr. Res.* **2022**, *153*, 106693. [CrossRef]
23. Snellings, R.; Salze, A.; Scrivener, K.L. Use of X-ray diffraction to quantify amorphous supplementary cementitious materials in anhydrous and hydrated blended cements. *Cem. Concr. Res.* **2014**, *64*, 89–98. [CrossRef]
24. Bullard, J.W.; Jennings, H.M.; Livingston, R.A.; Nonat, A.; Scherer, G.W.; Schweitzer, J.S.; Scrivener, K.L.; Thomas, J.J. Mechanisms of cement hydration. *Cem. Concr. Res.* **2011**, *41*, 1208–1223. [CrossRef]
25. Schöler, A.; Lothenbach, B.; Winnefeld, F.; Haha, M.B.; Zajac, M.; Ludwig, H.-M. Early hydration of SCM-blended Portland cements: A pore solution and isothermal calorimetry study. *Cem. Concr. Res.* **2017**, *93*, 71–82. [CrossRef]
26. Liu, X.; Chen, Q.; Song, F.; Zhang, Q.; Guo, S.; Qiao, Y.; Jiang, Z.; Zhu, H. Research on limestone calcined clay cement-based ultra-high performance concrete with high cement substitution: Mixture design and engineering properties. *Constr. Build. Mater.* **2023**, *409*, 133913. [CrossRef]
27. Wang, Y.; Tan, Y.; Wang, Y.; Liu, C. Mechanical properties and chloride permeability of green concrete mixed with fly ash and coal gangue. *Constr. Build. Mater.* **2020**, *233*, 117166. [CrossRef]
28. Wang, Y.; Shui, Z.; Gao, X.; Huang, Y.; Yu, R.; Li, X.; Yang, R. Utilizing coral waste and metakaolin to produce eco-friendly marine mortar: Hydration, mechanical properties and durability. *J. Clean. Prod.* **2019**, *219*, 763–774. [CrossRef]
29. Tang, J.; Ma, S.H.; Li, W.F.; Yang, H.; Shen, X.D. Research Progress on the Hydration of Portland Cement with Calcined Clay and Limestone. *Mater. Sci. Forum* **2021**, *1036*, 240–246. [CrossRef]
30. Avet, F.; Scrivener, K. Investigation of the calcined kaolinite content on the hydration of Limestone Calcined Clay Cement (LC^3). *Cem. Concr. Res.* **2018**, *107*, 124–135. [CrossRef]
31. Avet, F.; Li, X.; Scrivener, K. Determination of the amount of reacted metakaolin in calcined clay blends. *Cem. Concr. Res.* **2018**, *106*, 40–48. [CrossRef]
32. Grzegorz, L.G. Enhancement fracture behavior of sustainable cementitious composites using synergy between fly ash (FA) and nanosilica (NS) in the assessment based on digital image processing procedure. *Theor. Appl. Fract. Mech.* **2024**, *131*, 104442.
33. Zhang, P.; Sun, X.Y.; Wang, F. Mechanical Properties and Durability of Geopolymer Recycled Aggregate Concrete: A Review. *Polymers* **2023**, *15*, 615. [CrossRef] [PubMed]

34. Tang, J.; Wei, S.; Li, W.; Ma, S.; Ji, P.; Shen, X. Synergistic effect of metakaolin and limestone on the hydration properties of Portland cement. *Constr. Build. Mater.* **2019**, *223*, 177–184. [CrossRef]
35. Zhang, W.; Tang, J.; Li, W.; Ma, S.; Sun, J. The properties and hydration of a calcined coal series metakaolin-limestone-Portland cement system. *J. Therm. Anal. Calorim.* **2024**, *149*, 5153–5163. [CrossRef]
36. Maddalena, R.; Roberts, J.J.; Hamilton, A. Can Portland cement be replaced by low-carbon alternative materials? A study on the thermal properties and carbon emissions of innovative cements. *J. Clean. Prod.* **2018**, *186*, 933–942. [CrossRef]
37. Kaminskas, R.; Monstvilaite, D.; Valanciene, V. Influence of low-pozzolanic activity calcined mica clay on hydration and hardening of Portland cement. *Adv. Cem. Res.* **2018**, *30*, 231–239. [CrossRef]
38. Li, X.; Cao, M. Recent Developments on the Effects of Micro- and Nano-Limestone on the Hydration Process, Products, and Kinetics of Cement. *Materials* **2024**, *17*, 2133. [CrossRef]
39. Yu, T.; Zhang, B.; Yuan, P.; Guo, H.; Liu, D.; Chen, J.; Liu, H.; Setti Belaroui, L. Optimization of mechanical performance of limestone calcined clay cement: Effects of calcination temperature of nanosized tubular halloysite, gypsum content, and water/binder ratio. *Constr. Build. Mater.* **2023**, *389*, 131709. [CrossRef]
40. Li, X.; Ouzia, A.; Scrivener, K. Laboratory synthesis of C3S on the kilogram scale. *Cem. Concr. Res.* **2018**, *108*, 201–207. [CrossRef]
41. Wu, M.; Zhang, Y.; Liu, Z.; She, W.; Liu, C.; Wang, D.; Sui, S. Study on the composition of calcium alumina silicate hydrate (CASH) in lime-activated low carbon cementitious materials: The influence of alkaline additives. *Constr. Build. Mater.* **2023**, *392*, 132089. [CrossRef]

materials

MDPI

Article

Effect of Stöber Nano-SiO_2 Particles on the Hydration Properties of Calcined Coal Gangue-Blended Cement

Nan Zhang [1,†], Hao Zhou [2,†], Yueyang Hu [3], Jiaqing Wang [4], Guihua Hou [1], Jian Ma [1,*] and Ruiyu Jiang [1,*]

1 Key Laboratory for Advanced Technology in Environmental Protection of Jiangsu Province, Yancheng Institute of Technology, Yancheng 224051, China; zhangnanpaper@163.com (N.Z.); houguihua@ycit.cn (G.H.)
2 Nanjing Institute of Environmental Sciences, Ministry of Ecology and Environment of the People's Republic of China, Nanjing 210042, China; zhouhao@nies.org
3 Yancheng Institute of Technology, College of Materials Science and Engineering, Yancheng 224051, China; huyueyang1989@163.com
4 College of Civil Engineering, Nanjing Forestry University, Nanjing 210037, China; jiaqingw@njfu.edu.cn
* Correspondence: jian_ma@njtech.edu.cn (J.M.); jiangry@ycit.edu.cn (R.J.)
† These authors contributed equally to this work.

Abstract: This study focuses on the calcined coal gangue (CCG)-blended cements containing Stöber nano-SiO_2 (SNS) particles. The effects of SNS particles on the workability, hydration behaviour, mechanical properties and microstructure evolution of the blended cements were comprehensively investigated at curing ages ranging from 1 to 28 d. The hydration behaviour was studied via isothermal calorimetry test, X-ray diffraction (XRD) and thermogravimetric (TG) tests. The microstructural evolution was studied using mercury intrusion porosimetry (MIP) and scanning electron microscopy (SEM). The results show that the incorporation of SNS led to a significant reduction in fluidity, particularly at an SNS content of 3%. The SNS significantly increased the compressive strength of the CCG-blended cement at all curing ages, and the optimum SNS content was found to be 2%. SNS significantly accelerated not only the early cement hydration but also the pozzolanic reaction of CCG at later curing ages, resulting in a decrease in portlandite, as evidenced by the isothermal calorimetry, XRD and TG analysis. Microstructural analysis shows that the incorporation of SNS effectively refined the pore structure of the CCG-blended cement, resulting in the formation of a dense microstructure. All these beneficial effects of SNS provides advantages in the development of the compressive strength of the CCG-blended cement at all curing ages.

Keywords: cement; calcined coal gangue; Stöber nano-SiO_2; hydration; compressive strength

1. Introduction

The production of Portland cement (PC) results in high consumption of natural resources and CO_2 emissions. As a rough estimate, almost 0.930 tonnes of CO_2 is produced for every tonne of cement clinker. The cement industry alone is responsible for about 8% of all the CO_2 emissions worldwide, causing serious ecological and economic problems [1]. To achieve sustainable development in the cement industry, supplementary cementitious materials (SCMs) [2,3], including fly ash (FA), silica fume (SF) and granulated blast furnace slag (GBFS), are often blended to replace part of the cement clinker to reduce clinker factors, which is a common practice nowadays.

Coal gangue (CG) is a solid waste produced during coal mining and production [4]. The main mineral component of many CGs is kaolinite, which can be converted to metakaolin (MK) with high pozzolanic reactivity after high temperature activation [5]. This is commonly referred to as calcined coal gangue (CCG). As a potential pozzolanic material, CCG has been intensively studied for partial replacement of PC [6]. In this way, environmental problems such as the land occupation of CG and CO_2 emissions from cement production can be alleviated to some extent [7].

The current literature reports that CCG can react with portlandite (CH) from cement hydration to produce additional calcium silicate hydrate (C-S-H) and calcium alumina silicate hydrate (C-A-S-H) gels [8,9]. These products are beneficial for refining pore structures [10], resulting in improved mechanical properties and durability of the mortar or concrete within an appropriate substitution range of 10–20% [5,11]. Yang et al. [10] investigated the compressive strength of mortars in which cement was replaced by different levels of CCG. Compared with the reference mortar, the 90-day compressive strength of the mortars prepared with 10% and 20% of CCG increased by 11.7% and 6.2% and the 180-day compressive strength by 16.9% and 11.4%, respectively. However, the corresponding early compressive strength was found to respectively decrease by 14.4% and 23.1% at 3 d and by 6.2% and 16.2% at 7 d. Furthermore, the authors [10] also claimed that further increases in the level of CCG substitution in cement would significantly weaken the strength development of the corresponding mortar. Similar observations were also reported in the work of Guo et al. [12], who studied the compressive strength of mortar prepared from 30 wt.% CCG with an optimal calcination condition of CG at 800 °C and a holding time of 2 h. They found that the compressive strength of the CCG-prepared mortar at 3, 7 and 28 d was 30.9%, 22.8% and 21.1% lower than that of the reference mortar, respectively. The current research indicates that although the inclusion of CCG up to 10–20% can ensure a satisfactory compressive strength at later ages, its strength development at early ages is much lower than that of the reference sample, which is very unfavourable for its popularization in engineering.

The cement paste is composed of calcium silicate hydrate (C-S-H) gel, portlandite, ettringite, numerous pores, etc., of which C-S-H accounts for about 70% of the cement paste [13]. The particle size of the C-S-H gel is about 10 nm, indicating that the hardened cement paste is a nanomaterial dominated by C-S-H. However, the microstructure of the hardened cement paste is very rough. In the past few years, nanomaterials have been applied in various scientific fields of engineering [14] and physics [15,16] to improve the micro and macro performance of materials. Due to their special ultrafine structures, nanomaterials exhibit many excellent properties such as grain small size effect, interfacial effect and quantum size effect [17]. Extensive research indicates that nano-SiO_2 has been proven to effectively improve the performance of cement-based materials, especially in overcoming the problems of early slow performance development. In addition to providing nucleation sites for cement hydration, nano-SiO_2 [18,19] can also react with $Ca(OH)_2$ to form more C-S-H gel. The gel is conducive to the densification of the microstructure. Ultimately, the mechanical properties of cement-based materials are improved.

Of all the different forms of nano-SiO_2 used to modify the properties of cement-based materials, Stöber nano-SiO_2 (SNS) synthesized via the traditional Stöber method has a typical characterization of uniform and monodispersed spheres with high purity [20]. In the preparation process, controllable particle size and narrow size distribution can also be easily achieved [21]. Singh et al. [22–24] found that the addition of 5 wt.% SNS can not only increase the compressive strength of the control cement paste by 64% at 1 d and 35% at 28 d, but also reduce Ca leaching by 60% at 28 d. Oertel et al. [25–27] compared the effect of SNS and silica fume on the properties of ultra-high performance concrete (UHPC) and found that SNS can maintain its primary size with smallest agglomerates in UHPC. These monodispersed SNS particles were more conducive to achieving higher strength in UHPC. Similar findings were also found in the work of Sun et al. [28,29], where the calcined SNS can improve the compressive strength of PC and high-volume fly ash mortars better than the initial SNS. Although the relevant study of the effect of SNS on cement-based materials is limited in current research, it is still a promising method to keep nano-particles in a highly dispersed state and maximize the beneficial effect on the performance of cement-based materials.

Based on this, this study first investigates the effects of SNS content (from 0% to 3%) on the fluidity and strength development trends of Portland cement with 20 wt.% of CCG (PC-CCG). It then proceeds to discuss the potential mechanism of hydration in the SNS-added PC-CCG system using isothermal calorimetry, X-ray diffraction (XRD), thermogravimetric (TG), mercury intrusion porosimetry (MIP) and scanning electron microscopy (SEM). The knowledge gained from this research is expected to provide a new and in-depth understanding of the effect of SNS on the early age hydration performance of blended cement pastes containing CCG, thereby facilitating the practical application of CCG and promoting the sustainable development of the coal and cement industries.

2. Materials and Methods

2.1. Materials

Portland cement (PC, Chinese P·I 42.5) and CCG were used for the preparation of blended cement pastes. CG sourced from Guizhou Province was subjected to drying, crushing, ball-milling and calcination processes to obtain CCG. The milling time was set at 15 min and the calcination was set at 800 °C for 1 h. Figure 1 displays the chemical and mineralogical compositions of PC and CCG. It can be seen from Figure 1b that tricalcium silicate (C_3S), dicalcium silicate (C_2S), tricalcium aluminate (C_3A) and ferrite (F) are the main mineral phases in PC. Their particle size distributions are summarized in Figure 2. It can be seen that the D50 of CCG is smaller than that of PC, and their Blaine specific surface areas (SSAs) are 288 and 442 m^2/kg, respectively.

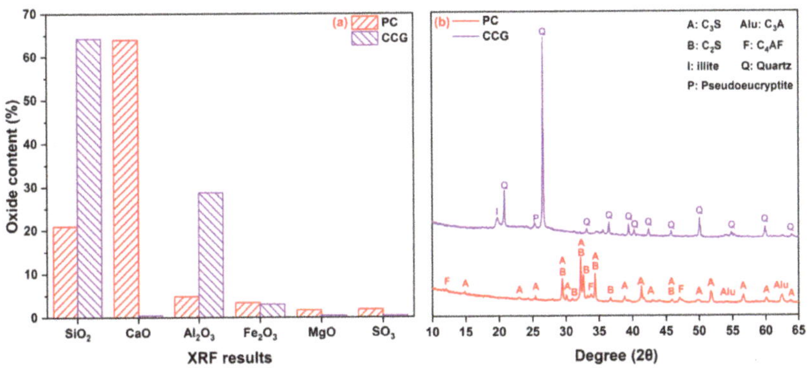

Figure 1. (**a**) Chemical composition and (**b**) mineralogical composition of raw materials.

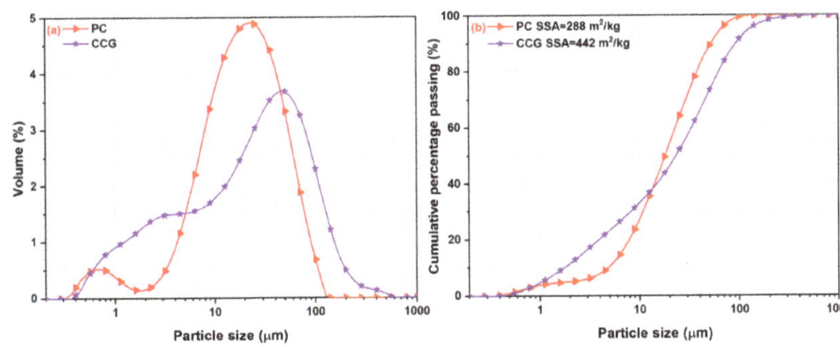

Figure 2. Particle size distributions of raw materials.

Stöber nano-SiO_2 particles (SNS) were first synthesized in the lab according to a classical Stöber method [20]. Deionized water, ammonium hydroxide and ethanol were mixed in a ratio using a magnetic stirrer followed by adding tetraethyl orthosilicate drop by drop. In particular, the experimental temperature and the mixing speed were set at 20 ± 1 °C and 800 rounds/minute, respectively. After synthesis, the above solution was centrifuged and the resulting solid calcined at 500 °C for 2 h [28]. Figure 3a shows the particle size and morphology of SNS. As observed, SNS particles show a uniform and monodisperse spherical shape and their average sizes are around 200 nm.

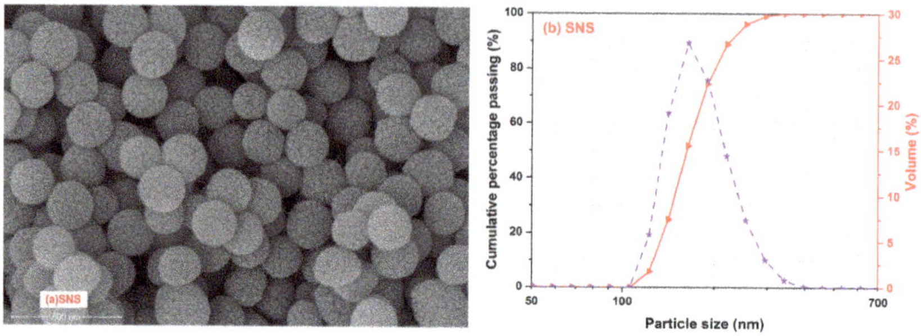

Figure 3. (**a**) Morphology and (**b**) particle size distributions of SNS.

2.2. Samples Preparations

Table 1 shows the detailed mix proportions for the mortar samples in this experiment. As shown in Table 1, the blended cement consists of 80 wt.% PC and 20 wt.% CCG. In the process of using nan-SiO_2 to modify cement-based materials, the amount of nan-SiO_2 is usually controlled within 3% [18]. Therefore, in order to investigate the effect of SNS on the compressive strength of this blended cement, the replacement levels of SNS to PC in this experiment are set to 1 wt.%, 2 wt.%, and 3 wt.%, respectively. In addition, the water/solid (cement + CCG + SNS) ratio is kept at 0.5 throughout the experiment.

Table 1. Detailed mixture proportions for the mortar samples (g, using 100 g cement as the basis).

Code	Cement	CCG	SNS	Sand	Water
PC	100	0	0	300	50
C20	80	20	0	300	50
1SNS	79	20	1	300	50
2SNS	79	20	2	300	50
3SNS	79	20	3	300	50

The raw materials weighed according to Table 1 were poured into the mortar mixing pot for low-speed mixing for 30 s, then the standard sand was added for low-speed mixing for 30 s, followed by high-speed mixing for a further 90 s. The prepared mortar was first tested for fluidity using the jumping table test, then poured into moulds of 40 mm × 40 mm × 160 mm and vibrated on the shaking table for about 1 min. Finally, the fresh mortar and moulds were placed in a standard curing room at 20 ± 1 °C for 24 h. After demoulding, the mortar specimens were immersed in water at 20 ± 1 °C until the compressive strength test at 1, 3, 7 and 28 d in accordance with GB/T17671-2021 "Test Method for Cement Mortar Strength" [30].

The preparation of paste samples was carried out to elucidate the underlying mechanism behind the variations in compressive strength of PC-CCG mortar resulting from the addition of SNS. The same mixture proportion to the mortar was used without the addition of sand. The weighed PC, CCG, SNS and water were mixed in a blender at 1600 rpm for 2 min, put into flat plastic tubes, gently shaken to shake out the potential bubbles introduced during the mixing process and covered with lids. Afterwards, the fresh paste and tubes were placed in a standard curing room at 20 ± 1 °C and ≥95% humidity until the microstructural analysis. At the time of testing, the cured paste was removed from the tubes, broken into small pieces and immersed in anhydrous ethanol to stop hydration (the anhydrous ethanol was updated once after 12 h). After 7 d of immersion, the piece samples were removed from the ethanol and dried in a vacuum drying oven at 40 °C for 1 d. Part of the small pieces was finely ground with an agate mortar until all powders could pass through an 80 μm square sieve for XRD and TG tests. Another part of the pieces was kept in the oven for MIP and SEM tests.

2.3. Methods

2.3.1. Characterization of Mortar

The fluidity of the mortar was tested in accordance with GB/T 2419-2005 "Method for Determination of flow of cement mortar" [31]. The compressive strength was determined in accordance with GB/T17671-2021 "Test Method for Cement Mortar Strength" [30].

2.3.2. Characterization of Paste

A TAM Air 8-channel microcalorimeter (Thermometric AB, Jarfalla, Sweden) was used to monitor the hydration heat behaviour of the hydrated paste. Approximately 4 g of fresh paste prepared in the paste sample preparation section was filled into the ampoule, sealed with a cap and placed in the calorimeter. Data were collected at a constant temperature of 20 °C for 3 d.

A MiniFlex 600 X-ray powder diffractometer (XRD, Rigaku, Tokyo, Japan) with a copper target (CuKα, $\lambda = 0.154$ nm) was used to determine the phase compositions of the hydrated paste. The working parameters were an accelerating voltage of 40 kV, accelerating current of 15 mA, scanning range of 5–65° and scanning speed of 5°/min. For qualitative analysis, the phases were identified with Search-Match using ICDD PDF-2.

A Mettler Toledo TGA/DSC1 synchronous thermal analyser (NETZSCH, Selb, Germany) with an accuracy of 0.1 mg was used to qualitatively and quantitatively determine the phase compositions of the hydrated paste by measuring the relative mass loss over a specified temperature range. The heating temperature range was 30–1000 °C, and the heating rate was 10 °C/min. The N_2 atmosphere was used, and the inlet flow rate was 30 mL/min.

A PoreMaster 60 GT mercury injection apparatus (Quantachrome instruments, Boynton Beach, FL, USA) was used to evaluate the pore structures of the hydrated cement paste. The low pressure range was 0.2–50 psia and the high pressure range was 20–60,000 psia. Pore sizes ranging from 3 nm to 10.8 μm were able to be detected.

A Nova NanoSEM 450 scanning electron microscope (SEM, FEI Company, Hillsboro, OR, USA) was used to observe the morphological characteristics of the hydrated paste. Images were taken at an acceleration voltage of 15 kV, a working distance of 7–9 mm and a magnification of 2–20 k times. Prior to SEM analysis, the samples were coated with a layer of gold to improve their conductivity.

3. Results and Discussion

3.1. Fluidity

Figure 4 shows the fluidity of the fresh mortar samples. As shown in Figure 4, the fluidity of C20 was only slightly lower than that of the reference PC mortar, possibly due to the comparable particle size distributions of CCG and PC, which did not show much effect on the fluidity of PC, even though 20% of the cement was replaced by CCG. However, the fluidity of C20 was significantly reduced from 219 mm to 209 mm with the addition of 1% SNS. This is because the average particle sizes of the SNS particles are much smaller than those of the PC and CCG particles (Figures 2 and 3), which contributes to the increased specific surface area and water needed to form lubricating layers around each particle. As a result, the C20 mortar with SNS experienced a reduced workability, particularly at 3% SNS, where the fluidity of C20 was reduced by 15.5%. It is also recognized by the current research that the introduction of nano-particles usually resulted in a sharp decrease in the workability of mortar or concrete [32,33].

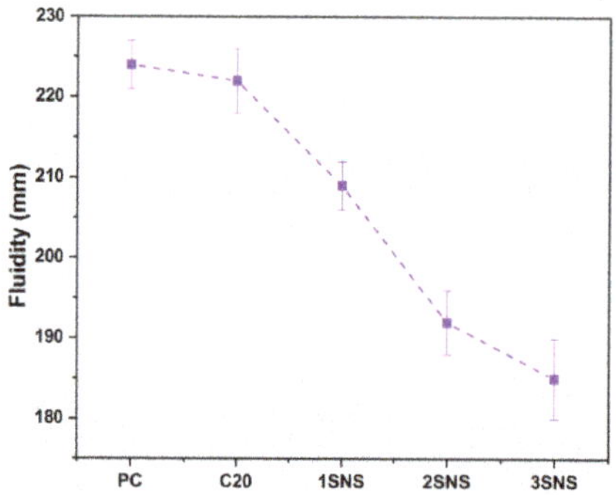

Figure 4. Fluidity results of all fresh mortars.

3.2. Compressive Strength

Figure 5 shows the variation in compressive strength of all mortars developed from 1 to 28 d. As shown in Figure 5, the introduction of CCG decreased the compressive strength of PC mortar by 4.8 MPa after 1 d due to the much lower reactivity of CCG than that of PC. This is also consistent with other reports that when the CCG replacement level exceeded 10%, the strength development of PC-CCG mortar was severely inhibited within the first 1 d [34]. After 3 d, the compressive strength of PC-CCG mortar increased faster than that of PC mortar, and the difference in the strength values between PC-CCG and PC mortars was 4.3, 3.1 and 2.5 MPa at 3, 7 and 28 d, respectively. This can be explained by the starting reaction of CCG with CH from 3 d and the fact that more hydrates would be generated, especially at later ages. However, the increased quantity of hydrates in the PC-CCG system cannot fully compensate for the strength loss caused by the 20 wt.% reduction in the cement content.

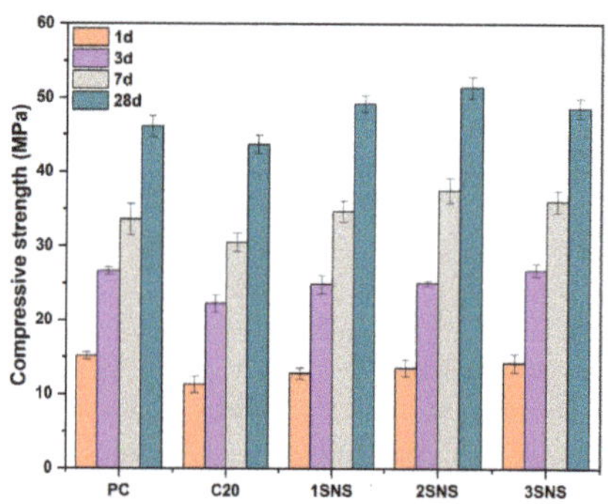

Figure 5. Compressive strength development of all mortars at 1, 3, 7 and 28 d.

When SNS was added, a clear trend was observed where increasing the SNS content significantly improved the compressive strength at 1 d. Compared to C20, the compressive strength of 1SNS, 2SNS and 3SNS improved by 12.9%, 19.4% and 25.9%, respectively. This is associated with the highly dispersed SNS particles, which can well exert the nucleation site effect that greatly accelerates cement hydration (as also will be discussed in Section 3.3), thereby effectively increasing the compressive strength of PC-CCG mortar [28]. Similar findings can also be observed at 3 d. However, it seems that the contribution of SNS gradually decreased, as reflected by the negligible strength differences between 2SNS and 3SNS mortars with the increase in the amount of SNS from 2% to 3% during this period. This observation is also consistent with the influence of other nanomaterials on the compressive strength of cement-based materials [35]. That is, the later the age of curing, the less the strengthening effect of the nanomaterials on the strength development.

From 7 d onwards, all the SNS-added mortars showed higher compressive strength than the reference mortar. As shown in Figure 5, the compressive strength of 1SNS, 2SNS and 3SNS mortars reached 34.6, 37.5 and 36.1 MPa after 7 d, which were 3.1%, 11.7% and 7.3% higher than those of the reference PC mortar, respectively. At 28 d, the corresponding increases were still as high as 6.7%, 11.5% and 5.4%, respectively. These values indicate that the addition of SNS can fully compensate for the loss of strength caused by the addition of 20 wt.% CCG from 7 to 28 d, the reason for which will be discussed in the following section. In addition, it is important to note that the SNS content should not exceed 2%, as this can lead to a stagnation of the strength growth in the medium-to-late stages of the mortar.

In order to better understand the mechanism of the introduction of SNS to improve the compressive strength of the PC-CCG system and to simplify the research scheme, the systems of PC, C20 and 2SNS were selected for further hydration study.

3.3. Isothermal Calorimetry

Figure 6 shows the hydration heat evolution of PC, C20 and 2SNS pastes, normalized per gram of cement. As shown in Figure 6a, the heat flow curve of PC paste had two exothermic peaks [36]. The first peak was mainly associated with hydration of alite and the second one was related to the renewed formation of ettringite (AFt), appearing at 7.2 h and 8.8 h, respectively. The addition of CCG exhibited a hydration heat curve with a trend comparable to that of PC during the acceleration stage, and the first peak was observed at about 6.9 h. The C20 sample, although containing only 80 wt.% cement, exhibited a heat release rate comparable to that of the PC sample, indicating that CCG greatly accelerated cement hydration. When 2 wt.% SNS was added, the corresponding paste showed a noticeably shorter induction period compared to the PC and C20 pastes, with the first peak appearing at 4.1 h. In addition, the slope of the C_3S exothermic peak increased in the SNS-added samples, and this peak also occurred at an earlier stage and had a higher heat flow. All these observations suggest that the addition of SNS significantly enhances cement hydration.

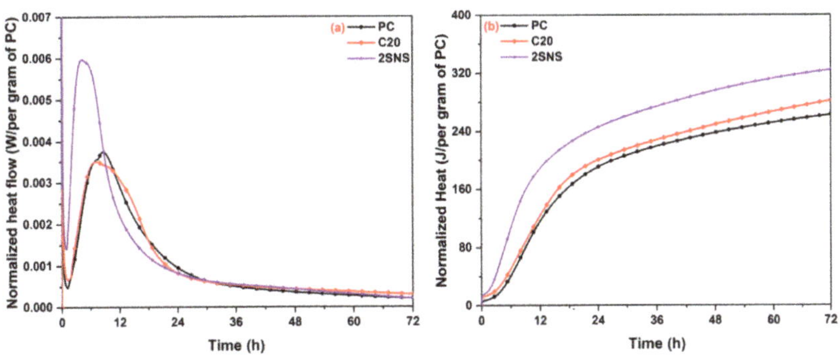

Figure 6. (**a**) Heat flow and (**b**) cumulative heat release of the selected pastes.

Figure 6b presents the cumulative hydration heat curves of samples within 72 h. The total heat release of PC was 262 J/g. For the C20 and 2SNS samples, the total heat of samples experienced an increase of 7.6% and 23.7%, respectively. The results suggest that the presence of SNS can significantly enhance the hydration process of cement. This acceleration effect is attributed to a nucleation effect of SNS [28] caused by its significantly higher surface area than PC and CCG. In addition, the acceleration of cement hydration can result in the rapid formation of a dense microstructure, thereby improving the compressive strength of the cement. This is in good agreement with the strength data, as evidenced by the significant increase in the 1- and 3-day compressive strength data of the CCG-prepared mortar after the addition of SNS in Figure 5.

3.4. XRD Analysis

Figure 7 shows the XRD patterns of hydrated cement pastes at different curing ages. The main hydration products of PC, as shown in Figure 7, are portlandite (CH) and ettringite. CH is formed by the hydration of the silicate phase in PC. It can be seen that its peak intensity (2θ 18°, oval area in Figure 7) showed an increasing trend with increasing curing age due to the continuous hydration of the silicate phase. Ettringite is formed by the hydration of the aluminium phase. After the incorporation of CCG, portlandite and ettringite remained as the main hydration products. This indicates that the presence of CCG did not alter the type of hydration products of PC. At 1 d, the characteristic peak intensity of CH in the C20 sample increased compared to that in PC, indicating that the addition of CCG promoted cement hydration, which is consistent with the results of the hydration heat observation in Figure 6a. After 3 d, there was a decrease in the intensity of the characteristic peak of CH in the C20 sample, which can be attributed to the pozzolanic reaction of CCG resulting in the consumption of portlandite, particularly after 7 d (Figure 7b). This reaction between CCG and CH produced additional C-A-S-H that can support the strength development of C20, aligning well with the analysis of a narrowing strength gap between C20 and PC (Figure 5).

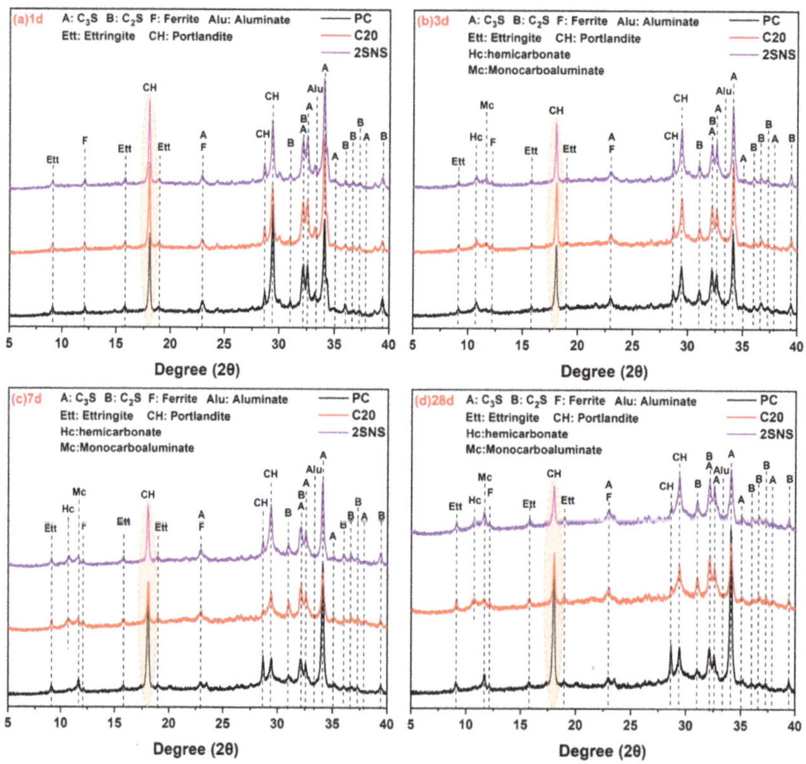

Figure 7. XRD patterns of the selected pastes at (**a**) 1 d, (**b**) 3 d, (**c**) 7 d and (**d**) 28 d.

In the presence of SNS, the primary hydration products consisted of portlandite and ettringite, which were identical to those observed in PC. The peak intensity of CH showed an increasing trend and the peak intensity of the mineral phase (2θ 18°) of the clinker showed a decreasing trend after the addition of SNS during early curing ages. This can be attributed to the accelerating effect of SNS, which is consistent with the hydration heat data. Compared with C20, the peak intensity of CH decreased significantly at later curing ages (especially at 28 d in Figure 7d), which can be attributed to the acceleration of the pozzolanic reaction by SNS.

3.5. TG Analysis

The TG−DTG results of the selected pastes after 3, 7 and 28 d of hydration are shown in Figure 8. From the DTG curves, it can be seen that four major mass loss peaks were observed during tests [37]. The first mass loss peak between 50 and 100 °C was mainly associated with the dehydration of calcium silicate hydrate (C−S−H) and ettringite. The second mass loss peak between 130 and 200 °C was related to the dehydration of carboaluminates (Hc and Mc) and AFm phase. The third peak between 400 and 500 °C was due to the decomposition of the CH. The mass loss in the range 600−800 °C was mainly due to the decarbonation of $CaCO_3$. The DTG analysis showed that C-S-H, AFt, Mc, carboaluminates and CH were the common hydrates present in all samples. The crystalline hydrates observed from DTG curves were found to be in good agreement with the results obtained from XRD analysis (Figure 7).

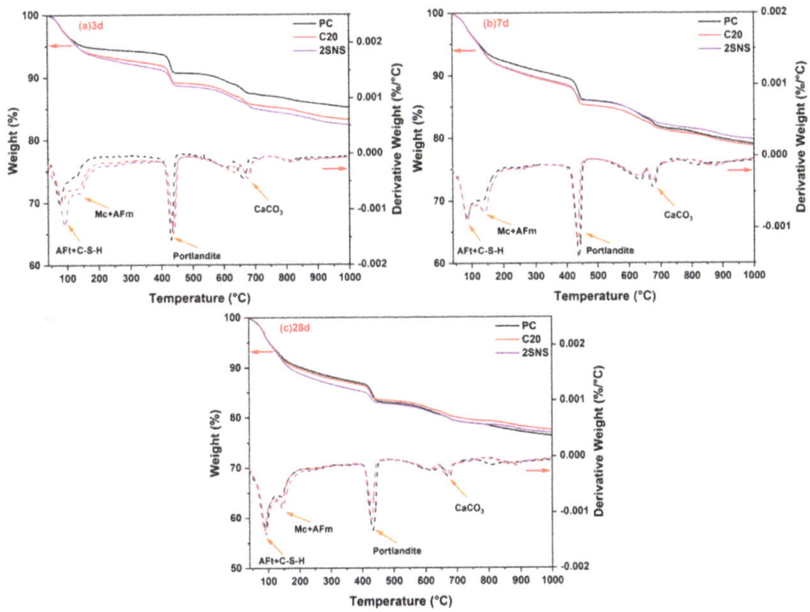

Figure 8. TG−DTG curves of selected pastes at (**a**) 3 d, (**b**) 7 d and (**c**) 28 d.

The area of the weight loss peak in the DTG curve reflected the weight of the hydration products. From 3 d to 28 d, the weight loss peak area in the range 50−200 °C of the SNS sample was observed to be the largest among the selected pastes (Figure 8a), indicating that it generated the highest quantity of hydration products. This is due to the fact that the addition of SNS resulted in the acceleration of hydration. However, at 7 d and 28 d, the weight loss peak area in the range 400−500 °C of the SNS sample was observed to be the smallest among the selected pastes. At later curing ages, the pozzolanic reaction of CCG consumed CH, leading to a reduction in its weight loss. When SNS was added, the pozzolanic reaction of CCG was accelerated. Consequently, it can be observed that the selected pastes exhibit the smallest weight loss peak area for CH (Figure 8b,c). The observation is also in line with the XRD analysis (Figure 7c,d).

3.6. Pore Structure Analysis

The pore structure, including porosity and pore size distribution, is crucial to support the performance of cement-based materials. Figure 9a shows the cumulative pore volume curves of PC, C20 and 2SNS pastes at 28 d. It can be seen from Figure 9a that both C20 and SNS had much higher porosities than PC at 28 d due to the significantly decreased gel-to-space ratio when 20 wt.% cement was replaced by CCG. In addition, the total porosity of C20 and SNS was 17.1% and 15.9%, respectively. This indicated that SNS effectively reduced the porosity of C20 at 28 d.

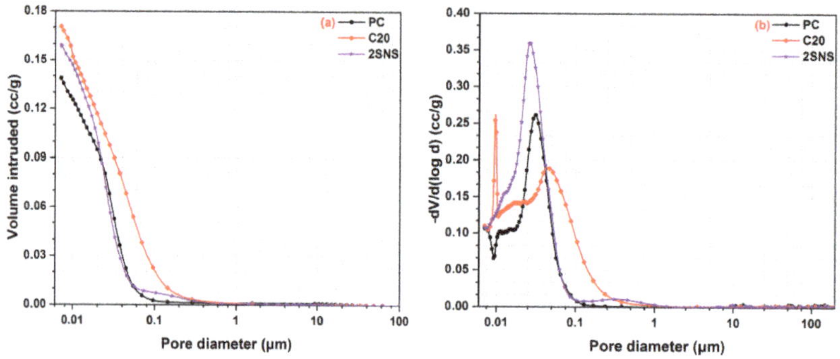

Figure 9. (**a**) Cumulative pore volume and (**b**) differential pore size distribution curves of the selected pastes at 28 d.

As shown in Figure 9b, the pores in these pastes can be classified into three types: gel pores (<10 nm), medium capillary pores (10–50 nm), and large capillary pores (>50 nm). The medium capillary pores were beneficial to the strength and permeability of cement-based materials, while the large capillary pores could deteriorate the relevant properties [38]. According to the classification of these types of pores, the volume fractions of different types of pores were calculated and shown in Figure 10.

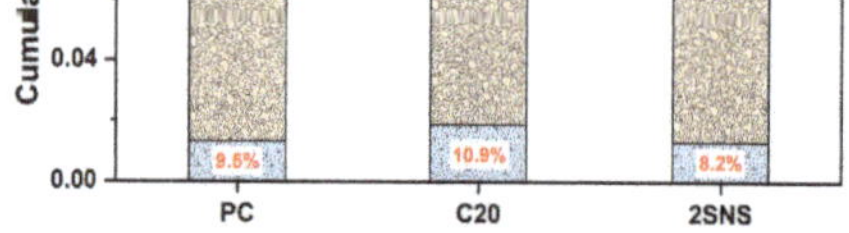

Figure 10. Volume fractions of different types of pores in the selected pastes.

It can be seen that the volume fractions of large capillary pores in C20 and SNS-added paste were 35.2% and 8.7%, while the volume fractions of the medium capillary pores in these two pastes was 54.0% and 83.2% at 28 d, respectively. This can be explained by the promotion effect of SNS on the cement hydration and the CCG reaction that produced more hydrates than that in C20, which filled the large capillary pores. Through the analysis of the pore structure, it was found that the addition of SNS could refine the pore structure of C20, as evidenced by the decreased porosity and large capillary pores and the increased medium capillary pores. This refining effect of SNS on the pore structures of C20 also well explains the improved compressive strength development of C20 with the addition of SNS at 28 d.

3.7. SEM Analysis

The microscopic morphology of selected samples at different curing ages are shown in Figure 11. At 3 d, AFt, CH and C-S-H gels were produced in hydrated PC (Figure 11a). In addition, large quantities of pores were also noticeable, which were left by the free water. As shown in Figure 11b, the morphology of the hydrates changed significantly in C20, where some irregular metakaolin particles, more pores and a looser structure than that in PC can be observed. This also helps to explain the increased porosity (Figure 9) and decreased compressive strength (Figure 5) of C20 when CCG was added to cement at 3 d. Compared with the SEM image of C20, the microstructure of 2SNS (Figure 11c) was much denser due to the increased precipitation of hydrates caused by the addition of 2% SNS as discussed above, which also accounted for its much higher compressive strength at 3 d (Figure 5).

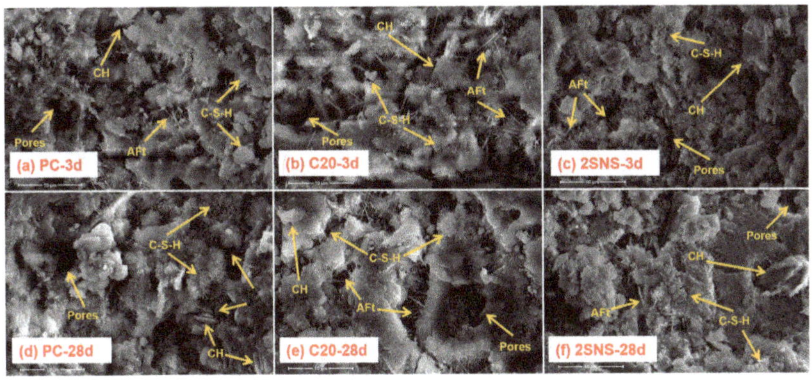

Figure 11. Microscopic image of the selected pastes after 1 and 28 d of curing.

As hydration progressed up to 28 d (Figure 11d–f), the structural compactness of each sample increased significantly, as reflected by the reduction in the number of pores and overall porosity. This is because continuous reactions in these samples would produce more hydrates, such as carboaluminates and C-A-S-H gels. These hydrates would become closer, interact with each other and form a dense microstructure. Compared with PC (Figure 11d), a more compact morphology can be observed in 2SNS (Figure 11f), where the hydrates were closely linked together. However, the microstructure of C20 (Figure 11e) was still less compact than that of PC. These findings are well in line with the XRD, TG, MIP and compressive strength results above.

4. Conclusions

This paper investigated the effect of SNS particles on the workability, compressive strength, hydration performance and microstructural development of CCG-blended cements. Based on the experimental results above, the following conclusions can be drawn:

(1) The incorporation of SNS resulted in a significant reduction in the fluidity of CCG-blended mortar. This is due to the increased specific surface area of SNS, which needs more water to form lubricating layers around each particle and consequently decreased the fluidity of the mortar.

(2) SNS significantly improved the compressive strength of the CCG-blended cement at all curing ages, even compensating for the strength loss caused by the 20 wt.% replacement of PC by CCG. Furthermore, the optimum SNS content was found to be 2%, beyond which the compressive strength of CCG-blended cement would decrease.

(3) The addition of SNS promoted not only cement hydration at the early stage but also the reaction between CCG and CH at the late stage, which produced more hydrates than that in the CCG-blended paste.

(4) The addition of SNS could refine the pore structure of the CCG-blended paste, as evidenced by the decreased porosity and large capillary pores and the increased medium capillary pores. This refinement also contributed to the formation of a dense microstructure of the CCG-blended paste, offering advantages in the development of the compressive strength of the CCG-blended cement at all curing ages.

The findings of this study have important implications, as they suggest a new strategy to efficiently improve the early compressive strength of PC with CCG. However, further research and investigation into durability, including shrinkage, carbonation and chloride ion or sulphate attack, is also required to fully understand the benefits of SNS on the performance of PC with CCG.

Author Contributions: N.Z.: Investigation, Data curation, Writing—original draft. H.Z.: Investigation, Writing—original draft. Y.H.: Formal analysis, Editing. J.W.: Writing—review and editing. G.H.: Resource, Conceptualization. J.M.: Writing—review and editing, Supervision, Funding acquisition. R.J.: Resource, Conceptualization, Data curation, Supervision. All authors have read and agreed to the published version of the manuscript.

Funding: The authors are grateful for CNBM Technology Innovation Academy, the Key Laboratory of Inorganic Functional Materials and Intelligent Manufacturing of Shandong Province (2021SFGC0903), the Key Research and Development Plan Science and Technology Demonstration Project of Shandong Province "Integrated Innovation and demonstration of new green building materials" (2021SFGC0903), the funding for school-level research projects of Yancheng Institute of Technology (XJR2023053), the special fund project for the Special Fund of Chinese Central Government for Basic Scientific Research Operations in the commonwealth Research Institute (GYZX220301), and the Special Fund of Carbon Peak and Carbon Neutrality Research Institute supported by the Nanjing Institute of Environmental Sciences, Ministry of Ecology and Environment (ZX2023SZY060). The work was supported by the high-level training talents of '333' project in Jiangsu province.

Institutional Review Board Statement: Not applicable.

Informed Consent Statement: Not applicable.

Data Availability Statement: The data used in this study can be required from the corresponding author. The data are not publicly available due to information that could compromise the research participants' privacy.

Conflicts of Interest: The authors declare no conflicts of interest.

References

1. Plaza, M.G.; Martinez, S.; Rubiera, F. CO_2 capture, use, and storage in the cement industry: State of the art and expectations. *Energies* **2020**, *13*, 21. [CrossRef]
2. Behera, S.K.; Mishra, D.P.; Singh, P.; Mishra, K.; Mandal, S.K.; Ghosh, C.N.; Kumar, R.; Mandal, P.K. Utilization of mill tailings, fly ash and slag as mine paste backfill material: Review and future perspective. *Constr. Build. Mater.* **2021**, *309*, 125120. [CrossRef]
3. Haha, M.B.; Termkhajornkit, P.; Ouzia, A.; Uppalapati, S.; Huet, B. Low clinker systems—Towards a rational use of SCMs for optimal performance. *Cem. Concr. Res.* **2023**, *174*, 107312. [CrossRef]
4. Xue, B.; Zhang, Y.; Li, J.; Wang, Y.Y. A review of coal gangue identification research-application to China's top coal release process. *Environ. Sci. Pollut. Res.* **2023**, *30*, 14091–14103. [CrossRef]
5. Zhang, Y.L.; Ling, T.C. Reactivity activation of waste coal gangue and its impact on the properties of cement-based materials—A review. *Constr. Build. Mater.* **2020**, *234*, 117424. [CrossRef]
6. Li, J.Y.; Wang, J.M. Comprehensive utilization and environmental risks of coal gangue: A review. *J. Clean. Prod.* **2019**, *239*, 117946. [CrossRef]
7. Li, L.; Shao, X.; Ling, T.-C. Life cycle assessment of coal gangue composite cements: From sole OPC towards low-carbon quaternary binder. *J. Clean. Prod.* **2023**, *414*, 137674. [CrossRef]
8. Kalpokaitė-Dičkuvienė, R.; Baltušnikas, A.; Levinskas, R.; Česnienė, J. Incinerator residual ash—Metakaolin blended cements: Effect on cement hydration and properties. *Constr. Build. Mater.* **2019**, *206*, 297–306. [CrossRef]
9. Cai, R.; He, Z.; Tang, S.; Wu, T.; Chen, E. The early hydration of metakaolin blended cements by non-contact impedance measurement. *Cem. Concr. Comp.* **2018**, *92*, 70–81. [CrossRef]
10. Yang, J.; Su, Y.; He, X.; Tan, H.; Jiang, Y.; Zeng, L.; Strnadel, B. Pore structure evaluation of cementing composites blended with coal by-products: Calcined coal gangue and coal fly ash. *Fuel Process. Technol.* **2018**, *181*, 75–90. [CrossRef]

11. Zhao, Y.; Qiu, J.; Ma, Z.; Sun, X. Eco-friendly treatment of coal gangue for its utilization as supplementary cementitious materials. *J. Clean. Prod.* **2021**, *285*, 124834. [CrossRef]
12. Guo, Z.H.; Xu, J.J.; Xu, Z.H.; Gao, J.M.; Zhu, X.L. Performance of cement-based materials containing calcined coal gangue with different calcination regimes. *J. Build. Eng.* **2022**, *56*, 104821. [CrossRef]
13. Scrivener, K.L.; Juilland, P.; Monteiro, P.J.M. Advances in understanding hydration of Portland cement. *Cem. Concr. Res.* **2015**, *78*, 38–56. [CrossRef]
14. Onaizi, A.M.; Huseien, G.F.; Lim, N.H.A.S.; Amran, M.; Samadi, M. Effect of nanomaterials inclusion on sustainability of cement-based concretes: A comprehensive review. *Constr. Build. Mater.* **2021**, *306*, 124850. [CrossRef]
15. Ishibe, T.; Tomeda, A.; Watanabe, K.; Kamakura, Y.; Mori, N.; Naruse, N.; Mera, Y.; Yamashita, Y.; Nakamura, Y. Methodology of thermoelectric power factor enhancement by controlling nanowire interface. *Acs Appl. Mater. Interfaces* **2018**, *10*, 37709–37716. [CrossRef] [PubMed]
16. Xiong, R.; Yu, S.; Smith, M.J.; Zhou, J.; Krecker, M.; Zhang, L.; Nepal, D.; Bunning, T.J.; Tsukruk, V.V. Self-assembly of emissive nanocellulose/quantum dot nanostructures for chiral fluorescent materials. *ACS Nano* **2019**, *13*, 9074–9081. [CrossRef]
17. Sanchez, F.; Sobolev, K. Nanotechnology in concrete—A review. *Constr. Build. Mater.* **2010**, *24*, 2060–2071. [CrossRef]
18. Abhilash, P.P.; Nayak, D.K.; Sangoju, B.; Kumar, R.; Kumar, V. Effect of nano-silica in concrete; a review. *Constr. Build. Mater.* **2021**, *278*, 122347.
19. Zhuang, C.L.; Chen, Y. The effect of nano-SiO_2 on concrete properties: A review. *Nanotechnol. Rev.* **2019**, *8*, 562–572. [CrossRef]
20. Stöber, W.; Fink, A.; Bohn, E. Controlled growth of monodisperse silica spheres in the micron size range. *J. Colloid Interface Sci.* **1968**, *26*, 62–69. [CrossRef]
21. Liang, X.; Lian, L.X.; Liu, Y.; Kong, Q.Q.; Wang, L. Controlled synthesis of monodisperse silica particles. *Micro Nano Lett.* **2016**, *11*, 532–534. [CrossRef]
22. Singh, L.P.; Bhattacharyya, S.K.; Singh, P.; Ahalawat, S. Granulometric synthesis and characterisation of dispersed nanosilica powder and its application in cementitious system. *Adv. Appl. Ceram.* **2012**, *111*, 220–227. [CrossRef]
23. Singh, L.P.; Bhattacharyya, S.K.; Ahalawat, S. Preparation of size controlled silica nano particles and its functional role in cementitious system. *J. Adv. Concr. Technol.* **2012**, *10*, 345–352. [CrossRef]
24. Singh, L.; Agarwal, S.; Bhattacharyya, S.; Sharma, U.; Ahalawat, S. Preparation of silica nanoparticles and its beneficial role in cementitious materials. *Nanomater. Nanotechnol.* **2011**, *1*, 44–51. [CrossRef]
25. Oertel, T.; Helbig, U.; Hutter, F.; Kletti, H.; Sextl, G. Influence of amorphous silica on the hydration in ultra-high performance concrete. *Cem. Concr. Res.* **2014**, *58*, 121–130. [CrossRef]
26. Oertel, T.; Hutter, F.; Tänzer, R.; Helbig, U.; Sextl, G. Primary particle size and agglomerate size effects of amorphous silica in ultra-high performance concrete. *Cem. Concr. Comp.* **2013**, *37*, 61–67.
27. Oertel, T.; Hutter, F.; Helbig, U.; Sextl, G. Amorphous silica in ultra-high performance concrete: First hour of hydration. *Cem. Concr. Res.* **2014**, *58*, 131–142. [CrossRef]
28. Sun, J.; Shi, Z.; Dai, J.; Song, X.; Hou, G. Early hydration properties of Portland cement with lab-synthetic calcined stöber nano-SiO_2 particles as modifier. *Cem. Concr. Comp.* **2022**, *132*, 104622.
29. Tian, L.; Zhang, F.; Chen, J.; Sun, J.; Zhu, H. Modification of high volume fly ash composites containing calcined stöber nano-SiO_2 particles. *J. Build. Eng.* **2022**, *51*, 104272.
30. *GB/T17671-2021*; Method of Testing Cements-Determination of Strength. State Administration for Market Regulation: Beijing, China, 2021.
31. *GB/T2419-2005*; Test Method for Fluidity of Cement Mortarx. The Standardization Administration of China: Beijing, China, 2005.
32. Liu, J.; Li, Q.; Xu, S. Influence of nanoparticles on fluidity and mechanical properties of cement mortar. *Constr. Build. Mater.* **2015**, *101*, 892–901. [CrossRef]
33. Quercia, G.; Hüsken, G.; Brouwers, H. Water demand of amorphous nano silica and its impact on the workability of cement paste. *Cem. Concr. Res.* **2012**, *42*, 344–357. [CrossRef]
34. Wu, D.; Chen, T.; Hou, D.S.; Zhang, X.X.; Wang, M.H.; Wang, X.P. Utilization of coal gangue powder to improve the sustainability of ultra-high performance concrete. *Constr. Build. Mater.* **2023**, *385*, 131482. [CrossRef]
35. Praveenkumar, T.R.; Vijayalakshmi, M.M.; Meddah, M.S. Strengths and durability performances of blended cement concrete with TiO_2 nanoparticles and rice husk ash. *Constr. Build. Mater.* **2019**, *217*, 343–351. [CrossRef]
36. Sun, J.; Zunino, F.; Scrivener, K. Hydration and phase assemblage of limestone calcined clay cements (LC^3) with clinker content below 50%. *Cem. Concr. Res.* **2024**, *177*, 107417. [CrossRef]
37. Deboucha, W.; Leklou, N.; Khelidj, A.; Oudjit, M.N. Hydration development of mineral additives blended cement using thermogravimetric analysis (TGA): Methodology of calculating the degree of hydration. *Constr. Build. Mater.* **2017**, *146*, 687–701. [CrossRef]
38. Wang, P.M.; Liu, X.P. Effect of temperature on the hydration process and strength development in blends of Portland cement and activated coal gangue or fly ash. *J. Zhejiang Univ. Sci. A* **2011**, *12*, 162–170. [CrossRef]

materials

MDPI

Article

A Comparative Study on the Texture of Exposed Aggregate Concrete (EAC) Pavements Using Different Measurement Techniques

Pawel Gierasimiuk *, Marta Wasilewska and Wladyslaw Gardziejczyk

Faculty of Civil Engineering and Environmental Sciences, Bialystok University of Technology, Wiejska 45E Street, 15-351 Bialystok, Poland; marta.wasilewska@pb.edu.pl (M.W.); w.gardziejczyk@pb.edu.pl (W.G.)
* Correspondence: p.gierasimiuk@pb.edu.pl

Abstract: This paper presents issues related to the assessment of the texture of aggregate concrete (EAC) surfaces using various methods for its verification. Microtexture was assessed using the British Pendulum Tester (BPT) and Dynamic Friction Tester (DFT). Two laser profilometers were used to assess macrotexture, circular texture meter (CTM) and stationary laser profilograph (SPL), as well as the commonly known volumetric method. Measurements were carried out on left and right tracks and in between them on five test sections of expressways. Based on the analyses performed, it was found that the results obtained by the DFT were less sensitive to changes in microtexture between individual tracks compared to the results obtained by the BPT. The BPN values in the left track were lower than those in the right track. However, the difference between the DFT20 results in these spots was insignificant. Both MPD and MTD values did not show significant differences between the right and left tracks. However, some differences were observed between the MPD parameters obtained using the CTM and SPL. This resulted from the different frequency and length of the scanned surface profile. However, the differences were at an acceptable level. A very high linear correlation was obtained in the case of BPN and DFT20 values (r – 0.719), and in the case of MPD and MTD values, the correlation was almost certain (r above 0.900). Based on a comparative analysis of the models estimating mean texture depth (MTD/ETD), a significant difference was observed between models based on EAC pavement results and those based on asphalt surfaces.

Keywords: exposed aggregate concrete; macrotexture; microtexture; mean profile depth; mean texture depth; friction coefficient

1. Introduction

The texture of a pavement's upper layer has a significant impact on the value of the friction force between the tire and the road surface, which plays a key role in both the dynamics and kinematic reactions of vehicles, as well as the stability and maneuverability of a vehicle in critical driving conditions. According to the classification system contained in [1,2], texture is divided into megatexture, macrotexture and microtexture, according to the wavelength and the amplitude. Megatexture is unevenness corresponding to wavelengths above 50 mm, which adversely affect driving comfort, rolling resistance, interior noise and road user safety. They are caused by damage to the pavement, i.e., cracks, defects, etc. The macrotexture corresponds to unevenness with a wavelength from 0.5 mm to 50 mm. It affects not only noise but also the drainage of water from contact between the tire and the road pavement, thus reducing the risk of aquaplaning [3,4]. The level of this unevenness results, in the case of concrete pavements, in the texturing method [5,6]. Microtexture is characterized by unevenness with a wavelength of less than 0.5 mm. It is responsible for breaking the water film to create a dry surface to generate the required friction force. It depends primarily on the polishing resistance of the coarse aggregate and the content of fine aggregate in the wearing course [7–9]. Due to the role of texture in ensuring the appropriate level of skid resistance, in many countries, it is mandatory to check it during the period of use of the pavement.

In practice, texture in terms of macrotexture unevenness is assessed using the volumetric method based on the MTD (Mean Texture Depth) parameter determined in accordance with EN 13036-1 [10] or the profilometric method based on the MPD (Mean Profile Depth) parameter in accordance with EN ISO 13473-1 [11]. Both methods are very well described in the literature [12,13]. Although the volumetric method is a very cheap and easy solution for obtaining information about texture, it does not enable a continuous registration of changes on the tested surface. Additionally, the measurement requires time and disabling the tested section from traffic. The development of contactless pavement scanning techniques made it possible to replace the volumetric method with parameters based on profile registration. The parameters determined on their basis may not only be MPD but also Ra (Arithmetic mean deviation), Rms (Root Mean Square), Rsk (Skewness) and Rku (Kurtosis). Many stationary and dynamic laser profilometers have been developed, and they must meet the requirements of the standard EN ISO 13473-1. Stationary devices enable measurements both in laboratory and in situ. However, similarly to the volumetric method, these measurements require road closure at some point. Dynamic profilometers are mounted on a vehicle; therefore, they have better performance compared to stationary profilometers. This enables continuous data recording. However, they cannot perform measurements in difficult-to-reach places, e.g., roundabouts, intersections, road junctions or during road construction. It is thus recommended to use stationary profilometers [14,15].

The registration of microtexture unevenness requires specialized profilometers. For this purpose, scanning images taken under a microscope or 3D scanners are primarily used [16]. Images of the upper layer surface of the pavement are also used to assess the macrotexture, and the development of technology in recent years makes it also possible to determine the parameters characterizing the pavement texture in 3D format [17–20]. Using such equipment in road conditions is very difficult or even impossible. Therefore, studies on the assessment of microtexture are carried out in the laboratory on samples cut from the pavement or samples prepared in the laboratory and simulated phenomena occurring in real traffic conditions. However, an indirect measure for assessing microtexture in situ are the results obtained from stationary or dynamic devices simulating braking conditions with full wheel lock at low slip speed (10–20 kph). It should be noted that the advantage of stationary devices for both macrotexture and microtexture assessment is the ability to use calibration plates to check the reliability of the obtained results. In the case of dynamic devices, quality control procedures are more complex.

Parameters describing the texture on both the macro and micro scale are taken into account in numerous models that have been developed to predict parameters related to the friction generated at the tire–pavement interface [21]. Some researchers emphasize that the division of the texture scale presented in EN ISO 13473-1 [11] is simplified and has limitations [22,23]. The use of decomposition methods, i.e., wavelet transform, Hilbert–Huang transform or Fourier transform, to identify individual scales showed that simplified division between microtexture and macrotexture does not represent their complex impact on the level of friction force values. An alternative approach to describing the texture is to analyze pavement unevenness through parameters closely related to the height, shape and density of the irregularities. However, it has been proven that the characterization of textures largely depends on the resolution of the recorded surface topography and specialized measurement techniques [24]. This study makes a significant contribution to understanding the mechanisms related to texture and its impact on the phenomena occurring at the interface between the tire and the road pavement. In practice, dynamic or stationary laser profilometers and devices reflecting full wheel lock braking are used to monitor texture.

Texture is closely related to the technology of construction of the pavement's upper layer. Consequently, this means that the proper selection of materials, determination of the composition of mixtures, texturing techniques, etc. will allow for maintaining the required level of skid resistance during the pavement life cycle.

EAC pavements are not as popular a technology used in road construction as asphalt pavements. This type of pavement is primarily used for roads with heavy traffic—mainly expressways and highways. Properly selected proportions between individual aggregate fractions, the maximum size of aggregate grains, and the texturing process allow for the arrangement of coarse aggregate to be such that a positive texture profile is created [25–27]. It should be noted that EAC pavements are characterized by a very good durability of the macrotexture during road use, unlike concrete pavements textured using burlap drag, brushing or grooving of fresh concrete mix [28,29]. It should be noted that EAC is a very difficult technology for constructing upper layers of road pavements. One of the difficulties in the construction of EAC pavements is the requirement to obtain a homogeneous texture [3]. This is due to the difficulty in determining the appropriate time from the moment of spraying the compacted concrete mixture with a set-retarding agent to the beginning of the removal of unbound cement mortar and exposure of the aggregate, taking into account numerous environmental factors such as wind velocity, air temperature and humidity [30]. To obtain the proper texture, it is very important to comply with the regimes during its construction, especially at the stage of its texturing and curing. If a mistake is made, repair is very time-consuming and expensive compared to layers of asphalt mixtures. Therefore, controlling texture conditions is very important because it affects not only skid resistance but also rolling resistance and tire/road noise [5,31,32].

Some researchers make use of a given texture assessment method depending on the type of road pavement. For example, assessing the texture of open (PA) or semi-open (SMA LA, BBTM) pavements is debatable. Profilometric methods do not accurately reflect the shape of the surface irregularities of free spaces due to open texture. However, in the case of the volumetric method, it is possible that the glass beads or sand used for the test penetrate into the free spaces of the layer structure, thus not reflecting the texture surface. In such cases, determining the relationship between profilometric and volumetric methods is inadvisable [33]. In the case of concrete pavements, when choosing a method for assessing texture, attention should be paid to the surface texturing technique. On surfaces textured with burlap drag or brush, spreading glass beads or calibrated sand is very difficult due to the low texture depth (below 0.5 mm). Therefore, it is recommended to use profilometric methods. However, in the case of longitudinally grooved concrete pavements, when using a dynamic laser profilometer recording a 2D profile, there is a risk of underestimating the depth of the pavement texture. The laser beam can record a profile in the groove, i.e., a flat surface. Therefore, it is recommended to use a 3D laser that will fully reflect pavement irregularities. However, when using stationary devices, they must be properly positioned on the pavement during measurement. The texture of EAC pavements can be assessed by both profilometric and volumetric methods, as they are multi-directional. There are numerous regression models available in the literature that were developed on the basis of results obtained using volumetric and profilometric methods using both dynamic and stationary devices [17,19,34–36]. These models are dedicated to estimating parameters describing the texture or friction coefficient. They were mainly developed based on measurements of asphalt mixture layers. This is due to the fact that this technology for pavements' upper layers is the most frequently used in the world and very well described in the technical literature. Concrete pavements are an alternative to building roads using asphalt mixtures. They are mostly preferred for roads with very heavy traffic.

The aim of this article is to provide a comparative analysis of parameters describing texture using various stationary devices, in terms of assessing Exposed Aggregate Concrete pavement in real conditions on the road. This study includes an analysis of texture variability in the cross-section of the road lane, taking into account the left, right and center tracks. Additionally, the verification of regression models developed by other researchers for estimating parameters describing texture was presented, using the results of this study recorded in actual road conditions.

2. Research Program

Five test EAC sections (TS−1, TS−2, TS−3, TS−4, TS−5; TS−Test Section) with a length of 100.0 m located on expressways were selected for testing. The test sections were selected on the basis of preliminary measurements of the MPD parameter, which were used to determine its impact on road surface noise presented in the article [5]. Sections were selected that varied in texture level. Figure 1 shows images of the tested pavements and samples of their unevenness profiles.

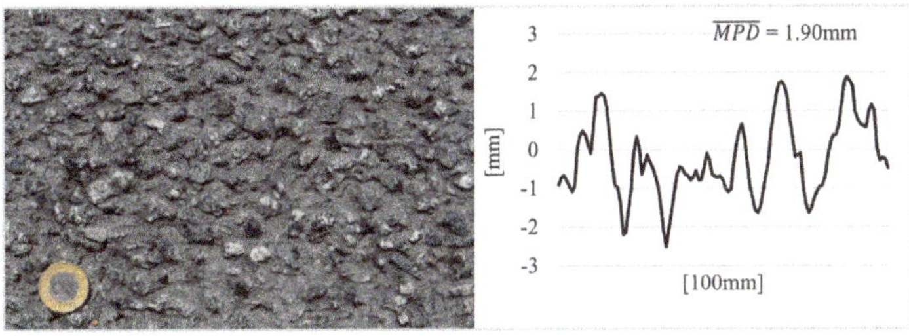

(a)

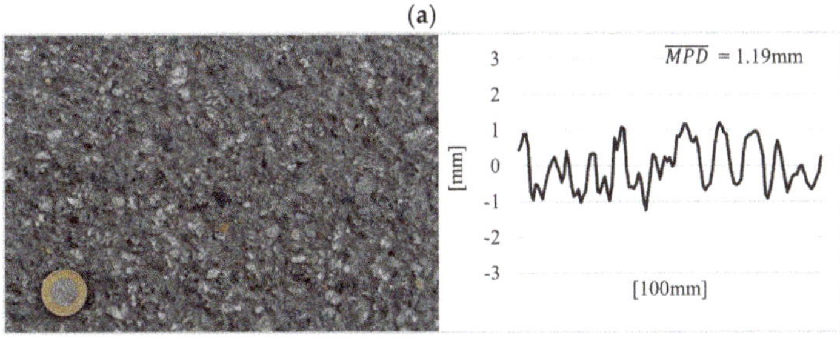

(b)

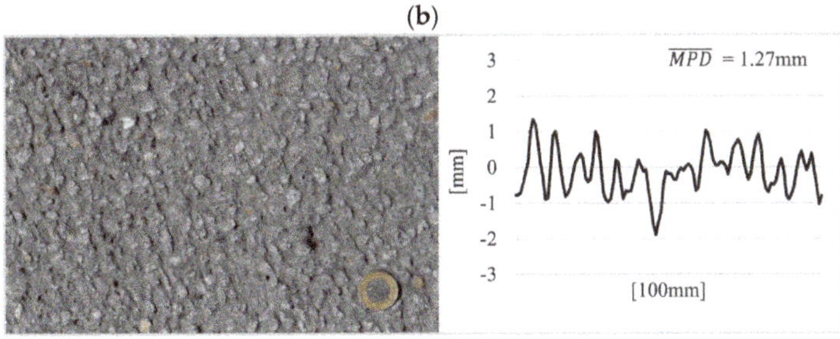

(c)

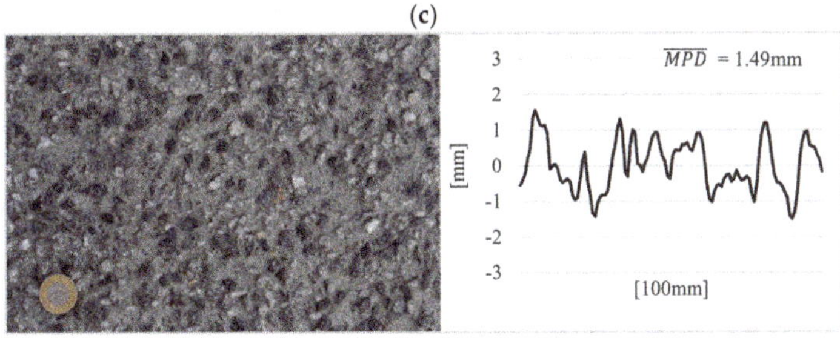

(d)

Figure 1. *Cont.*

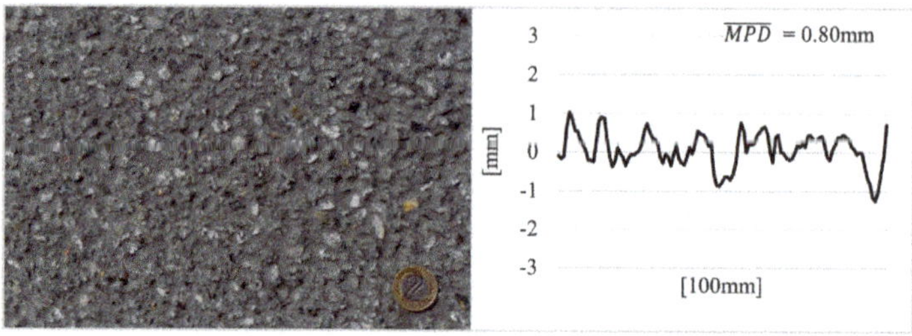

(e)

Figure 1. Pavements of test sections with samples of unevenness profiles (the coin has a diameter of 21.5 mm): (**a**) TS−1, (**b**) TS−2, (**c**) TS−3, (**d**) TS−4 and (**e**) TS−5.

The scope of this research included the following measurements:

- British Pendulum Number (BPN) using the British Pendulum Tester (BPT) according to ASTM 303 [37];
- Friction coefficient DFT20 using Dynamic Friction Tester (DFT) according to ASTM E1911 [38];
- MPD values using Circular Texture Meter (CTM) according to ASTM E2157-01 [39];
- MPD values using Stationary Laser Profilograph (SLP) according to EN ISO 13473-1 [11];
- MTD values according to EN 13036-1 [10].

Five measurement cross-sections were marked every 25.0 m on the right lane of the road on each test section. Measurements were carried out in the left wheel track (L—left), in the right wheel track (R—right) and between the wheel tracks (C—center) (Figure 1). Three measurements of each parameter were made in each track (Figure 2). The total number of test spots on the section was 45.

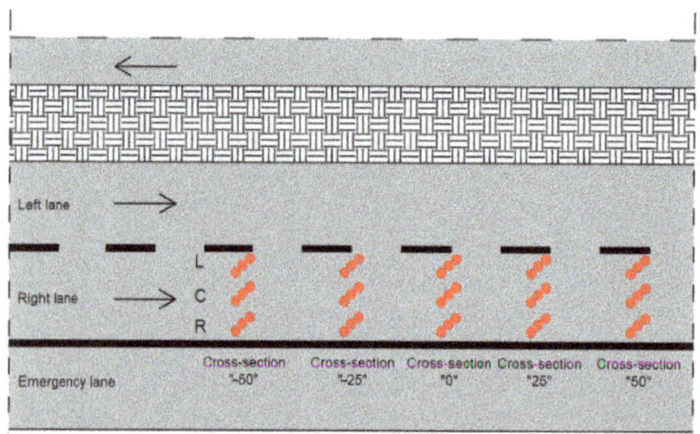

Figure 2. Location of measurement cross-sections and test spots on the test section.

The test sections were marked on the expressway. In order to ensure optimal safety during the measurements, the right lane of the road was closed for approximately 500 m in each section (Figure 3).

Figure 3. Example of a test section.

3. Description of Measurement Procedures

3.1. Measurement of the BPN

The BPN value was determined by measuring the friction between the rubber slider of the British Pendulum Tester and the pavement (Figure 4). Before the measurement, the pendulum was calibrated and then adjusted so that the length of the slide was 126 mm. The surface and the slider of the pendulum were then wetted with water. The pendulum arm with the slider was released, and the friction value was read from the large scale. This operation was repeated 5 times. The BPN value was calculated from the last three readings.

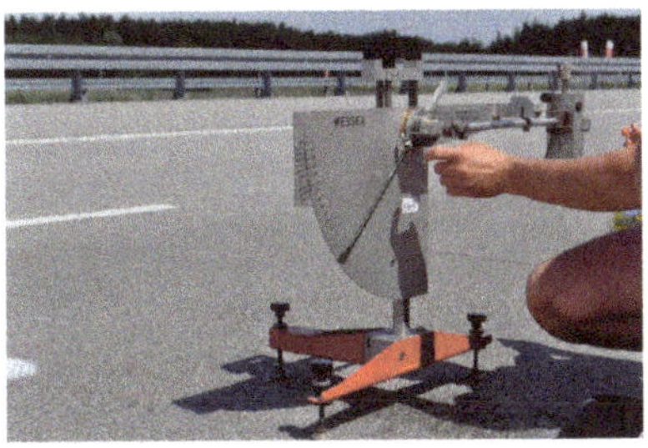

Figure 4. Measurement of British Pendulum Tester.

3.2. Measurement of the DFT20

The Dynamic Friction Tester is a stationary device in which the friction coefficient is measured in the slip speed ranging from 0 kph to 80 kph (Figure 5). The measurement principles correspond to mobile locked wheel devices. The DFT consists of two disks that are connected to each other by a spring and a sensor. There are three rubber sliders mounted on the lower disk. Discs rotating at 80 kph are lowered to the pavement, and the friction coefficient is measured. The software allows for determining the friction coefficient DFT20 at a slip speed of 20 kph, DFT40 at 40 kph and DFT60 at 60 kph. The DFT20 parameter was adopted for the analysis.

Figure 5. DFT device during field measurements.

3.3. Measurement of the MTD Using Volumetric Patch Method (VPM)

Measurement involves spreading glass beads with dimensions of 180/250 μm and a volume of 25,000 mm^3 with a rubber disk so that they form a circle on the tested pavement (Figure 6). This operation was repeated in four places in the measurement area. The average texture depth—MTD (Mean Texture Depth)—was calculated based on the measured circle diameter.

Figure 6. Macrotexture measurements using the volumetric patch method.

3.4. Measurement of the MPD Using the CTM

The Circular Texture Meter (CTM) (Figure 7) determines the unevenness profile based on which parameters characterizing the macrotexture are calculated—MPD and RMS. The profile is measured by a CCD laser displacement sensor that moves around the circumference of a circle with a radius of 142 mm. The resolution of the device is 0.87 mm. The MPD_{CTM} parameter was adopted for the analysis. Before the measurement, the correct operation of the sensor was checked on a calibration board with a known profile.

Figure 7. CTM during measurements.

3.5. Measurement of the MPD Using Stationary Laser Profilometer (SLP)

The SLP device was built at the Bialystok University of Technology in Bialystok (Figure 8). The measuring part is 1.7 m long. The SLP was equipped with a laser sensor with parameters similar to the laser sensor from the CTM device. As a result of the measurement, a surface texture profile was obtained; based on this, the MPD value was determined. The sampling resolution could be 0.86 mm or 0.43 mm. The MPD_{SPL} results obtained at a resolution of 0.43 mm were taken into account for the analysis.

Figure 8. Stationary Laser Profilometer.

4. Results and Discussion

4.1. Microtexture Assessment Using a Method Based on DFT and BPT Measurements

Figure 9 shows the results of the DFT20 and BPN friction coefficients in the left (L), and right (R) track and between the wheel tracks (C) on individual test sections. Table 1 summarizes their descriptive statistics, i.e., average values. STD represents standard deviations, and V represents coefficients of variation.

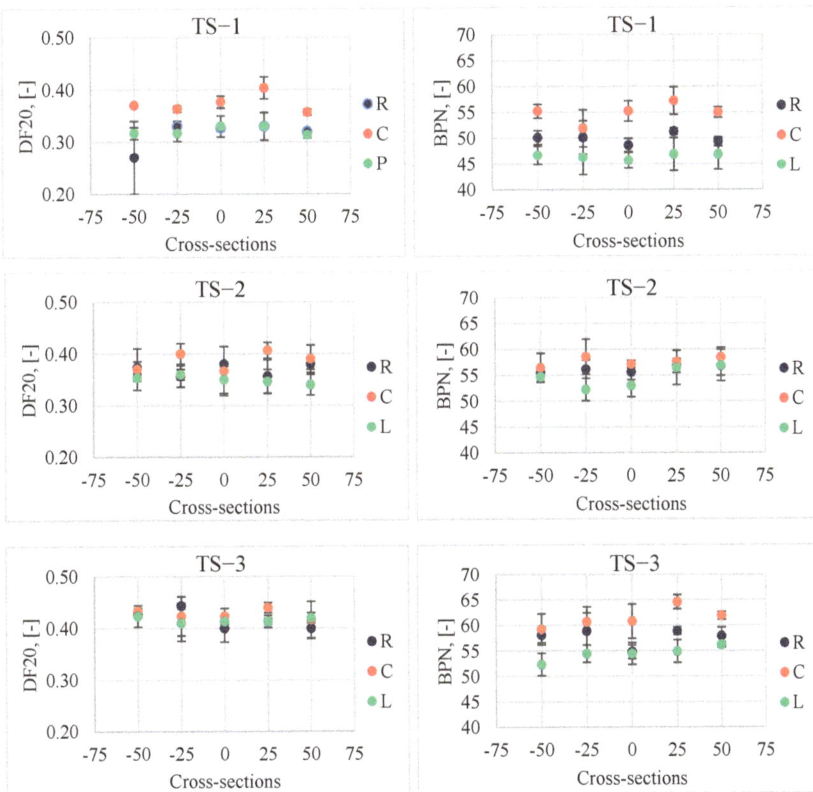

Figure 9. *Cont.*

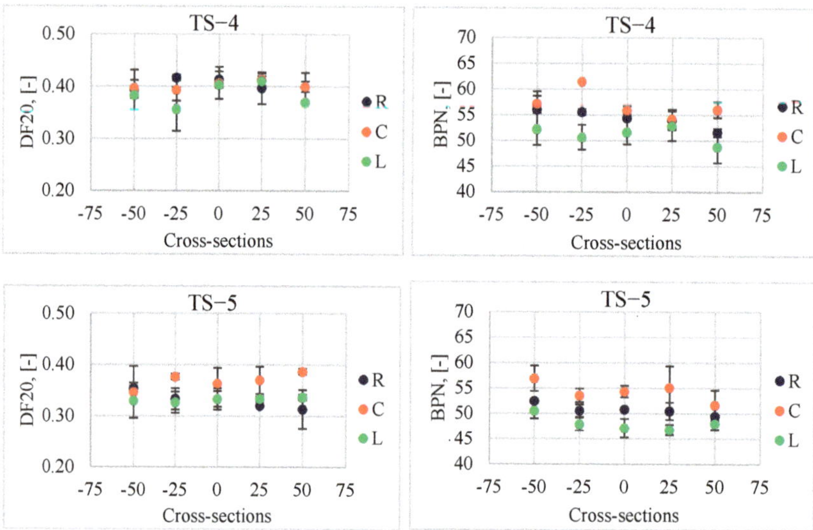

(**a**) (**b**)

Figure 9. Average values of DFT20 (**a**) and BPN (**b**) with standard deviations on individual test sections.

Table 1. Descriptive statistics of the DFT20 and BPN parameters.

	TS−1			TS−2			TS−3			TS−4			TS−5		
	R	C	L	R	C	L	R	C	L	R	C	L	R	C	L
$\overline{DFT20}$ [-]	0.32	0.37	0.32	0.37	0.39	0.35	0.42	0.43	0.42	0.40	0.40	0.38	0.33	0.37	0.33
STD [-]	0.04	0.02	0.02	0.02	0.03	0.02	0.02	0.02	0.02	0.02	0.02	0.03	0.02	0.03	0.02
V [%]	11.8	5.1	5.1	5.2	8.2	4.9	5.3	5.3	4.1	5.4	5.0	6.9	6.8	7.7	5.4
$\overline{BPN}$ [-]	50	55	46	56	58	55	58	61	54	54	57	51	51	54	48
STD [-]	1.73	2.64	2.31	1.39	2.15	2.81	2.60	2.67	2.04	2.23	2.76	2.78	1.29	2.89	1.82
V [%]	3.5	4.8	5.0	2.5	3.7	5.1	4.5	4.3	3.7	4.1	4.8	5.4	2.6	5.3	3.8

The difference between the DFT20 results recorded in individual tracks is only noticeable on sections TS-1 and TS-5. In the remaining sections, the differences are insignificant. The DFT device is clearly less sensitive to microtexture variations compared to the British Pendulum Tester. This is evidenced by slight differences in the DFT20 values between the left and right track. The most polished part of the lane is the left traffic track. This is due to the higher frequency and higher speed of crossings resulting from overtaking maneuvers. This is clearly confirmed by the results of tests conducted by the BPT on the EAC pavements. There is a significant difference between the BPN values depending on the location of the analyzed tracks on the tested sections; the highest values were obtained in the middle (C), then in the right (R) and left track (L).

The coefficients of variation indicate that the surfaces are homogeneous in terms of microtexture in individual tracks (C, L, R). The values are in a range of 3 ÷ 8%. A larger scatter was only recorded when measured with the DFT device on the TS-1 section in the right track.

4.2. *Macrotexture Assessment by Volumetric and Profilometric Methods*

The measurement results of the MPD and MTD parameters recorded on individual test sections in the left wheel track (L), in the right wheel track (R) and between the wheel tracks (C) are presented in Figures 10 and 11. Table 2 provides descriptive statistics, i.e., average values, standard deviations (STD) and coefficients of variation (V).

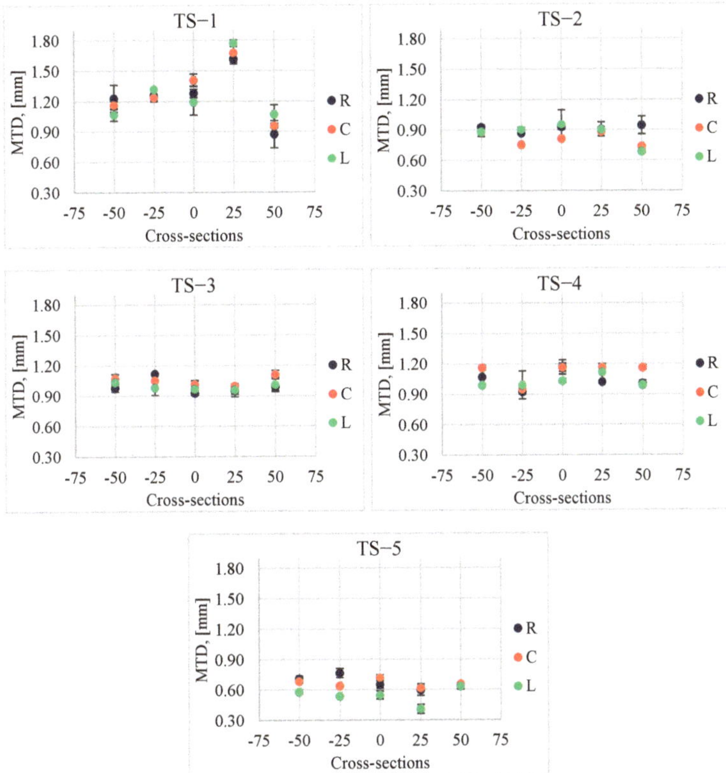

Figure 10. Average MTD values with standard deviations on individual test sections.

Table 2. Descriptive statistics of MTD, MPD_{CTM} and MPD_{SPL} parameters on test sections.

	TS−1			TS−2			TS−3			TS−4			TS−5		
	SP	**SS**	**SL**	**SP**	**SS**	**SL**	**SP**	**SS**	**SL**	**SP**	**SS**	**SL**	**SP**	**SS**	**SL**
$\overline{MTD}$ [mm]	1.25	1.29	1.28	0.91	0.81	0.86	0.99	1.05	0.99	1.04	1.12	1.03	0.67	0.66	0.54
STD [mm]	0.27	0.27	0.29	0.06	0.07	0.12	0.08	0.05	0.06	0.10	0.10	0.09	0.07	0.04	0.09
V [%]	21.3	21.1	22.3	7.0	8.2	14.1	7.9	5.1	6.3	9.8	8.5	8.9	11.1	6.2	15.9
$\overline{MPD_{CTM}}$ [mm]	1.63	1.69	1.57	1.18	1.10	1.12	1.26	1.33	1.21	1.33	1.39	1.24	0.89	0.83	0.76
STD [mm]	0.26	0.36	0.28	0.08	0.07	0.08	0.08	0.09	0.07	0.09	0.12	0.09	0.13	0.10	0.06
V [%]	16.0	21.1	17.6	6.6	6.1	6.7	6.6	7.0	6.0	6.5	8.9	7.5	14.7	11.6	7.3
$\overline{MPD_{SLP}}$ [mm]	1.69	1.81	1.66	1.20	1.21	1.18	1.33	1.40	1.36	1.38	1.52	1.39	0.90	0.91	0.77
STD [mm]	0.32	0.45	0.37	0.07	0.09	0.09	0.07	0.08	0.07	0.16	0.07	0.11	0.10	0.10	0.11
V [%]	19.1	25.0	22.6	6.2	7.7	7.7	5.4	5.8	4.8	11.8	4.9	8.0	11.7	11.1	14.3

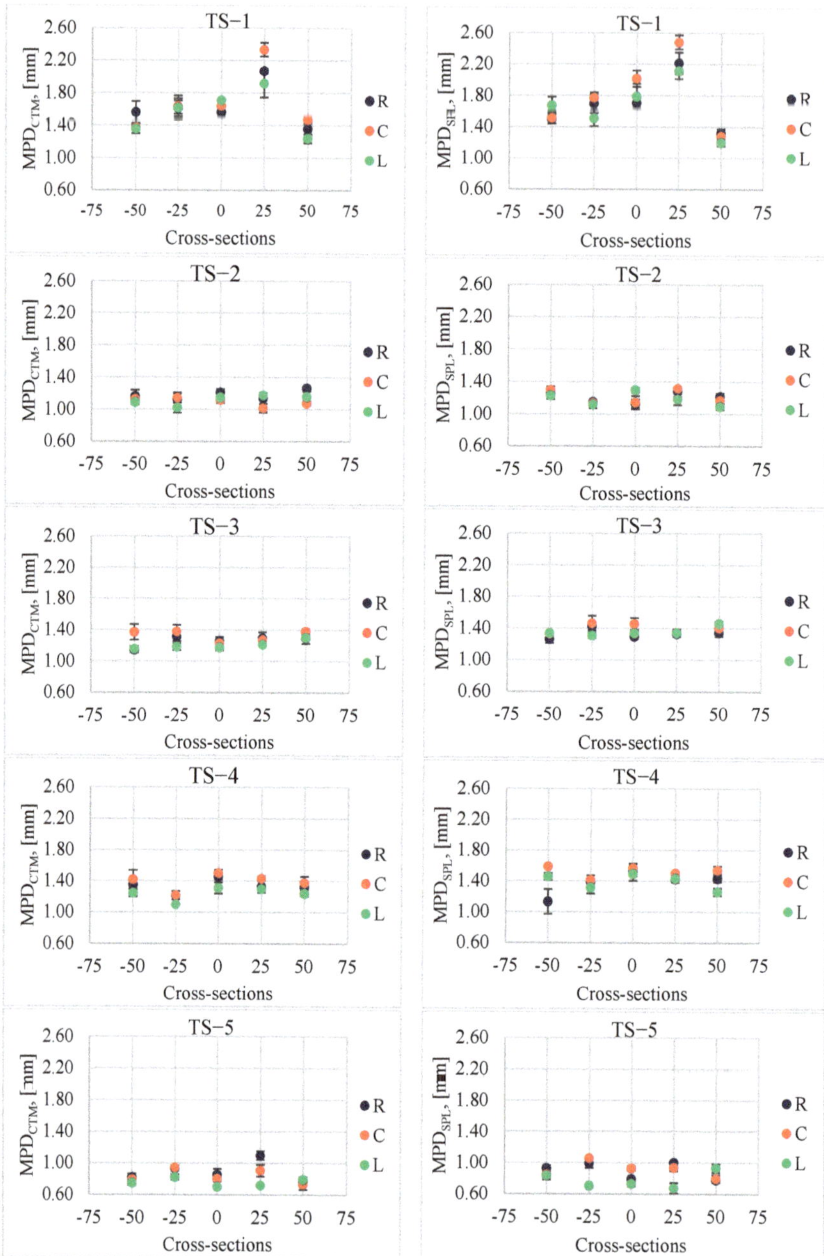

(a) (b)

Figure 11. Average MPD values with standard deviations on individual test sections obtained from (**a**) CTM and (**b**) SPL.

In some rare cases, slightly lower MPD and MTD values were observed in the left track compared to the values obtained in the right and the middle tracks of the lane. Generally, it was found that the differences in the parameters describing the macrotexture depending on the track (L, R or C) were insignificant. This is the opposite situation compared to changes in the values of friction coefficients, BPN and DFT20, where the track was important. In the case of macrotexture changes, the recorded differences were dictated by the differences in the macrotexture level of the selected test sections. However, when analyzing changes on a given section, the differences were visible in the cross-sections and not in the track of the vehicle's wheels. This may be related to the process of curing and exposing the coarse aggregate.

It was noticed that the results obtained from individual measurement techniques show the same trends in changes at individual measurement points on the test sections. This is especially visible on sections TS−1 and TS−4. In the case of TS−1, the highest values were obtained at locations "0" and "+25", which result from the loss of aggregates from the surface. In these sections, high values of all analyzed parameters were obtained, MPD_{CTM}, MPD_{SLP} and MTD.

When comparing the parameters obtained from volumetric and profilometric methods, it is evident that the MTD values are lower than MPD_{CTM} and MPD_{SLP}. However, there are slight differences between the MPD results from CTM and SLP. The methods for determining the average profile depth in both cases meet the requirements specified in EN-ISO 13473-1 [11]. However, the CTM device scans the surface profile in a circle, while the SLP device scans it in a straight line. Therefore, it is not possible for each of these devices to scan the profile in exactly the same place on the tested surface. This causes differences in the results. Figure 12 shows the deviation from the average MPD (1.27 mm) calculated from a set of data at 225 measurement points located in individual cross-sections on test sections, which were obtained from measurements with two repetitions (runs) with CTM and SPL devices.

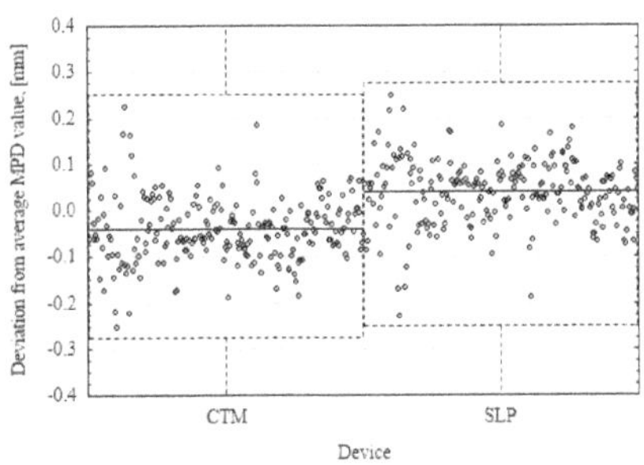

Figure 12. Deviation from the average MPD value for a particular profilometer.

The range of deviations calculated at individual measurement points for two devices is comparable. Therefore, differences are accepted.

4.3. Determining Linear Regression between Individual Parameters

A correlation analysis was performed to determine linear relationships between the analyzed parameters characterizing microtexture and macrotexture. The matrix scatterplot for each pair of variables obtained from microtexture and macrotexture and the linear regression lines are presented in Figure 13. Pearson correlation coefficients, r, are also shown in each scatter plot.

It was found that there was no correlation between the parameters describing the microtexture (BPN, DFT20) and the parameters describing the macrotexture (MTD, MPD). This is also confirmed by the results obtained by other researchers [9,23]. However, a very high correlation was obtained in the case of BPN and DFT20 (r − 0.719), and, in the case of MPD and MTD, the correlation was almost full (r above 0.900). Therefore, linear regression relationships were determined in these configurations of variables (Figure 14).

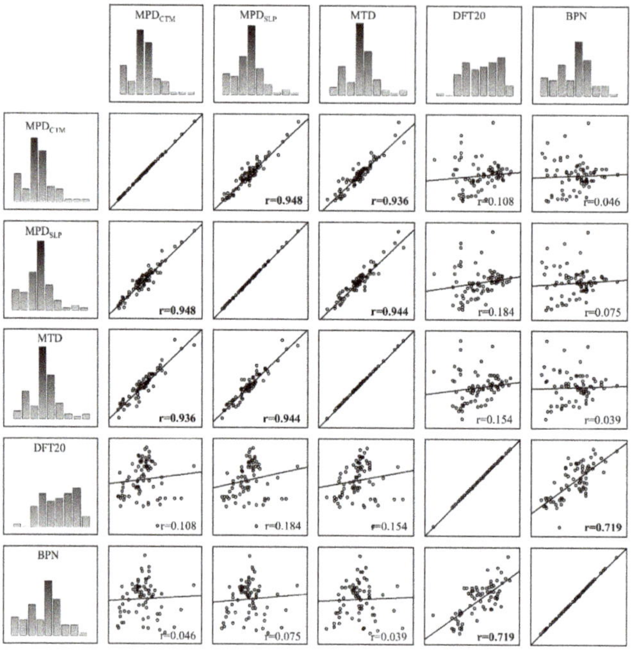

Figure 13. Matrix scatterplot.

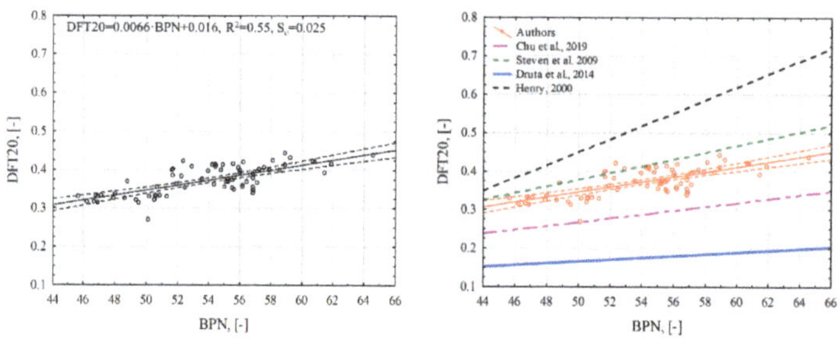

Figure 14. Regression relationships between DFT20 and BPN coefficients: (**a**) regression relationship with 95% confidence interval; (**b**) comparison of the regression relationship of the authors and other researchers [34–36,40].

The obtained model (Figure 14a) explains 55% of the DFT20 values through the values of the independent variable BPN. Table 3 lists the regression relationships established by individual researchers, and Figure 14b presents their graphical presentation. They were determined on the basis of measurements primarily made on wearing courses made from asphalt mixtures.

The relationships differ in the values of the regression coefficients b0 and b1 and the coefficient of determination R^2. They may result from the number and spread of results that were the basis for developing these models. BPT and DFT devices reproduce braking conditions with a locked wheel during measurement. The analysis of the results proved that both DFT and BPT are sensitive to changes in microtexture. However, differences between devices related to sliding speed, slider size and rubber hardness most likely influence the differences in the mechanism of generating friction force between DFT and BPN. Consequently, the results obtained from these two devices cannot be used interchangeably.

Table 3. Linear relationships between DFT20 and BPN.

Relationship	R^2	No. of Reference
DFT20 = 0.016 + 0.0066BPN	0.55	Authors
DFT20 = −0.400 + 0.017BPN	0.86	[34]
DFT20 = 0.018 + 0.0051BPN	0.39	[35]
DFT20 = −0.0668 + 0.0089BPN	0.51	[36]
DFT20 = 0.0517 + 0.0023BPN	0.36	[40]

In the case of the relationship developed on the basis of the results of parameters describing the macrotexture profile, it was found that the MPD_{SLP} values explain approximately 90% of the variability of the MPD_{CTM} results (Figure 15a). Differences in surface profile scanning and laser sampling frequency contribute to some scatter in the results. However, statistical analysis has proven that these two devices can be used interchangeably when assessing macrotexture in a range from 0.40 mm to 2.00 mm, as the differences in results are approximately 10%. However, the relationships presented in Figure 15b,c, which were obtained using individual laser profilometers, explain 87% (MPD_{CTM}) and 89% (MPD_{SLP}) of the variability of the MTD parameter. Dynamic laser profilometers are usually used to diagnose the condition of the road surface. Dependencies that allow for MTD estimation based on MPD results or vice versa are practiced quite often. The reasons may be the desire to refer to archived data, hardware limitations or limitations in performing measurements with a specific device. The standard EN ISO 13473-1 [11], both the 2005 and 2019 versions, presents functional dependencies for estimating ETD (Estimated Texture Depth). However, many other linear relationships that use MTD and MPD results can be found in the technical literature. Table 4 shows dependencies, and Figure 15d shows their graphical presentation.

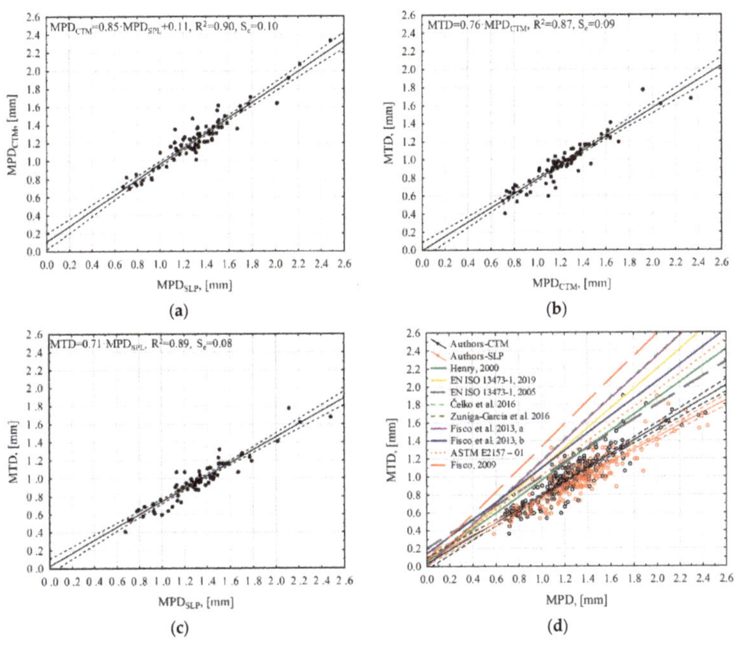

Figure 15. Regression relationships between macrotexture coefficients: (**a**) MPD_{CTM}—MPD_{SLP}, (**b**) MTD—MPD_{CTM}, (**c**) MTD—MPD_{SLP} and (**d**) MTD—MPD comparison of the results of the authors and other researchers [11,15,17,19,20,34,39,41].

Table 4. Regression relationship between MTD and MPD.

Relationship	R^2	Measurement Method of MPD	No. of Reference
$MTD = 0.76MPD_{CTM}$	0.87	Circular Texture Meter	Authors
$MTD = 0.71MPD_{SPL}$	0.89	Stationary Laser Profilograph	Authors
$ETD = 1.1MPD$	-	-	[11], 2019
$ETD = 0.2 + 0.8MPD$	-	-	[11], 2005
$MTD = 0.0762 + 1.2587MPD$	0.89	Circular Texture Meter	[15]
$MTD = 0.016 + 1.167MPD_{ZS}$	0.94	ZScanner® 800	[17]
$MTD = -0.12 + 1.10MPD_{CTM}$	0.94	Circular Texture Meter	[19]
$MTD = 0.053 + 0.91MPD_{CTM}$	0.98	Circular Texture Meter	[34]
$MTD = 0.069 + 0.947MPD$	-	Circular Texture Meter	[39]
$MTD = 1.17MPD$	0.98	Laser scanner	[41], a
$MTD = 0.139 + 0.96MPD$	0.91	Laser profiler	[41], b

The values of the slope coefficients of the regression Equation (b1), determined by the authors, are much smaller than 1. However, in the remaining analyzed cases, most of the equations have a value around 1 or above this value. This may be due to the profile characteristics of the tested surface. To explain the differences, other parameters that can be determined from the profile should be analyzed. EAC pavements are characterized by a positive texture, and wearing courses made of asphalt mixtures, i.e., SMA, BBTM, PA, AC, are characterized by a negative texture. In the case of the analyzed EAC pavements, the Rsk value is close to or above 0, which indicates the positive nature of the texture (the Rsk parameter evaluates the symmetry of the texture curve with the baseline) [25,42].

Due to the positive texture of the EAC, during macrotexture measurements using the volumetric method, glass beads are distributed in the spaces between the grains in a different way than on layers of asphalt mixtures with a negative profile. There is also a difference when calculating the MPD parameter. The average value of the profile height is much lower than in the case of negatively textured surfaces. which results in higher maximum profile heights necessary to calculate the MPD (Figure 16). The described phenomena regarding measurements using the volumetric and profilometric methods of EAC pavements influence the values of the b1 coefficients of the linear regression between the MTD and MPD parameters.

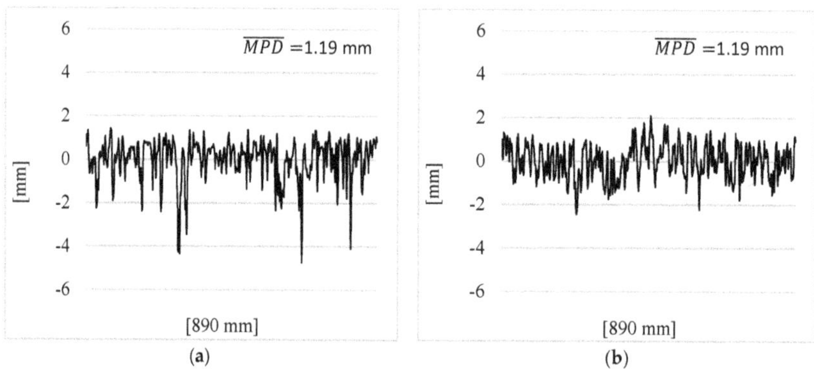

Figure 16. Examples of profiles obtained from measurement with the CTM device: (**a**) SMA11; (**b**) EAC8.

Comparative analyzes of the texture of various pavements are based on values recorded mainly on wearing courses of asphalt mixtures. Unlike EAC pavements, BBTM, SMA, PA and AC wearing courses have a negative profile. However, differences in the profile may significantly affect the interpretation of the results. Therefore, measurements on EAC pavements made it possible to verify the relationships between parameters characterizing the texture based on various measurement techniques and to refer to linear regressions developed by other researchers. The planned studies will concern the inclusion of the results obtained in detailed analyses of wearing layers of asphalt mixture types, varying in terms of the content of voids. This will allow us to develop regression relationships that will take into account the technological details necessary for use in practice.

5. Conclusions

Texture is crucial to user safety due to the role it plays in generating the friction force between tires and the pavement in slipping conditions. Texture is related to the technology of making the upper road layer. Therefore, at the design stage, it is important to consciously choose individual materials, the type of pavement or texturing technique in the case of concrete pavement.

Based on the analysis of the results of parameters describing microtexture and macrotexture recorded on five test sections with Exposed Aggregate Concrete surface and their comparison with data from the literature, the following conclusions were formulated:

1. Both DFT and BPT devices can be used to estimate the microtexture of EAC pavements. The DFT device is less sensitive to changes occurring in the microtexture compared to BPT. BPN values confirm that the left wheel track is most exposed to polishing factors.
2. Based on the developed linear regression relationship, it was found that DFT20 explains 55% of the values of the independent variable BPN. BPT and DFT20 device measurements allow for the ranking of test areas in terms of microtexture variability.
3. The EAC macrotexture can be verified by volumetric and profilometric methods using the CTM and SLP. The lower MTD values obtained compared to the MPD values are probably related to the positive EAC surface profile. The average difference between MTD and MPD_{SLP} values is approximately 0.34 mm, and between MTD and MPD_{CTM} approximately 0.27 mm.
4. Statistical analysis proved that there is almost certain correlation in the individual configuration of MPD_{CTM}, MPD_{SLP} and MTD parameter pairs (correlation coefficient r above 0.90).
5. The deviation from the average MPD value calculated from measurement data on five test sections is approximately 0.40 mm obtained for both the CTM and SLP devices. Slight differences in the results of MPD_{CTM} and MPD_{SLP} are caused by the sampling frequency of the profile scanning laser and the operating characteristics of the devices. However, this does not significantly affect the variability of the obtained results, and CTM and SLP profilometers can be used interchangeably.
6. During comparative analyzes of the results relating to the macrotexture obtained using different techniques, the technology of making the upper layers surface should be taken into account. Due to their characteristics, EAC pavements should be treated independently compared to asphalt pavements when estimating the MTD/ETD parameter. This is particularly important in the practical use of dependencies when controlling the quality of this type of surface.

Author Contributions: Conceptualization, W.G., P.G. and M.W.; methodology, W.G., P.G. and M.W.; validation, P.G., M.W. and W.G.; formal analysis, P.G. and M.W.; investigation, P.G. and M.W.; resources, M.W.; data curation, P.G.; writing—original draft preparation, P.G. and M.W.; writing—review and editing, P.G., M.W. and W.G.; visualization, P.G.; supervision, W.G.; project administration, P.G.; funding acquisition, M.W. and W.G. All authors have read and agreed to the published version of the manuscript.

Funding: This work was financially supported by a subsidy from the Ministry of Science and Higher Education to the Bialystok University of Technology (No. WZ/WB-IIL/7/2023).

Institutional Review Board Statement: Not applicable.

Informed Consent Statement: Not applicable.

Data Availability Statement: The raw data supporting the conclusions of this article will be made available by the authors on request.

Acknowledgments: Marek Motylewicz, Phd Eng. (Bialystok University of Technology), is the author of the field measurement photos presented in this paper.

Conflicts of Interest: The authors declare no conflicts of interest.

References

1. Wambold, J.C.; Antle, C.E.; Henry, J.J.; Rado, Z. *International PIARC Experiment to Compare and Harmonize Texture and Skid Resistance Measurements*; PIARC: Paris, France, 1995; p. 346.
2. Hall, J.W.; Smith, K.L.; Titus-Glover, L.; Wambold, J.C.; Yager, T.J.; Rado, Z. *Guide for Pavement Friction*; NCHRP Web-only document 108; Transportation Research Board: Washington, DC, USA, 2009; p. 257.
3. Haider, M.; Wehr, R.; Conter, M.; Kriegisch, M.; Gasparoni, S. Texture and noise characteristics of exposed aggregate concrete road surfaces. In Proceedings of the 12th International Symposium on Concrete Roads, Prague, Czech Republic, 23–26 September 2014.
4. Hablovicova, B.; Krivanek, V.; Markova, P. Comparison of Exposed Aggregate Cement Concrete Surface and Stone Mastic Asphalt Surface Noise Emissions by Close-Proximity Method. *Appl. Sci.* **2021**, *11*, 10359. [CrossRef]
5. Gardziejczyk, W.; Gierasimiuk, P. Influence of texturing method on tyre/road noise of cement concrete pavement. *Int. J. Pavement Eng.* **2018**, *19*, 1061–1076. [CrossRef]
6. Leng, Z.; Fan, Z.; Liu, P.; Kollmann, J.; Oeser, M.; Wang, D.; Jiang, X. Texturing and evaluation of concrete pavement surface: A state-of-the-art review. *J. Road Eng.* **2023**, *3*, 252–265. [CrossRef]
7. Zong, Y.; Li, S.; Zhang, J.; Zhai, J.; Li, C.; Ji, K.; Feng, B.; Zhao, H.; Guan, B.; Xiong, R. Effect of aggregate type and polishing level on the long-term skid resistance of thin friction course. *Constr. Build. Mater.* **2021**, *282*, 122730. [CrossRef]
8. Qi, L.; Liu, Y.; Liu, Z.; Zhang, C.; Chen, Z.; Lv, J.; Wan, H. Skid Resistance Attenuation of Asphalt Pavement Based on Multifactor Accelerated Wear Test. *Coatings* **2023**, *13*, 717. [CrossRef]
9. Liu, Y.; Cheng, X.; Yang, Z. Effect of Mixture Design Parameters of Stone Mastic Asphalt Pavement on Its Skid Resistance. *Appl. Sci.* **2019**, *9*, 5171. [CrossRef]
10. *EN 13036-1*; Road and Airfield Surface Characteristics—Test Methods—Part 1: Measurement of Pavement Surface Macrotexture Depth Using a Volumetric Patch Technique. European Committee for Standardization (CEN): Brussels, Belgium, 2010.
11. *EN ISO 13473-1*; Characterization of Pavement Texture by Use of Surface Profiles. Part 1: Determination of Mean Profile Depth. International Organization for Standardization: Geneva, Switzerland, 2019.
12. Praticò, F.G.; Astolfi, A. A new and simplified approach to assess the pavement surface micro- and macrotexture. *Constr. Build. Mater.* **2017**, *148*, 476–483. [CrossRef]
13. Flintsch, G.W.; de León, E.; McGhee, K.K.; AI-Qadi, I.L. Pavement Surface Macrotexture Measurement and Applications. *Transp. Res. Rec.* **2003**, *1860*, 168–177. [CrossRef]
14. D'Apuzzo, M.; Evangelisti, A.; Flintsch, G.W.; Izeppi, E.d.L.; Mogrovejo, D.E.; Nicolosi, V. Evaluation of Variability of Macrotexture Measurement with Different Laser-Based Devices. In Proceedings of the Airfield and Highway Pavements 2015, Miami, FL, USA, 7–10 June 2015. [CrossRef]
15. Fisco, N.R. Comparison of Macrotexture Measurement Methods. Master's Thesis, The Ohio State University, Columbus, OH, USA, 2009.
16. Lin, Y.; Dong, C.; Wu, D.; Jiang, S.; Xiang, H.; Weng, Z. Study of Pavement Macro- and Micro-Texture Evolution Law during Compaction Using 3D Laser Scanning Technology. *Appl. Sci.* **2023**, *13*, 5736. [CrossRef]
17. Čelko, J.; Kováč, M.; Kotek, P. Analysis of the pavement surface texture by 3D scanner. *Transp. Res. Proc.* **2016**, *14*, 2994–3003. [CrossRef]
18. Kane, M.; Lim, M.; Do, M.T.; Edmondson, V. A new predictive skid resistance model (PSRM) for pavement evolution due to texture polishing by traffic. *Constr. Build. Mater.* **2022**, *342*, 128052. [CrossRef]
19. Zuniga-Garcia, N.; Prozzi, J.A. *Contribution of Micro- and Macro-Texture for Predicting Friction on Pavement Surfaces*; U.S. Department of Transportation's University Transportation Centers Program: Washington, DC, USA, 2016; p. 106.
20. Pérez-Acebo, H.; Gonzalo-Orden, H.; Findley, D.J.; Rojí, E. A skid resistance prediction model for an entire road network. *Constr. Build. Mater.* **2020**, *262*, 120041. [CrossRef]
21. Kane, M.; Edmondson, V. Skid resistance: Understanding the role of road texture scales using a signal decomposition technique and a friction model. *Int. J. Pavement Eng.* **2020**, *23*, 499–513. [CrossRef]
22. Brown, C.A.; Hansen, H.A.; Jiang, X.J.; Blateyron, F.; Berglund, J.; Senin, N.; Bartkowiak, T.; Dixon, B.; Le Goïc, G.; Quinsat, Y.; et al. Multiscale analyses and characterizations of surface topographies. *CIRP Ann.* **2018**, *67*, 839–862. [CrossRef]
23. Torbruegge, S.; Wies, B. Characterization of pavement texture by means of height difference correlation and relation to wet skid resistance. *J. Traffic Trans. Eng.* **2015**, *2*, 59–67. [CrossRef]
24. Kane, M.; Do, M.T.; Cerezo, V. Optimising pavement texture measurement resolution for accurate friction determination: A preliminary investigation. *Road Mater. Pavement Des.* **2023**, *25*, 1894–1909. [CrossRef]
25. Vieira, T.; Sandberg, U.; Erlingsson, S. Negative texture, positive for the environment: Effects of horizontal grinding of asphalt pavements. *Road Mater. Pavement Des.* **2021**, *22*, 1–22. [CrossRef]

26. Šernas, O.; Zofka, A.; Vaitkus, A.; Gražulytė, J. The effect of exposed aggregate concrete gradation on the texture characteristics and durability. *Constr. Build. Mater.* **2020**, *261*, 119921. [CrossRef]
27. Wasilewska, M.; Gardziejczyk, W.; Gierasimiuk, P. Effect of aggregate graining compositions on skid resistance of Exposed Aggregate Concrete pavement. *IOP Conf. Ser. Mater. Sci. Eng.* **2018**, *356*, 012001. [CrossRef]
28. Rith, M.; Kim, Y.K.; Lee, S.W. Characterization of long-term skid resistance in exposed aggregate concrete pavement. *Constr. Build. Mater.* **2020**, *256*, 119423. [CrossRef]
29. Vaitkus, A.; Andriejauskas, T.; Šernas, O.; Čygas, D.; Laurinavičius, A. Definition of concrete and composite precast concrete pavements texture. *Transport* **2019**, *34*, 404–414. [CrossRef]
30. Dąbrowski, M.; Brachaczek, A.; Bogusz, K.; Glinicki, M.A. Experimental assessment of appropriate time for aggregate exposure at the surface of cement concrete pavement. *Int. J. Pavement Eng.* **2024**, *25*, 2318607. [CrossRef]
31. Gardziejczyk, W.; Gierasimiuk, P.; Motylewicz, M.; Wasilewska, M. Evaluation of noisiness of exposed aggregate cement concrete pavement. *Road Mater. Pavement Des.* **2021**, *22*, 2352–2368. [CrossRef]
32. Sun, Z.; Premarathna, W.A.A.S.; Anupam, K.; Kasbergen, C.; Erkens, S.M.J.G. A state-of-the-art review on rolling resistance of asphalt pavements and its environmental impact. *Constr. Build. Mater.* **2024**, *411*, 133589. [CrossRef]
33. Praticò, F.G.; Vaiana, R. A study on the relationship between Mean Texture Depth and Mean Profile Depth of asphalt pavements. *Constr. Build. Mater.* **2015**, *101*, 72–79. [CrossRef]
34. Henry, J.J. *Evaluation of Pavement Friction Characteristics. Synthesis of Highway Practice No. 291*; Transportation Research Board: Washington, DC, USA, 2000.
35. Chu, L.; Cui, X.; Zhang, K.; Fwa, T.F.; Han, S. Directional Skid Resistance Characteristics of Road Pavement: Implications for Friction Measurements by British Pendulum Tester and Dynamic Friction Tester. *Transp. Res. Rec.* **2019**, *2673*, 793–803. [CrossRef]
36. Steven, B.; Monismith, C.L.; Tsai, B.-W. *Friction Testing of Pavement Preservation Treatments: Friction Measurements on Fog Seal Trials Using Six Rejuvenators. State Route KER58*; California Department of Transportation: Sacramento, CA, USA, 2009.
37. *ASTM E303-22*; Standard Test Method for Measuring Surface Frictional Properties Using the British Pendulum Tester. ASTM: West Conshohocken, PA, USA, 2022.
38. *ASTM E1911*; Standard Test Method for Measuring Paved Surface Frictional Properties Using the Dynamic Friction Tester. ASTM: West Conshohocken, PA, USA, 2019.
39. *ASTM E2157–01*; Standard Test Method for Measuring Pavement Macrotexture Properties Using the Circular Track Meter. ASTM: West Conshohocken, PA, USA, 2017.
40. Druta, C.; Wang, L.; Lane, D.S. *Evaluation of the MMLS3for Accelerated Wearing of Asphalt Pavement Mixtures Containing Carbonate Aggregates*; Report No.: FHWA/VCTIR 14-R17; Virginia Center for Transportation Innovation and Research: Charlottesville, VA, USA, 2014.
41. Fisco, N.; Sezen, H. Comparison of surface macrotexture measurement methods. *J. Civ. Eng. Manag.* **2013**, *19*, 153–160. [CrossRef]
42. Chen, S.; Liu, X.; Luo, H.; Yu, J.; Chen, F.; Zhang, Y.; Ma, T.; Huang, X. A state-of-the-art review of asphalt pavement surface texture and its measurement techniques. *J. Road Eng.* **2022**, *2*, 156–180. [CrossRef]

materials

MDPI

Article

Optimization of Foams—Polypropylene Fiber-Reinforced Concrete Mixtures Dedicated for 3D Printing

Magdalena Rudziewicz, Marcin Maroszek, Kinga Setlak, Mateusz Góra and Marek Hebda *

Faculty of Materials Engineering and Physics, Cracow University of Technology, Warszawska 24, 31-155 Kraków, Poland; magdalena.rudziewicz@doktorant.pk.edu.pl (M.R.); marcin.maroszek@doktorant.pk.edu.pl (M.M.); kinga.plawecka@pk.edu.pl (K.S.); mateusz.gora@doktorant.pk.edu.pl (M.G.)
* Correspondence: marek.hebda@pk.edu.pl

Abstract: The continued global urbanization of the world is driving the development of the construction industry. In order to protect the environment, intensive research has been carried out in recent years on the development of sustainable materials and ecological construction methods. Scientific research often focuses on developing building materials that are renewable, energy-efficient, and have minimal impact on the environment throughout their life cycle. Therefore, this article presents research results aimed at developing a concrete mixture using cement with reduced CO_2 emissions. In the context of increasing ecological awareness and in line with European Union policy, the development of a mixture based on environmentally friendly cement is of key importance for the future development of the construction industry. The article compares the physical properties of two mixtures, their foaming possibilities, and the influence of the added polypropylene (PP) fibers on the strength properties of the produced composites. It was found that bending strength and compressive strength were highest in the material with silica fume and aluminum powder at 5.36 MPa and 28.76 MPa, respectively. Microscopic analysis revealed significant pore structure differences, with aluminum foamed samples having regular pores and hydrogen peroxide foamed samples having irregular pores. Optimizing aluminum powder and water content improved the materials' strength, crucial for maintaining usability and achieving effective 3D printing. The obtained results are important in the development of research focused on the optimization of 3D printing technology using concrete.

Keywords: foamed concrete; additive manufacturing; fiber reinforcement

1. Introduction

1.1. Aerated Concrete

Foamed concrete is gaining prominence as an increasingly favored alternative to conventional concrete in residential construction applications. This material is utilized for its acoustic and thermal insulating properties, alongside its notably high fire resistance. Furthermore, aerated concrete demonstrates potential in reducing the weight of prefabricated building elements and the mitigation of seismic impact on structural integrity. The utilization of prefabricated elements fabricated from foamed concrete presents a cost-effective solution compared to those constructed from traditional concrete, thereby contributing to overall construction cost reduction. Attributable to its advantageous properties, foamed concrete stands as a promising candidate for application in 3D concrete printing endeavors. In contrast to conventional lightweight concrete with densities ranging from 1000 to 2000 kg/m^3 and compressive strengths between 1 to 100 MPa [1,2] foamed concrete exhibits notably lower densities, potentially reaching up to 400 kg/m^3 [3]. Elements fabricated from foamed materials are distinguished by notably lower density compared to conventional solid counterparts. Moreover, it is possible to use various raw materials and foaming methods [4,5]. One manufacturing approach involves incorporating pre-prepared foam into cement paste to introduce porosity within the composite. Alternatively, aerated concrete can be generated by incorporating chemical foaming agents into the concrete matrix [6]. During the reaction between aluminum powder (as the foaming agent) and the calcium hydroxide generated during cement hydration, hydrogen gas bubbles are produced [7]. Achieving a stable aerated concrete (utilizing aluminum powder or hydrogen peroxide as the foaming agent) amenable to pre-processing, pumping, and 3D printing poses a formidable challenge, with scant research hitherto addressing this exigency. Alternatively, traditional concrete formulation has undergone extensive scrutiny regarding the impact of the aluminum (Al) content on the compressive strength of aerated concrete [8–12]. Shabbar et al. demonstrated a decrease in the compressive strength of aerated concrete (AC) specimens with escalating aluminum powder content across samples containing 0.25%, 0.5%, 0.75%, and 1% Al. The internal surfaces of specimens featuring aluminum (Al) powder concentrations ranging from 0.5 to 1% exhibited irregularly shaped pores of greater dimensions than those apparent in specimens with 0.25% content [2]. Guglielmi et al. observed an initial marginal decline in compressive strength followed by augmentation as porosity diminished. Moreover, it was discerned that excessive aluminum powder hindered pore formation due to interactions with hydrogen peroxide [13]. Additionally, aerated concrete, characterized by a density of 686 kg/m^3 and a compressive strength of approximately 6.3 MPa, was successfully produced [14].

In the realm of advanced building materials, aerated concrete finds extensive application in external building insulation, internal gap filling, and fire protection systems. A judicious selection of the foaming agent and cement matrix profoundly impacts the properties of foam concrete. Hydrogen peroxide (H_2O_2) emerges as a particularly advantageous blowing agent owing to its controlled decomposition kinetics, rendering it a cost-effective and environmentally friendly alternative to aluminum powder and conventional blowing agents. Hydrogen peroxide, an inorganic liquid, exhibits complete miscibility with water in any proportion. Its decomposition rate can be effectively regulated using catalysts, such as manganese dioxide and permanganate. As a fully ecological product, its decomposition yields water and oxygen as by-products. Several researchers, including Pan et al. [15], Rooyen et al. [3], and Wu et al. [16], have demonstrated the use of H_2O_2 as a foaming agent for conventional concrete based on Portland cement, resulting in concrete foam with densities ranging from 150 kg/m^3 to 300 kg/m^3 [15]. However, it is essential to consider potential challenges when utilizing foamed concrete in additive manufacturing technology. Foamed concrete is prone to deformation in the lower layers of prints due to the weight exerted by subsequent layers [17,18]. Therefore, the initial steps in designing a foamed material tailored for 3D printing and developing production methods should involve verifying the stability and consistency of the prepared foamed materials and assessing their mechanical properties. The development of foam materials in concrete technology has been limited due to their brittleness and excessive susceptibility to cracks. One method to limit crack propagation in concrete is the incorporation of fibers, which serve as bridging mechanisms and enhance the material's resistance to cracking.

1.2. Fibers

In order to enhance the viscosity and consistency of the material for 3D concrete printing (3DCP), fibers can be introduced as additives. Their incorporation aims to improve properties related to maintaining layer shape after extrusion and to limit deformation resulting from the deposition of subsequent layers [19]. The addition of fibers plays a crucial role in preserving the geometry of the concrete throughout the printing process, thereby increasing the accuracy and precision of the resulting prints [20]. In traditional concrete structures, fibers are commonly used to improve the performance of manufactured elements. The most frequently used types of fibers include small synthetic fibers (up to 75 mm in length) known for their resistance to various factors, including oxidation. Carbon [21], basalt [22], glass [23], polypropylene [13,24], steel [25], and polyvinyl alcohol (PVA) [26] fibers are among the popular options. Fibers can be categorized into microfibers (with a diameter of less than 0.32 mm and a length of less than 30 mm) and macrofibers (with a diameter of more than 0.32 mm and a length of more than 30 mm) [27]. The introduction of fibers can modify the flow characteristics of concrete, thereby directly influencing its pumpability during the 3D printing process. Research conducted thus far indicates that the addition of nanosilica (NS) and polypropylene (PP) fibers to gypsum formulations can significantly enhance printability and mitigate the observed anisotropy in printed samples. This improvement is attributed to the minimal quantity of NS required [28]. The incorporation of dispersed fibers into concrete also serves as an effective reinforcement method, particularly in experiments involving additive manufacturing technology [29–31]. For instance, studies on 3D printing with geopolymers reinforced with polypropylene (PP) fibers demonstrated enhanced resistance to deformation under a load [32]. Additionally, experiments with fiber-reinforced foamed concrete revealed that adding 0.4% PP fibers significantly improved the compressive strength of the material, increasing it by 35% [33]. Further research indicated that PP fibers also enhanced the bending strength of foamed concrete by 40% [34]. Consequently, the enhanced shape stability and construction possibilities render printed concrete a viable option for implementation in civil infrastructure construction.

The analysis of using fiber-reinforced foamed concrete, specifically with polypropylene fibers, in the context of 3D concrete printing is an innovative approach that could bring significant benefits to the construction industry. Before commencing research in this area, it was essential to thoroughly understand the behavior and consistency of the produced mix. Polypropylene fibers added to concrete can enhance its tensile strength and reduce the risk of cracking. A critical aspect of this research is determining the open time for specific mixes, as this is crucial for the 3D printing process. The open time defines the period during which the mix retains sufficient plasticity for shaping. Additionally, it was important to investigate how the material behaves during flow tests at different intervals after preparation. Flow tests allow the assessment of the mix's ability to maintain proper consistency and homogeneity during application. Polypropylene fibers can also improve the structural stability of the concrete during the printing process. The research aims to determine the optimal mix proportions that will provide the best mechanical and rheological properties. The results of these studies could lead to the development of new standards and guidelines for 3D concrete printing. The innovation of this approach lies in the synergy between advanced building materials and modern manufacturing technologies. This study provides a preliminary characterization of the constituents and methodologies employed in the preparation of foamed materials designed for 3D concrete printing (3DCP). A comparative analysis was performed to assess the stability and rheological consistency of foamed mixtures in comparison to those incorporating polypropylene fibers, forming the basis for the development of a foamed mixture with dispersed reinforcement optimized for 3DCP. The judicious selection of binders and additives, combined with a comprehensive understanding of the material's behavior, is essential for establishing the new trajectory required to advance innovations in 3DCP technology.

2. Materials and Methods

2.1. Tested Materials

The research was carried out using two base materials. The first of them was Binder 1 (B1), which, in addition to Portland cement, contains silica fly dust up to 50%. Binder 2 (B2) contained only standard Portland cement. Foaming agents, such as aluminum powder and hydrogen peroxide H_2O_2 at a concentration of 30%, were used to produce foamed concrete. PP-12 polypropylene fiber with a length of 12 mm was also used in the study. The physical properties of the ingredients used in accordance with the data sheets issued by the manufacturers are presented in Tables 1–3.

Table 1. Physical properties of the aluminum powder [35].

Active Aluminum Content [%]	Water Coverage [cm^2/g]	Particle Size Distribution [μm]	Bulk Density [g/cm^3]
≥93	ca. 7300	102–119	0.16

Table 2. Physical properties of the hydrogen peroxide H_2O_2 [36].

Solubility in Water	Boiling Point	Melting Point	Density [g/cm^3]
Miscible	150 °C	−0.43 °C	1.13

Table 3. Physical properties of the polypropylene fibers [37].

Water Absorption (%)	Tensile Strength [N/mm^2]	Young's Modulus [N/mm^2]	Density [g/cm^3]
0	420	3800	0.91

The above-mentioned materials were used to prepare 20 mixtures, which were poured into molds measuring 40 × 40 × 160 mm and 100 × 100 × 100 mm. After demolding, the samples were inspected and found to be free of any visible cracks, pores, or defects. The reference samples were made only from a mixture of base binder (30% for B1 and 25% for B2), river sand (70% for B1 and 65% for B2), and water. The precise values for these parameters are provided in Table 4. To evaluate the impact of foaming agents on materials B1 and B2, both aluminum powder (Al) and hydrogen peroxide (H_2O_2) were analyzed. Additionally, samples of foamed concrete containing 0.4% polypropylene (PP) fibers were produced. The composition and proportions of the prepared samples are detailed in Table 4.

Table 4. Designation of the investigated samples depending on the composition and the foaming agent used.

No.	Designation of the Mixtures	Cementitious Binder [%]		PP-12 [%]	Foaming Admixture [%]		Water [%]
1	B1/H2O2	B1	84.53	-	H_2O_2	2.86	12.61
2	B1/AL		87.00	-	Al	0.15	12.85
3	B1/PP		86.62	0.40	-	-	12.98
4	B1		87.23	-	-	-	12.77
5	B1/AL/PP		86.60	0.40	Al	0.15	12.85
6	B1/H2O2/PP		84.10	0.40	H_2O_2	3.30	12.20
7	B1/05AL/PP		86.68	0.40	Al	0.075	12.85
8	B1/15AL/PP		80.85	0.40	Al	0.15	18.60
9	B1/05H2O2/PP		85.90	0.40	H_2O_2	1.50	12.20
10	B1/15H2O2/PP		80.03	0.40	H_2O_2	3.30	16.27
11	B2/H2O2	B2	80.09	-	H_2O_2	3.51	16.40
12	B2/AL		82.37	-	Al	0.15	17.48
13	B2/PP		82.12	0.40	-	-	17.48
14	B2		83.10	-	-	-	16.90
15	B2/AL/PP		82.15	0.40	Al	0.15	17.60
16	B2/H2O2/PP		79.70	0.40	H_2O_2	3.10	16.80
17	B2/05AL/PP		81.93	0.40	Al	0.075	17.60
18	B2/15AL/PP		79.29	0.40	Al	0.15	20.16
19	B2/05H2O2/PP		81.85	0.40	H_2O_2	0.15	17.60
20	B2/15H2O2/PP		79.50	0.40	H_2O_2	3.10	17.00

2.2. Methods

2.2.1. Slump Flow Test

Consistency tests were conducted using the falling cone method according to the PN-EN 12350-4:2019-08 standard [38]. The method employed involves measuring the flow diameter of the concrete mixture under its own weight after the removal of the mold. The concrete mixture is initially placed in a mold (an Abrams cone) on a flat surface in a single layer, without additional compaction (see Figure 1). Following the removal of the mold, the flow diameter is measured in two perpendicular directions, and the result is reported as the average of these two measurements. This procedure is straightforward to execute both at construction sites and during 3D concrete printing operations. The results obtained from this method enable an assessment of the concrete's consistency when incorporating either a solid or liquid foaming agent. Moreover, it allows for a quick assessment of the potential benefits of introducing polypropylene fibers into the mixture to improve the rheology and stability of the printing process.

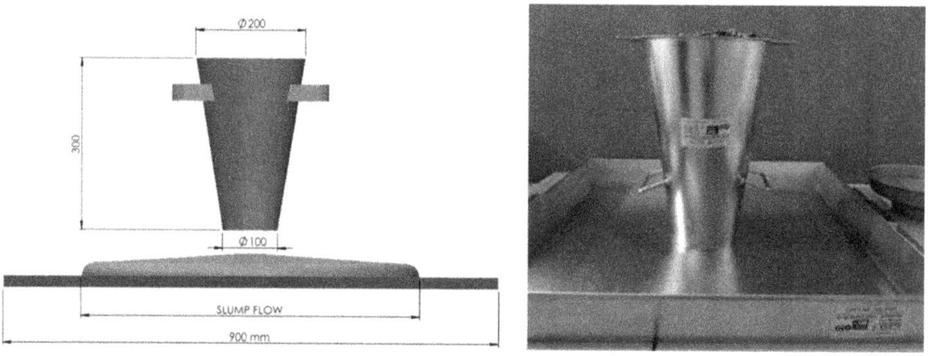

Figure 1. Consistency testing using the falling cone method—free flow. The measurement diagram is one the left, and a photo of the measuring set is on the right.

2.2.2. Compressive Strength Tests

Compressive strength tests were carried out in accordance with the PN-EN 12390-3:2019-07 standard [39], using a Matest 3000 kN universal testing machine (Matest, Treviolo, Italy). The dimensions of the prepared samples were 40 × 40 × 40 mm, and the linear speed of force increase of 0.5 MPa/sec increased until the sample was destroyed (Figure 2). The result was estimated from six measurement repetitions for each material variant.

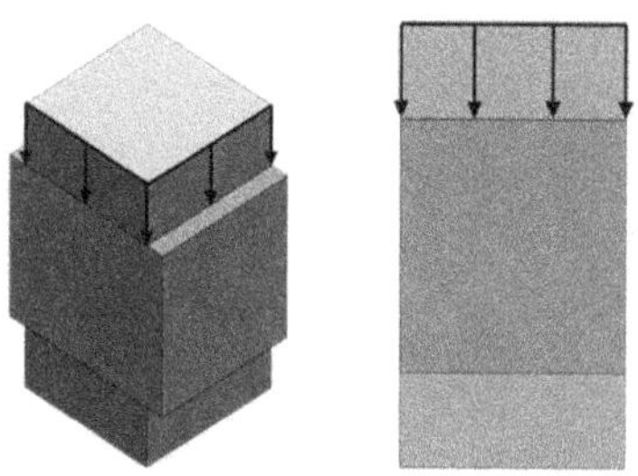

(a) (b)

Figure 2. Scheme showing the compressive strength test: (a) three-dimensional image; (b) cross-section.

2.2.3. Bending Strength Test

Flexural strength tests were carried out using a Matest 3000 kN testing machine on 40 × 40 × 160 mm specimens, based on EN 12390-5:2019-08 [40]. The linear speed of the force increase was 0.5 MPa/sec, with distances between supports of 150 mm (Figure 3). The result was estimated from three measurement repetitions for each material variant.

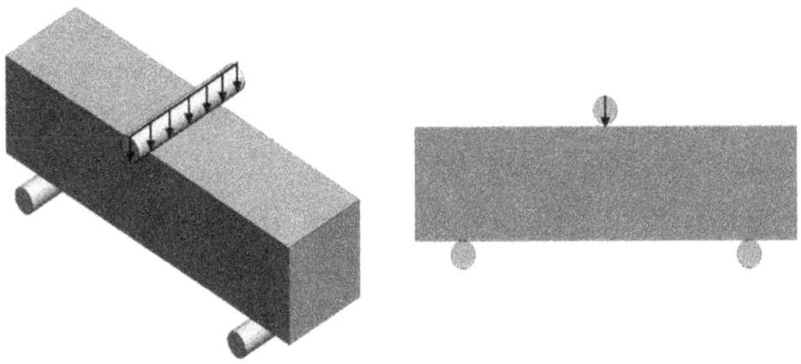

(a) (b)

Figure 3. Scheme showing the three-point bending strength test: (**a**) three-dimensional image; (**b**) cross-section.

2.2.4. Microscopy Observation

A Techrebal Banito B2920 digital microscope with 100× magnification, integrated with a Techrebal 48 MP camera (Techrebal, Wilczyce, Poland), was used for microscopy observation.

3. Results and Discussion

The concrete mix must maintain sufficient plasticity over an extended period, known as the open time, to allow for continuous extrusion without clogging or hardening prematurely. This is essential for the layer-by-layer deposition process. The mix needs to be fluid enough to be extruded through the printer nozzle yet stiff enough to hold its shape once deposited to prevent collapse or deformation of the printed layers. The material must have good buildability, meaning it can support subsequent layers without excessive deformation or collapse. This requires a balance between early strength gain and workable flowability [41]. The mix must be homogeneous and free from segregation or bleeding. For the material to be used in 3D printing, it must be pumpable through the delivery system without causing blockages. This requires the mix to have a low enough viscosity to be pumped but not so low that it affects the buildability. Each printed layer must adhere well to the previous one to ensure structural integrity. This requires the material to retain some level of tackiness or bonding capability during the printing process. The material must exhibit low shrinkage and high resistance to cracking, particularly during the early stages of setting [42,43]. The printed concrete must meet long-term durability requirements, including resistance to environmental factors, such as freeze–thaw cycles, chemical exposure, and moisture ingress. In fiber-reinforced mixes, it is essential to ensure an even distribution of fibers throughout the concrete to avoid weak spots and to enhance mechanical properties, such as tensile and flexural strength [44]. The mix should develop sufficient green strength (initial strength) quickly after extrusion to support the weight of additional layers and to resist deformation under its own weight.

During the 3D printing process using concrete, it is crucial to ensure the stability and durability of individual layers. Therefore, the research focused on verifying the consistency of the base mixture and modifying its composition. The flow measurement of concrete mixtures was carried out 1 min, 20 min, and 40 min after the mixture preparation [45–47]. The time interval between the individual tests is due to the fact that, during large-scale production, the interval between layers is significant. Additionally, because of the characteristics of the material, which must be sufficiently cohesive but not too fluid, it is necessary to examine whether printing with the given mixture at these intervals is feasible. It is crucial to verify this information to determine whether it is possible to mix a batch of material for one layer and if it will still have the appropriate consistency to apply a second layer with the same mixture. Similarly, the information obtained from these tests helps confirm whether, after 40 min, the consistency will change enough to allow for the application of another layer. Furthermore, this is particularly important in the context of interlayer forces and their strength [48]. The stability of the foaming process was also assessed depending on the type of foaming agent used. Table 5 shows the results of the slump flow test depending on the composition of the analyzed mixtures. Generally, slightly higher values were measured for samples based on raw material B1 than B2. Concrete mixtures modified with the addition of fibers showed the highest compactness, regardless of the mixing time and type of binder. It was also found that the addition of hydrogen peroxide to the mixture made it impossible to measure its consistency. This effect was independent of the type of raw material used. Comparison of mixtures with similar compositions after identical mixing times reveals clear differences in their properties. Lower flow values were measured each time for mixtures made from B2 raw material. To improve the consistency results of materials made of B1 material, further modifications to the composition of the mixtures are necessary. This result encourages further tests and the process of modifying the composition of the mixture, especially in the context of its suitability in 3D printing processes. In 3DCP technology, efficient mixing and pumping of the prepared mixture is crucial because the material must maintain a balance between being strong enough to maintain the integrity of the layer after printing, yet fluid enough to ensure smooth transport. Concomitantly, it is particularly important to prepare a material with a homogeneous structure; in the event of heterogeneity of the mixture or delamination of the concrete mixture after it dries, gaps and cracks may occur [49]. When segregation occurs and a significant amount of foam remains in the upper part of the layer, insufficient bonding between the layers may result. Cracking may also occur as the material dries and shrinks, often exacerbated by rapid drying [50]. Similarly, structural cracks may appear as a result of excessive loads or damage to the structure. Uneven or excessive settling of material from successive layers can affect long-term exposure to stress, causing permanent deformation. Therefore, achieving the appropriate consistency and stability of the foamed mixture is crucial.

Table 5. Slump flow test results depending on the composition and the foaming agent used.

No.	Designation of the Mixtures	Slump Flow Test		
		1 min [mm]	20 min [mm]	40 min [mm]
1	B1/H2O2	nd *	nd *	nd *
2	B1/AL	455	432	420
3	B1/PP	340	315	300
4	B1	458	435	420
5	B1/AL/PP	340	nd *	nd *
6	B1/H2O2/PP	465	nd *	nd *
7	B1/05AL/PP	445	370	nd *
8	B1/15AL/PP	415	nd *	nd *
9	B1/05H2O2/PP	full	460	nd *
10	B1/15H2O2/PP	480	nd *	nd *
11	B2/H2O2	nd *	nd*	nd *
12	B2/AL	400	320	290
13	B2/PP	370	345	320
14	B2	400	375	375
15	B2/AL/PP	325	nd *	nd *
16	B2/H2O2/PP	440	nd *	nd *
17	B2/05AL/PP	390	nd *	nd *
18	B2/15AL/PP	385	nd *	nd *
19	B2/05H2O2/PP	330	nd *	nd *
20	B2/15H2O2/PP	430	nd *	nd *

* The result cannot be presented due to the lack of fluidity of the mixture.

Figures 4 and 5 present a representative picture of the consistency changes in the tested materials depending on their composition and mixing time. By comparing the recorded photographic evidence, a notable distinction was observed between the mixtures prepared with raw materials B1 and B2. Mixtures formulated with material B2 demonstrated superior stability and a homogeneous consistency throughout the entire volume. Each measurement revealed a uniformly compact structure with consistent and well-defined flow edges. Moreover, the layers formed during the removal of the Abrams cone when the material with the addition of polypropylene fibers was tested were repeatable. Conversely, the material prepared based on B1 behaved differently each time. Concrete without additives had an uneven flow edge and water precipitation. The addition of aluminum powder initially only slightly changed the consistency of the material, but after 40 min of mixing the material noticeably foamed. By contrast, the addition of hydrogen peroxide to the tested materials caused such a quick reaction with the concrete mixture (without fibers) that it was not possible to conduct a consistency test. Moreover, with the extension of the mixing time and the introduction of additives in the form of polypropylene fibers, the consistency of the mixtures became more and more compact (Table 5, Figures 4 and 5). This effect was independent of the type of base raw material used.

It was observed that the B1/AL/PP and B1/H2O2/PP mixtures initially exhibited appropriate consistency post-mixing. However, conducting consistency tests after 20 min proved impractical due to the mixtures becoming excessively brittle, a condition deemed unacceptable. Furthermore, formulations containing 0.5% H_2O_2 exhibited excessive fluidity within the first minute of mixing, rendering them unsuitable for 3D concrete printing. After the first minute, B1/05H2O2/PP was excessively fluid, flowing through the cone in just 0.8 s. However, after 20 min of mixing, the flow time increased to 6.68 s. This rapid transformation in consistency suggests that the material undergoes swift alterations in its properties, making it unsuitable for additive manufacturing applications. Furthermore, the cone walls were heavily soiled by the mixture after the 20-min flow test. This indicates poor flow stability and residue formation, which would likely hinder the printing process and compromise the quality of the printed elements. The B1/05H2O2/PP mixture exhibited significantly different behavior compared to B1/05AL/PP. Conversely, the mixture incorporating 0.5% Al showed promising consistency both immediately after mixing and after the 20 min mark. The foamed concrete mixture B1/05AL/PP, reinforced with fibers, has demonstrated the most promising properties among the tested formulations. The flow time of this mixture through a cone was recorded at 4.43 s after 1 min and 5.78 s after 20 min, as illustrated in Figure 5. Although there is a noticeable increase, it is not significantly large, suggesting the mixture's stability over time. The consistency of the mixture after 20 min is markedly different from that after 1 min, likely due to hydration processes and structural changes within the concrete (Figure 4). Moreover, after the mixture flows through the cone, no larger material elements, such as clumped fibers or layers of concrete, remain on the cone walls. This observation indicates that the material can seamlessly pass through the transportation system in the printer and through the nozzle during the printing process. In summary, the B1/05AL/PP mixture exhibits advantageous characteristics for applications in concrete printing. Its stable flow, marked by a minimal increase in flow time, and the absence of residues on the cone walls, highlight its potential to achieve a uniform structure in printed elements. These properties may enhance the efficiency and quality of concrete printing, making the B1/05AL/PP mixture a promising material for modern construction applications involving 3D printing technology. The realm of 3D concrete printing necessitates materials with extended workability periods. Considering the potential future applications of these materials, the findings are encouraging. Nevertheless, further investigations are warranted into the optimal proportions of the frother and fiber, as well as the printing window of the mixture. The tested materials were subjected to strength tests. The results of compressive and flexural strength depending on the type of tested materials were presented in Tables 6 and 7, respectively.

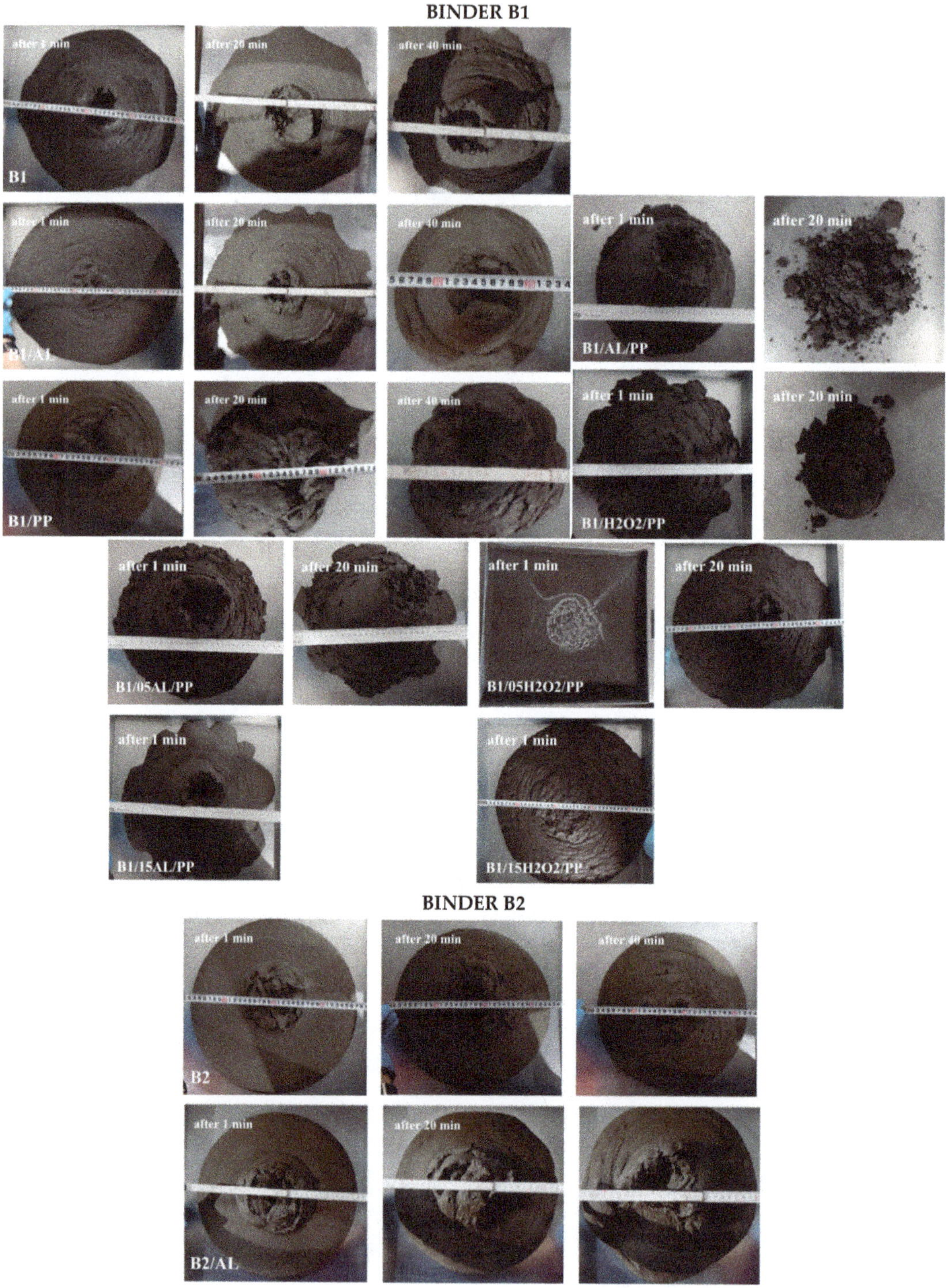

Figure 4. *Cont.*

Figure 4. Testing the consistency of the mixtures and their stability and buildability.

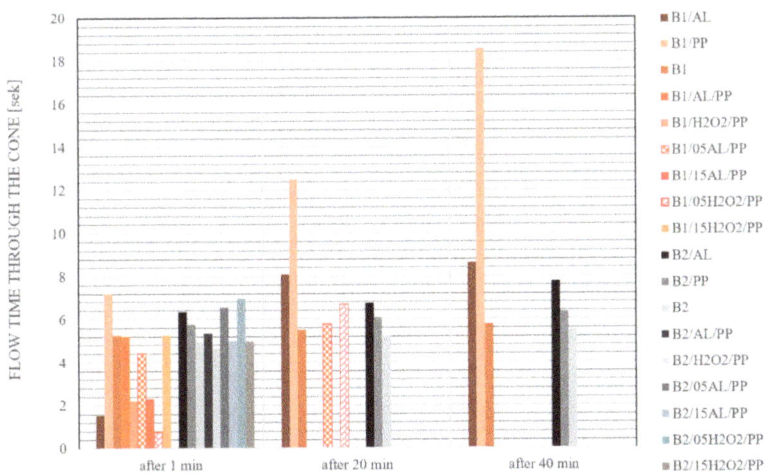

Figure 5. Flow time of the tested mixtures through the Abrams cone.

Analysis of the compressive strength results of the samples made of the B1 and B2 mixtures showed differences in values depending on their composition (Table 6).

The reference samples of the B1 material containing silica fly dust exhibited an average compressive strength of 21.55 MPa. The addition of polypropylene fibers (B1/PP) resulted in significantly enhanced strength properties, with values peaking at 33.12 MPa. Furthermore, it was found that the choice of foaming agent distinctly influenced the compressive strength of the samples. Samples with the addition of aluminum powder as the foaming agent (B1/AL, B1/05AL/PP, B1/15AL/PP) demonstrated compressive strengths of 28.76 MPa, 10.77 MPa, and 3.43 MPa, respectively.

Table 6. Compressive strength of samples based on the B1 and B2 materials depending on the composition and the foaming agent used.

No.	Designation of the Mixtures	Compressive Strength [MPa]	Standard Deviation [MPa]
1	B1/H2O2	10.37	0.49
2	B1/AL	28.76	2.17
3	B1/PP	33.12	1.47
4	B1	21.55	2.68
5	B1/AL/PP	1.99	0.80
6	B1/H2O2/PP	1.02	0.22
7	B1/05AL/PP	10.77	0.56
8	B1/15AL/PP	3.43	0.67
9	B1/05H2O2/PP	4.98	0.32
10	B1/15H2O2/PP	5.14	0.36
11	B2/H2O2	1.58	0.18
12	B2/AL	1.19	0.24
13	B2/PP	10.89	1.46
14	B2	13.87	0.25
15	B2/AL/PP	1.35	0.12
16	B2/H2O2/PP	1.85	0.21
17	B2/05AL/PP	2.59	0.21
18	B2/15AL/PP	0.72	0.09
19	B2/05H2O2/PP	4.62	0.29
20	B2/15H2O2/PP	4.04	0.26

Table 7. Flexural strength of samples based on the B1 and B2 materials depending on the composition and the foaming agent used.

No.	Designation of the Mixtures	Flexural Strength [MPa]	Standard Deviation [MPa]
1	B1/H2O2	3.34	0.18
2	B1/AL	5.36	0.96
3	B1/PP	0.67	0.15
4	B1	4.72	0.49
5	B1/AL/PP	0.79	0.05
6	B1/H2O2/PP	0.72	0.11
7	B1/05AL/PP	1.79	0.45
8	B1/15AL/PP	1.07	0.12
9	B1/05H2O2/PP	0.98	0.14
10	B1/15H2O2/PP	1.77	0.12
11	B2/H2O2	0.74	0.18
12	B2/AL	1.13	0.15
13	B2/PP	3.92	0.23
14	B2	2.93	0.47
15	B2/AL/PP	0.57	0.18
16	B2/H2O2/PP	0.73	0.23
17	B2/05AL/PP	1.16	0.16
18	B2/15AL/PP	0.50	0.11
19	B2/05H2O2/PP	0.82	0.18
20	B2/15H2O2/PP	1.32	0.24

Conversely, the use of H_2O_2 instead of Al powder as a foaming agent significantly reduced the compressive strength of materials with similar compositions. For the B1/05H2O2/PP sample, almost half the compressive strength value was measured compared to its counterpart foamed with Al powder. Moreover, the B1/H2O2 sample had even greater differences, with almost three times lower properties, compared to the B1/Al sample. It is worth noting that the B1/AL/PP and B1/H2O2/PP mixtures exhibited the lowest compressive strengths among the tested formulations based on B1, with values of 1.99 MPa and 1.02 MPa, respectively. These results were consistent with the observed brittleness of the materials 20 min after preparation. A similar trend was noted in the bending strength tests (Table 7).

The compressive strength test results of the B2 samples showed significantly lower values compared to samples made based on the B1 raw material. Any modification to the B2 composition resulted in a reduction in compressive strength. The introduction of polypropylene (PP) fibers had the least detrimental effect. The B2/PP mixture achieved 10.89 MPa. However, attempts to foam the material resulted in drastic reductions in strength properties. The compressive strength of the B2/AL and B2/AL/PP mixtures were similar, registering 1.19 MPa and 1.35 MPa, respectively. A notable difference was observed based on the type of foaming agent used. Samples foamed with hydrogen peroxide (H_2O_2) displayed better strength values compared to those foamed with aluminum powder. This effect was opposite to the previously described properties of materials based on raw material B1. Despite this, the compressive strengths were still insufficient for practical applications. Such low values restrict the potential use of these materials in additive manufacturing, as they would not provide the necessary compressive strength required for successive layers in 3D printing.

Table 7 presents the flexural strength of samples based on the B1 and B2 materials depending on the composition and the foaming agent used. Based on the recorded results, the three-point bending strength of the B1 material without additives reached 4.72 MPa. Lower values were observed for samples foamed with hydrogen peroxide (B1/H2O2), with an average strength of 3.34 MPa. For both foamed and fiber-reinforced samples, a significant trend was observed, showing lower values compared to the reference mixtures. Specifically, the B1/AL/PP and B1/H2O2/PP mixtures demonstrated significantly lower bending strengths than samples where only the amount of frother or water was modified. The three-point bending strength of B2 concrete samples was lower compared to the B1 material, with an average value of 2.93 MPa. The highest bending strength for B2-based mixtures was recorded for the mixture with added polypropylene fibers, reaching 3.92 MPa. Foamed samples, regardless of the additive type, exhibited low bending strength values, averaging around 1 MPa. Surprisingly low values were recorded for samples with added fibers (B1/PP).

The microscopic observation revealed significant differences in the structure, size, and edges of the pores between the B1 and B2 materials (Figures 6 and 7). In both cases, the pores created by foaming with aluminum powder were round. The B2 samples displayed small, regular pores almost uniformly distributed throughout the entire volume, contrasting with the B1 samples, which had larger and less evenly distributed pores, indicating a lower degree of foaming. The sample B1/AL exhibits the largest pores, with the highest average pore surface area among the tested group. The regularity of the pore shape enhances the uniformity of the mechanical properties, potentially benefiting the material's low thermal conductivity and overall mechanical performance. This pore configuration suggests a well-regulated production process. Similarly, samples B1/05AL/PP and B1/15AL/PP are characterized by a large number of regularly shaped pores. In the analysis of samples foamed with aluminum powder, it is noteworthy that only the sample B2/15AL/PP displays uneven edges of air bubbles. Samples foamed with hydrogen peroxide exhibited lower strength values and were characterized by small, densely packed pores with very irregular edges. Moreover, the B2-based materials showed a rougher microscopic structure. Additionally, cracks were visible in the structure of the B2/H2O2, B2/H2O2/PP, and B2/05H2O2/PP samples (marked with arrows). The B1/05H2O2/PP sample is characterized by the smallest average pore size and a high pore density. This phenomenon is likely associated with the relatively high density of the sample, which may be attributed to the fibers used during the production process, potentially leading to the penetration of air pores. A similar crack was observed in the B1-based material in the sample marked as B1/15H2O2/PP.

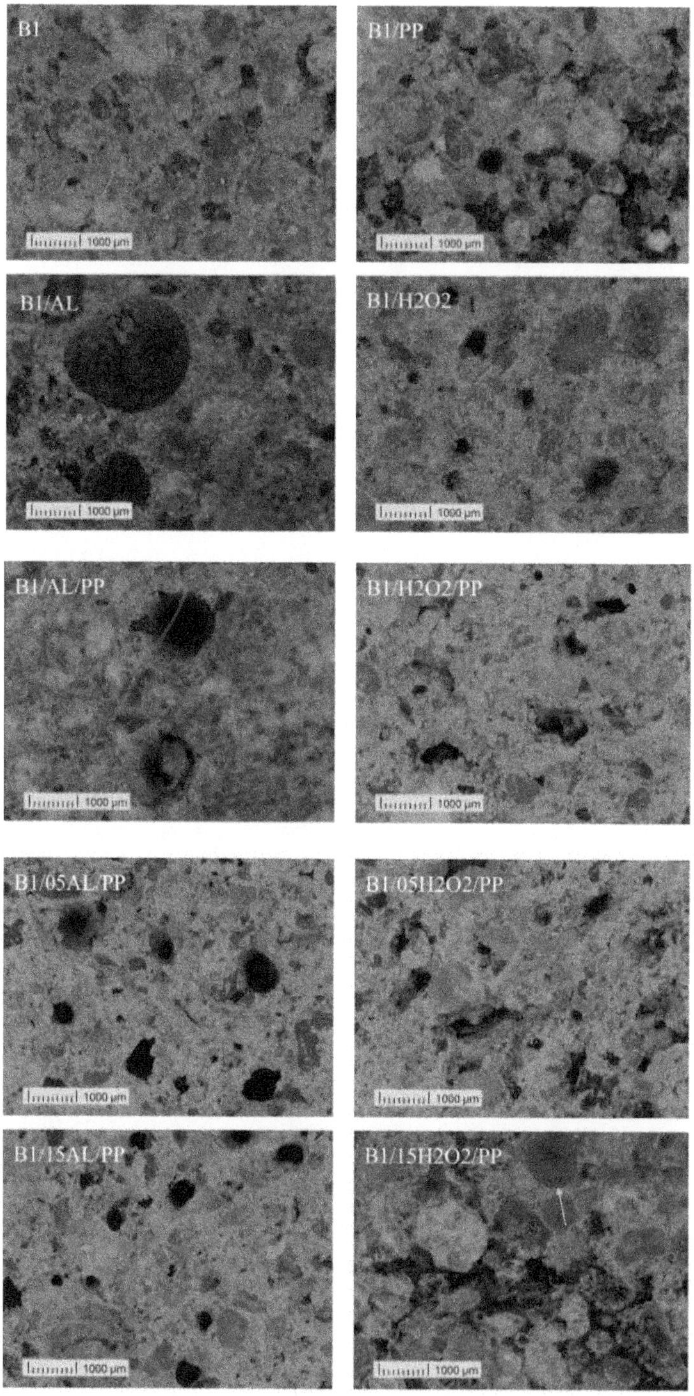

Figure 6. Representative microstructures of materials based on raw material B1: B1/AL, B1/H2O2, B1, B1/PP, B1/AL/PP, and B1/H2O2/PP. The arrow indicates a structural crack.

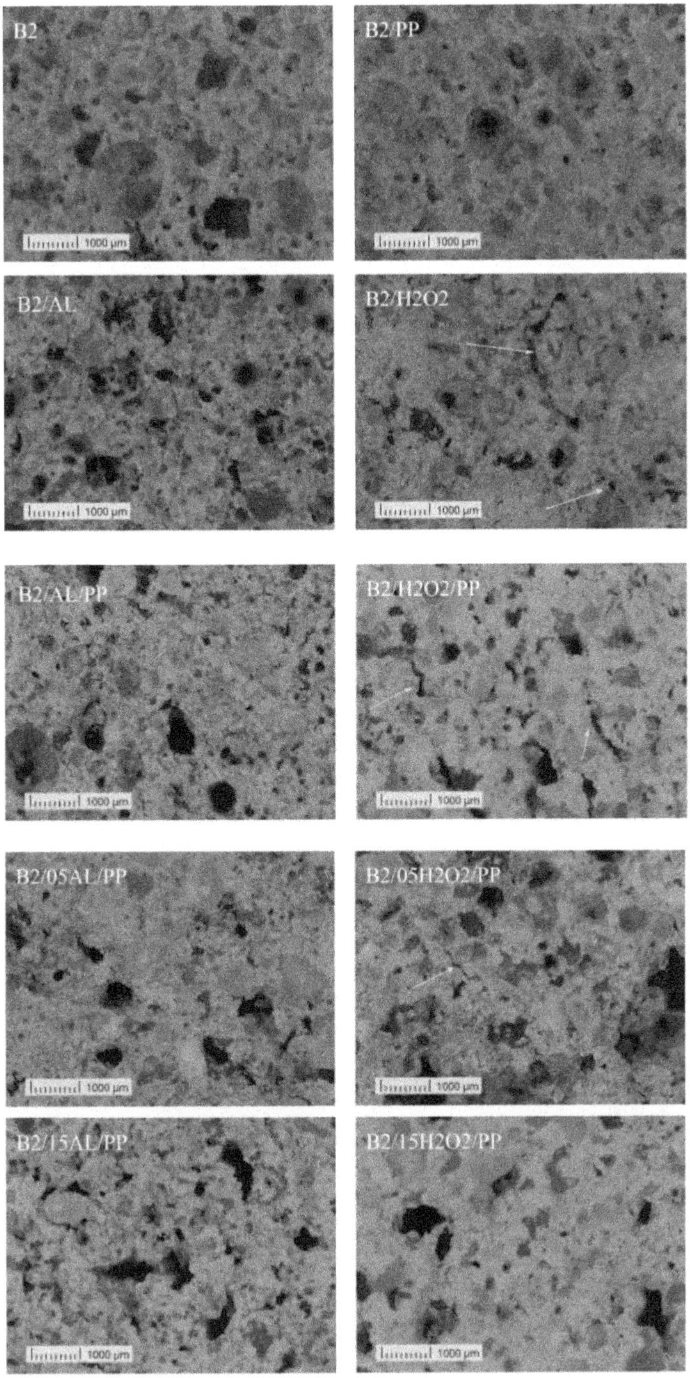

Figure 7. Representative microstructures of materials based on raw material B2: B2/AL, B2/H2O2, B2,B2/PP, B2/AL/PP, and B2/H2O2/PP. The arrows indicate structural cracks.

The density of samples made of the B2 material was characterized by significantly lower values compared to samples made of B1 raw material (Table 8). This difference highlights the importance of optimization and provides the ability to control the manufacturing process to tailor material properties to specific applications.

Table 8. Density of samples based on the B1 and B2 materials depending on the composition and the foaming agent used.

No.	Designation of the Mixtures	Density [kg/m^3]	Standard Deviation [kg/m^3]
1	B1/H2O2	1646.7	0.59
2	B1/AL	2009.9	0.54
3	B1/PP	2079.9	0.41
4	B1	2055.7	0.40
5	B1/AL/PP	2014.0	0.60
6	B1/H2O2/PP	1754.1	0.59
7	B1/05AL/PP	1828.6	0.40
8	B1/15AL/PP	1465.3	0.52
9	B1/05H2O2/PP	1794.1	0.46
10	B1/15H2O2/PP	1656.2	0.47
11	B2/H2O2	1312.1	0.60
12	B2/AL	1512.5	0.62
13	B2/PP	1541.9	0.55
14	B2	1607.5	0.39
15	B2/AL/PP	1514.6	0.54
16	B2/H2O2/PP	1502.3	0.52
17	B2/05AL/PP	1462.9	0.41
18	B2/15AL/PP	1375.7	0.42
19	B2/05H2O2/PP	1555.5	0.49
20	B2/15H2O2/PP	1552.2	0.59

The analysis of the literature indicates that the addition of foaming agents, such as aluminum powder and H_2O_2, significantly impacts cohesion and workability. The results obtained from observing the mixture flow through the Abrams cone and the concrete flow during testing reveal that, with minor exceptions (B1/05H_2O_2/PP and B1/15H_2O_2/PP), concrete foams initially exhibit lower consistency. Subsequently, their structure becomes significantly denser, and they demonstrate slower material spreading during tests, corroborating the findings of other researchers [51]. Furthermore, the comparative analysis of the mixtures in this study confirms that the addition of fibers in the tested materials increases water absorption [52]. The strength-to-density ratio of the tested samples is not always straightforward; higher material density does not consistently correlate with increased compressive or bending strength. An exemplary case is sample B1/H_2O_2, which, despite its low density, demonstrates the highest compressive and bending strength. Conversely, B1/H_2O_2/P and B1/AL/PP samples, which exhibit the highest density, do not achieve the highest strength values. The influence of additives on strength properties is substantial. Figures 8 and 9 illustrate the correlation between the mixture's density and its corresponding bending and compressive strengths.

The results of this study provide valuable insights into the structure and behavior of foamed concrete influenced by foaming agents, such as aluminum powder and hydrogen peroxide. The literature extensively describes the properties and outcomes achieved during experiments conducted by other researchers. However, there is a scarcity of studies detailing the behavior of the mixture itself during testing. The completed research demonstrates that, regardless of the amount of foaming chemicals added, a concrete mixture containing fibers is not feasible for use in 3D concrete printing. This finding aligns with numerous studies discussing 3D printing with foamed concrete using protein-based or synthetic foaming agents. The conducted studies demonstrate that 3D printing with foamed fiber-reinforced concrete is possible, but it requires a lot of research focused mainly on the behavior of the material in the context of 3D printing. Previous researchers have conducted experiments leading to the use of the material in additive technology by foaming concrete only with the use of foaming agents. This finding corroborates numerous studies discussing 3D printing with foamed concrete using protein-based or synthetic foaming agents. However, 3D printing with fiber-reinforced foamed concrete is theoretically possible, but it necessitates extensive research focused primarily on the behavior of the material specifically designed for 3D printing. A comprehensive understanding and optimization of the mixture's properties are essential to ensure adequate consistency, homogeneity, and strength during the printing process [4,6,33,49–51,53–55].

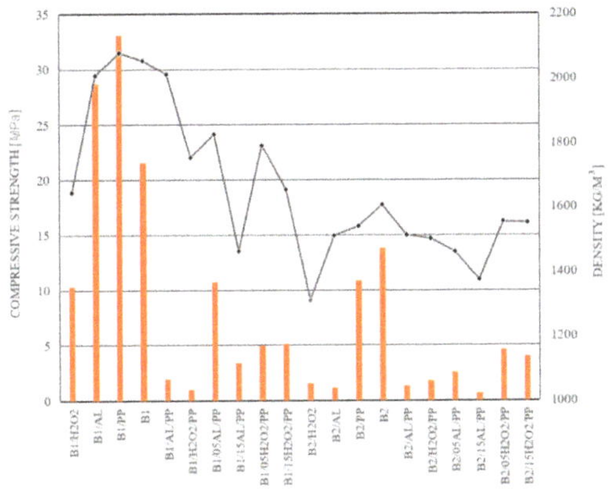

Figure 8. Characteristics of the tested materials in the context of the analysis of compressive strength and their density.

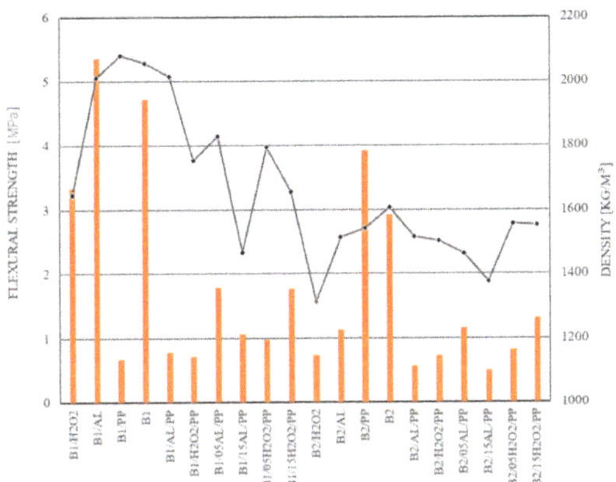

Figure 9. Characteristics of the tested materials in the context of the analysis of bending strength and their density.

4. Conclusions

In the consistency evaluation, by the slump flow method, compositions based on the B1 raw material demonstrated higher fluidity than those based on the B2 raw material. Modifying the composition of the mixtures by introducing fibers, the foaming process, and extending the mixing time influenced the change in consistency from liquid to dense plastic. This effect was independent of the type of raw material used. The content of foaming chemicals directly influences the bubble content in fresh foam mortar, as well as the pore structure of the hardened mortar, which is crucial for the rheological and hardening properties of 3D printed fiber-reinforced concrete. Mixtures with double the foaming agent content, i.e., 0.075% (B1/05AL/PP, B1/05H2O2/PP, B2/05AL/PP, B2/05H2O2/PP), demonstrated higher compressive and bending strengths with lower densities compared to the initial combined mixtures (B1/AL/PP, B1/H2O2/PP, B2/AL/PP, and B2/H2O2/PP). To utilize these materials effectively in 3D printing, additives should be incorporated to stabilize the mixture and extend its open time, ensuring better printability. Mixtures containing twice the amount of network water exhibited significantly altered rheological properties compared to fresh foam mortar. Despite their increased strength values, these mixtures proved to be very unstable for further testing in additive manufacturing technology. Moreover, B1/AL/PP and B1/H2O2/PP mixtures became too brittle after 20 min of mixing, rendering them unsuitable for 3D concrete printing technology. The B1/05H2O2/PP mixture initially exhibited excessive fluidity, flowing through the cone in just 0.8 s after the first minute. However, its flow time increased significantly to 6.68 s after 20 min of mixing. After the 20 min flow test, the cone walls were heavily soiled by the B1/05H2O2/PP mixture, indicating poor flow stability and residue formation. These characteristics suggest that the material has a narrow optimal application window, likely hindering the printing process and compromising the quality of the printed elements. Generally, the strength properties of samples from raw material B1 were significantly higher than those made from B2. Moreover, the foaming process further reduces the strength properties of the samples. The optimal results were achieved with the B1/H_2O_2 and B1/0.5AL/PP compositions. Microscopic analysis revealed significant differences in the pore structures of B1 and B2 materials. Samples foamed with aluminum powder had round, regular-shaped pores, while those foamed with hydrogen peroxide had small, densely packed, irregular pores. The irregular and densely packed pores in samples formed with hydrogen peroxide are consistent with previous findings by other researchers, who reported that hydrogen peroxide can produce a more irregular pore structure. The lower strength values observed in these samples align with the literature indicating that irregular pores can compromise the material's mechanical integrity [56,57]. Further continuation of research on the development of more ecological concrete mixtures, e.g., from the B1 raw material, is crucial for the possibility of their use in additive technologies. Specifically, the open time for printing must exceed 20 min, and the material must achieve the desired porosity while maintaining high compressive and bending strength. This optimization is essential for validating the material's performance and suitability for 3D printing applications.

Author Contributions: Conceptualization, M.R. and M.H.; Data curation, M.M., K.S. and M.G.; Formal analysis, M.R. and M.H.; Funding acquisition, M.H.; Investigation, M.R., M.M. and K.S.; Methodology, M.R., M.M., K.S., M.G. and M.H.; Resources, M.G. and M.H.; Supervision, M.H.; Visualization, M.R. and M.H.; Writing—original draft, M.R. and M.H.; Writing—review & editing, M.R. and M.H. All authors have read and agreed to the published version of the manuscript.

Funding: We would like to acknowledge the support of the Ministry of Education and Science within the program "Implementation Doctorate" and the Polish National Centre for Research and Development, grant number M-ERA.NET 3/2021/115/3D-FOAM/2022.

Institutional Review Board Statement: Not applicable.

Informed Consent Statement: Not applicable.

Data Availability Statement: No new data were created or analyzed in this study. Data sharing is not applicable to this article.

Conflicts of Interest: The authors declare no conflicts of interest.

References

1. Neville, A.M.; Brooks, J.J. *Concrete Technology*, 2nd ed.; Longman Scientific & Technical: Harlow, UK, 2010.
2. Shabbar, R.; Nedwell, P.; Wu, Z. Mechanical properties of lightweight aerated concrete with different aluminium powder content. *MATEC Web Conf.* **2017**, *120*, 02010. [CrossRef]

3. Van Rooyen, A.S. *Mechanics and Durability of Surface Treated Structural Foamed Concrete*; Stellenbosch University: Stellenbosch, South Africa, 2020.
4. Markin, V.; Sahmenko, G.; Nerella, V.N.; Nather, M.; Mechtcherine, V. Investigations on the foam concrete production techniques suitable for 3D-printing with foam concrete. *IOP Conf. Ser. Mater. Sci. Eng.* **2019**, *660*, 012039. [CrossRef]
5. Rudziewicz, M.; Maroszek, M.; Góra, M.; Dziura, P.; Mróz, K.; Hager, I.; Hebda, M. Feasibility Review of Aerated Materials Application in 3D Concrete Printing. *Materials* **2023**, *16*, 6032. [CrossRef] [PubMed]
6. Liu, C.; Chen, Y.; Zhang, Z.; Niu, G.; Xiong, Y.; Ma, L.; Fu, Q.; Chen, C.; Banthia, N. Study of the influence of sand on rheological properties, bubble features and buildability of fresh foamed concrete for 3D printing. *Constr. Build Mater.* **2022**, *356*, 129292. [CrossRef]
7. Wei, C.; Liu, X.; Zhang, Z.; Wu, P. Utilization of solid wastes for aerated concrete reparation: Mechanical properties and microstructural analysis. *J. Build. Eng.* **2024**, *82*, 108235. [CrossRef]
8. Van, L.T.; Kim, D.V.; Xuan, H.N.; Dinh, T.V.; Bulgakov, B.; Bazhenova, S. Effect of aluminium powder on light-weight aerated concrete properties. *E3S Web Conf.* **2019**, *97*, 02005. [CrossRef]
9. Paikara, R.K.; Gyawali, T.R. Influence of aluminum powder content and powder-to-sand ratio on the physical and mechanical properties of aerated lightweight mortar. *Clean. Mater.* **2023**, *10*, 100213. [CrossRef]
10. Ahmida, F.; Sayah, G.M.; Zineb, D.; Quéneudec-t'Kint, M. Experimental study on the effect of lime and aluminium content on porosity, introduced porosity, compressive strength and thermal conductivity of a lightweight cellular concrete based on limestone sand. *Constr. Build. Mater.* **2023**, *392*, 131552. [CrossRef]
11. Li, Z.; Gao, D.; Wu, C.; Lv, G.; Liu, X.; Zhai, H.; Huang, Z. Mechanical performance of aerated concrete and its bonding performance with glass fiber grille. *Nonlinear Eng.* **2021**, *10*, 240–244. [CrossRef]
12. Kurek, I.; Florek, E.; Gozdur, W.; Ziejewska, C.; Marczyk, J.; Łach, M.; Korniejenko, K.; Duży, P.; Choińska, M.; Szechyńska-Hebda, M.; et al. Foamed Eco-Geopolymer Modified by Perlite and Cellulose as a Construction Material for Energy-Efficient Buildings. *Energies* **2022**, *15*, 4297. [CrossRef]
13. Guglielmi, P.O.; Silva, W.R.L.; Repette, W.L.; Hotza, D. Porosity and mechanical strength of an autoclaved clayey cellular concrete. *Adv. Civ. Eng.* **2010**, *2010*, 194102. [CrossRef]
14. Yang, L.; Yan, Y.; Hu, Z. Utilization of phosphogypsum for the preparation of non-autoclaved aerated concrete. *Constr. Build. Mater.* **2013**, *44*, 600–606. [CrossRef]
15. Pan, Z.; Li, H.; Liu, W. Preparation and Characterization of Super Low Density Foamed Concrete from Portland Cement andAdmixtures. *Constr. Build. Mater.* **2014**, *72*, 256–261. [CrossRef]
16. Wu, J.; Zhang, Z.; Zhang, Y.; Li, L. Preparation and characterization of ultra-lightweight foamed geopolymer (UFG) based on fly ash-metakaolin blends. *Constr. Build. Mater.* **2018**, *168*, 771–779. [CrossRef]
17. Ziejewska, C.; Marczyk, J.; Korniejenko, K.; Bednarz, S.; Sroczyk, P.; Łach, M.; Mikuła, J.; Figiela, B.; Szechyńska-Hebda, M.; Hebda, M. 3D Printing of Concrete-Geopolymer Hybrids. *Materials* **2022**, *15*, 2819. [CrossRef] [PubMed]
18. Suiker, A.S.J.; Wolfs, R.J.M.; Lucas, S.M.; Salet, T.A.M. Elastic buckling and plastic collapse during 3D concrete printing. *Cem. Concr. Res.* **2020**, *135*, 106016. [CrossRef]
19. Tran, M.V.; Cu, Y.T.H.; Le, C.V.H. Rheology and shrinkage of concrete using polypropylene fiber for 3D concrete printing. *J. Build. Eng.* **2021**, *44*, 103400. [CrossRef]
20. Zhou, Y.; Althoey, F.; Alotaibi, B.S.; Gamil, Y.; Iftikhar, B. An overview of recent advancements in fibre-reinforced 3D printing concrete. *Front. Mater.* **2023**, *10*, 1289340. [CrossRef]
21. Zhao, J.; Trindade, A.C.C.; Liebscher, M.; Silva, F.d.S.; Mechtcherine, V. A review of the role of elevated temperatures on the mechanical properties of fiber-reinforced geopolymer (FRG) composites. *Cem. Concr. Compos.* **2023**, *137*, 104885. [CrossRef]
22. Liu, X.; Wang, S.; Han, F.; Qin, j.; Lu, L.; Xue, Q.; Ji, J. Mechanical relationship between compressive strength and sulfate erosion depth of basalt fiber reinforced concrete. *Constr. Build. Mater.* **2024**, *411*, 134412. [CrossRef]
23. Panda, B.; Chandra, P.S.; Tan, J.M. Anisotropic mechanical performance of 3D printed fiber reinforced sustainable construction material. *Mater. Lett.* **2017**, *209*, 146–149. [CrossRef]
24. Zhang, Y.; Aslani, F. Development of fibre reinforced engineered cementitious composite using polyvinyl alcohol fibre and activated carbon powder for 3D concrete printing. *Construct. Build. Mater.* **2021**, *303*, 124453. [CrossRef]
25. Chen, Y.; Zhang, Y.; Pang, B.; Wang, D.; Liu, Z.; Liu, G. Steel fiber orientational distribution and effects on 3D printed concrete with coarse aggregate. *Mater. Struct.* **2022**, *55*, 100. [CrossRef]
26. Sun, J.; Aslani, F.; Lu, J.; Wang, L.; Huang, Y.; Ma, G. Fibre-reinforced lightweight engineered cementitious composites for 3D concrete printing. *Ceram. Int.* **2021**, *47*, 27107–27121. [CrossRef]
27. Blazy, J.; Blazy, R. Polypropylene fiber reinforced concrete and its application in creating architectural forms of public spaces. *Case Stud. Constr. Mater.* **2021**, *14*, e00549. [CrossRef]
28. Jiang, Q.; Liu, Q.; Wu, S.; Zheng, H.; Sun, W. Modification effect of nanosilica and polypropylene fiber for extrusion-based 3D printing concrete: Printability and mechanical anisotropy. *Addit. Manuf.* **2022**, *56*, 102944. [CrossRef]
29. Raj, B.; Sathyan, D.; Madhavan, M.K.; Raj, A. Mechanical and durability. Properties of hybrid fiber reinforced foam concrete. *Constr. Build. Mater.* **2020**, *245*, 118373. [CrossRef]
30. Afraz, A.; Ali, M. Effect of Banana Fiber on Flexural Properties of Fiber Reinforced Concrete for Sustainable Construction. *Eng. Proc.* **2021**, *12*, 63. [CrossRef]

31. Awang, H.; Ahmad, M.H.; Al-Mulali, M.Z. Influence of kenaf and polypropylene fibers on mechanical and durability properties of fiber-reinforced lightweight foamed concrete. *J. Eng. Sci. Technol.* **2015**, *10*, 496–508.
32. Nematollahi, B.; Vijay, P.; Sanjayan, J.; Nazari, A.; Xia, M.; Naidu Nerella, V.; Mechtcherine, V. Effect of Polypropylene Fibre Addition on Properties of Geopolymers Made by 3D Printing for Digital Construction. *Materials* **2018**, *11*, 2352. [CrossRef]
33. Awang, H.; Mydin, M.A.O.; Roslan, A.F. Effects of fibre on drying shrinkage, compressive and flexural strength of lightweight foamed concrete. *Adv. Mat. Res.* **2012**, *587*, 144–149. [CrossRef]
34. Jaini, Z.M.; Rum, R.H.M.; Boon, K.H. Strength and fracture energy of foamed concrete incorporating rice husk ash and polypropylene mega-mesh 55. *IOP Conf. Ser. Mater. Sci. Eng.* **2017**, *248*, 012005. [CrossRef]
35. Available online: https://www.sunchemical.com/wp-content/uploads/2022/08/SunChemical-Benda-Lutz-Powders-Pastes-AAC-Brochure.pdf (accessed on 1 August 2022).
36. Available online: https://chempur.pl/pliki/karty_charakterystyk/wodoru_nadtlenek_30.pdf (accessed on 1 August 2004).
37. Available online: https://assets.ctfassets.net/xbq8te6l92zd/66mnjzCfMc9WpNJhAI2j5S/49b77e4594dc68a49a42f433f5d1915f/FIBRES_POLYPROPYLENE_adjuvant-beton.pdf (accessed on 26 March 2019).
38. Testing of Concrete Mixes—Part 4: Consistency Testing by Densitometry. Available online: https://sklep.pkn.pl/pn-en-12350-4-2019-08e.html (accessed on 30 July 2024).
39. Concrete Testing—Part 3: Compressive Strength of Test Specimens. Available online: https://sklep.pkn.pl/pn-en-12390-3-2019-07e.html (accessed on 30 July 2024).
40. Concrete Testing—Part 5: Flexural Strength of Test Specimens. Available online: https://sklep.pkn.pl/pn-en-12390-5-2019-08p.html (accessed on 30 July 2024).
41. Reiter, L.; Wangler, T.; Roussel, N.; Flatt, R.J. The role of early age structural build-up in digital fabrication with concrete. *Cem. Concr. Res.* **2018**, *112*, 86–95. [CrossRef]
42. Moelich, G.M.; Kruger, P.J.; Combrinck, R. Mitigating early age cracking in 3D printed concrete using fibres, superabsorbent polymers, shrinkage reducing admixtures, B-CSA cement and curing measures. *Cem. Concr. Res.* **2022**, *159*, 106862. [CrossRef]
43. Shahmirzadi, M.R.; Gholampour, A.; Kashani, A.; Ngo, T.D. Shrinkage behavior of cementitious 3D printing materials: Effect of temperature and relative humidity. *Cem. Concr. Compos.* **2021**, *124*, 104238. [CrossRef]
44. Warsi, S.B.F.; Panda, B.; Biswas, P. Exploring fibre addition methods and mechanical properties of fibre-reinforced 3D printed concrete: A review. *Dev. Built Environ.* **2023**, *16*, 100295. [CrossRef]
45. Pan, T.; Jiang, Y.; He, H.; Wang, Y.; Yin, K. Effect of Structural Build-Up on Interlayer Bond Strength of 3D Printed Cement Mortars. *Materials* **2021**, *14*, 236. [CrossRef] [PubMed]
46. Marczyk, J.; Ziejewska, C.; Gądek, S.; Korniejenko, K.; Łach, M.; Góra, M.; Kurek, I.; Doğan-Sağlamtimur, N.; Hebda, M.; Szechyńska-Hebda, M. Hybrid Materials Based on Fly Ash, Metakaolin, and Cement for 3D Printing. *Materials* **2021**, *14*, 6874. [CrossRef] [PubMed]
47. Vespalec, A.; Podroužek, J.; Boštík, J.; Míča, L.; Koutný, D. Experimental study on time dependent behaviour of coarse aggregate concrete mixture for 3D construction printing. *Constr. Build. Mater.* **2023**, *376*, 130999. [CrossRef]
48. Hager, I.; Maroszek, M.; Mróz, K.; Kesek, R.; Hebda, M.; Dvorkin, L.; Marchuk, V. Interlayer, Bond Strength Testing in 3D-Printed Mineral Materials for Construction Applications. *Materials* **2022**, *15*, 4112. [CrossRef] [PubMed]
49. Markin, V.; Mechtcherine, V. 3D Printing with Foam Concrete, Overview of the Ongoing Research. Available online: https://kniele.de/images/pdf-en/cpt-2021_01-3d-print-foam-concrete-kniele-kkm-tu-dresden.pdf (accessed on 30 July 2024).
50. Parmigiani, S.; Falliano, D.; Moro, S.; Ferro, G.A.; Restuccia, L. 3D-printed multi-functional foamed concrete building components: Material properties, component design, and 3D printing application. *Dev. Built Environ.* **2024**, 100483. [CrossRef]
51. Markin, V.; Nerella, V.N.; Schröfl, C.; Guseynova, G.; Mechtcherine, V. Material Design and Performance Evaluation of Foam Concrete for Digital Fabrication. *Materials* **2019**, *12*, 2433. [CrossRef] [PubMed]
52. Mohammed, M.; Jawad, A.J.M.; Mohammed, A.M.; Oleiwi, J.K.; Adam, T.; Osman, A.F.; Dahham, O.S.; Betar, B.O.; Gopinath, S.C.B.; Jaafar, M. Challenges and advancement in water absorption of natural fiber-reinforced polymer composites. *Polym. Test.* **2023**, *124*, 108083. [CrossRef]
53. Markin, V.; Krause, M.; Otto, J.; Schroefl, C.; Mechtcherine, V. 3D-printing with foam concrete: From material design and testing to application and sustainability. *J. Build. Eng.* **2021**, *43*, 102870. [CrossRef]
54. Cho, S.; Rooyen, A.; Kearsley, E.; Zijl, G. Foam stability of 3D printable foamed concrete. *J. Build. Eng.* **2022**, *47*, 103884. [CrossRef]
55. Gao, Y.; Hua, S.; Yue, H. Study on Preparation and Rheological Properties of 3D Printed Pre-Foaming Concrete. *Appl. Sci.* **2023**, *13*, 5303. [CrossRef]
56. Xiao, J.; Lv, Z.; Duan, Z.; Zhang, C. Pore structure characteristics, modulation and its effect on concrete properties: A review. *Constr. Build. Mater.* **2023**, *397*, 132430. [CrossRef]
57. Li, S.; Li, H.; Yan, C.; Ding, Y.; Zhang, X.; Zhao, J. Investigating the Mechanical and Durability Characteristics of Fly Ash Foam Concrete. *Materials* **2022**, *15*, 6077. [CrossRef] [PubMed]

materials

Article

Development of Negative-Temperature Cement Emulsified Asphalt Spraying Materials Based on Spraying Performance and Rheological Parameters

Yitong Hou [1,*], Kaimin Niu [1,2], Bo Tian [1,2], Junli Chen [3] and Xueyang Li [1]

1 School of Civil Engineering, Chongqing Jiaotong University, Chongqing 400074, China
2 Research Institute of Highway Ministry of Transport, Beijing 100088, China
3 Chongqing Communications Construction (Group) Co., Ltd., Chongqing 401120, China
* Correspondence: 611200111008@mails.cqjtu.edu.cn

Abstract: To develop a cement emulsified asphalt composite (CEAC) that can be sprayed under a plateau negative temperature environment, the effects of the water–solid ratio, calcium aluminate cement substitution rate, emulsified asphalt content, sand–binder ratio, and polyvinyl alcohol (PVA) fiber content on the spraying performance and rheological parameters of CEAC were explored through the controlled variable method. Additionally, the correlation between the spraying performance and rheological parameters of CEAC was established, and the optimal proportion of CEAC was determined. Then, the difference in frost resistance and pore structure between the cement slurry (CS) without emulsified asphalt and CEAC at the optimum proportion was analyzed. The results showed that the optimum proportions for sprayed CEAC were 0.14 water–solid ratio, 0.5 sand–binder ratio, 25% substitution of calcium aluminate cement, 5% emulsified asphalt content, and 1.5% PVA fiber volume mixing. The yield stress and plastic viscosity of CEAC were positively correlated with the build-up thickness, whereas the rebound rate and the latter showed a negative correlation. The spraying performance may be described by the rheological parameters; the ranges of yield stress and plastic viscosity of 2.37–3.95 Pa·s and 77.42–108.58 Pa, respectively, produced the best spray ability. After undergoing an equivalent number of freeze–thaw cycles, CEAC exhibited lower mass and strength loss rates compared to CS, thereby demonstrating superior frost resistance. In addition, the pore structure analysis showed that the difference in capillary and macropore contents was the main reason for the variability in frost resistance between CS and CEAC.

Keywords: negative-temperature environment; cement emulsified asphalt; spray performance; rheological properties; frost resistance; pore structure

1. Introduction

One type of quick-setting, hardened concrete is called shotcrete, for which a pressure gun is used to spray fine stone concrete on a target surface; it is widely used in tunnel lining structure support, a variety of repair and reinforcement projects, with good economic efficiency, short construction period, and so on, and the combination of reinforcement mesh, anchors, and other support methods can significantly improve the protective capacity of a structure [1–3]. The application technologies of shotcrete mainly include the dry-mix process and wet-mix process [4]. As the wet-mix process has better homogeneity and durability, and may successfully lower the dust concentration and rebound amount, the wet-mix process has become the development trend of shotcrete technology in the world [5–7]. However, due to the need to mix the aggregate, cement, and water in the designed proportion before construction, it is not possible to adequately ensure the quality of construction under adverse weather conditions, which limits this technique's application in the permafrost regions of high-altitude and cold regions [8]. Among the factors that influence shotcrete's performance, the properties of the binder hold the utmost significance [9].

Therefore, there is a need to develop high-quality sprayed materials suitable for permafrost regions. Although some scholars have successfully applied wet-sprayed concrete to high-altitude permafrost regions with negative temperatures all year round as well as to tunnels in cold regions by adding additives such as early-strength antifreeze binders, antifreeze, and accelerators to shotcrete, its performance is not effectively guaranteed [10–12]. In addition, due to the frequent alternation of positive and negative temperatures in perennial permafrost regions and tunnel entrance sections in cold regions, the internal moisture of shotcrete expands by freezing and thawing, so the frost resistance of shotcrete is also an important indicator of its durability performance [13]. Some academics have also conducted in-depth research on the resistance of shotcrete to frost. Haldane [14] believes that the frost resistance of shotcrete is influenced by the distribution of pores and is intricately linked to the water–cement ratio. Wan et al. [15] discovered that incorporating micro- and nano-bubbled water into shotcrete can improve its pore structure, thus improving the resistance of shotcrete to frost. Hu et al. [16] performed an investigation into the current status of shotcrete research in cold tunnels, summarized the various factors that influence shotcrete tunnels situated in cold regions, and forecasted the future trends in the development of such tunnels. Holter et al. [17], in response to the characteristics of shotcrete in a cold-region tunnel lining that is susceptible to freezing damage, investigated the freezing resistance of shotcrete employing a specific ratio of water to cement of 0.45–0.47 through experiments and put forward a new functional freeze–thaw cycle test method, which can more realistically simulate the conditions of moisture content and freezing gradients within a tunnel lining. Kalhori et al. [18] examined the influence of nanomaterials on the frost resistance properties of shotcrete, revealing that both nanoclay and nanosilica enhance its frost resistance capabilities. Furthermore, the study indicated that nanosilica exhibits a superior improvement effect compared to nanoclay.

Rheology pertains to the scientific discipline that investigates the deformation and flow behaviors of substances in response to external forces. For cement-based materials, which are non-homogeneous, multi-phase, and non-linear, their deformation and destructive behaviors under external forces are affected by a variety of factors. In cement-based materials, rheology mainly studies their deformation and flow characteristics, including properties such as fluidity, plastic viscosity, yield stress, creep, and relaxation. These characteristics play a crucial role in determining the spraying efficiency of cement-based materials. Yun et al. [19] conducted a study exploring the impact of mineral admixtures on the rheological properties of shotcrete by using Bingham's model, and based on the rheological properties, suggested a novel technique aimed at enhancing the pumping efficiency and spraying capabilities of shotcrete. Pan et al. [20] examined the rebound rate of sprayed concrete in relation to its rheological characteristics, including slump, yield stress, plastic viscosity, and others, discovering that these properties hold significant importance in the performance of shotcrete. Their findings revealed that variations in slump, yield stress, and water secretion rate exert a considerable influence on the rebound rate of concrete. Furthermore, they derived a multiple linear regression equation that can be utilized to predict the rebound rate. Feng et al. [21] formulated a sprayable ultra-high ductility magnesium cementitious composite (UHDMC) and established a correlation between sprayability factors and rheological parameters. This correlation enabled the determination of the optimal mixing ratio suitable for spraying UHDMC. In conclusion, to better predict the spraying performance of cementitious sprayed materials, the relationship between their sprayability and rheological properties needs to be investigated.

Based on this, in order to achieve shotcrete construction in negative-temperature environments and improve the frost resistance of concrete, the present study endeavors to devise a cement emulsified asphalt composite (CEAC) suitable for spraying in cold environments. Furthermore, it aims to explore the influence of various factors, including water–solid ratio, sand–binder ratio, calcium aluminate cement substitution rate, emulsified asphalt dosage, and PVA fiber dosage, on the sprayability characteristics and rheological parameters of CEAC. Moreover, utilizing identical mix parameters, this study conducted an analysis of data fitting to investigate the correlation between sprayability characteristics and rheological parameters, ultimately determining the optimal mix ratio. Finally, the difference in frost resistance and pore structure between cement slurry (CS) without emulsified asphalt and CEAC at the optimal mix ratio was determined by the MIP test, which revealed the intrinsic reasons for the difference in frost resistance. The results provide guidance for the application of CEAC in tunnels and other spraying projects in high-altitude permafrost regions. The flowchart of this study is shown in Figure 1.

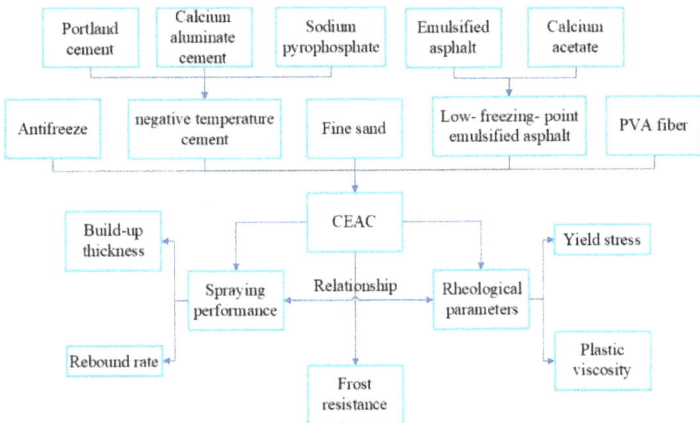

Figure 1. Experimental flowchart.

2. Experimental Program

2.1. Materials

According to the results of previous studies by our group, a mixture of ordinary silicate cement and calcium aluminate cement was used as the negative-temperature cement, and the antifreeze used was a 20 wt.% concentration of sodium nitrite solution, with 3% of the total cement mass of sodium pyrophosphate used as the accelerator [22]. Low-freezing-point emulsified asphalt was prepared by adding 8% calcium acetate by mass to emulsified asphalt, which was externally mixed. Two kinds of cement were provided by a cement factory in Beijing. The compositional and XRD patterns of the cement are shown in Table 1 and Figure 2. The emulsified asphalt used was obtained from Chongqing CNOOC Modern Transportation Oil Material Co., Ltd. (Chongqing, China). The sodium pyrophosphate, sodium nitrite, and calcium acetate were AR reagents produced by Shanghai Aladdin Biochemical Science and Technology Co., Ltd. (Shanghai, China). The water used was tap water.

Table 1. Composition of the cement.

Type of Cement	Mass Fraction (wt.%)									
	SiO_2	CaO	Al_2O_3	SO_3	MgO	Fe_2O_3	KO_2	Na_2O	TiO_2	LOI
Portland cement	21.50	64.20	4.14	2.89	2.57	2.40	0.84	0.67	0.32	0.40
Calcium aluminate cement	0.68	26.80	71.30	0.04	0.46	0.09	0.02	0.37	0.11	0.07

Quartz sand with a diameter of 0–0.1 mm was selected for the fine sand, which was provided by Beijing Construction Engineering Group Co., Ltd. (Beijing, China). The fiber material used was PVA fiber produced by Changzhou Tianyi Engineering Fiber Co., Ltd. (Changzhou, China). The performance factors of the PVA fiber are shown in Table 2. The water used was tap water.

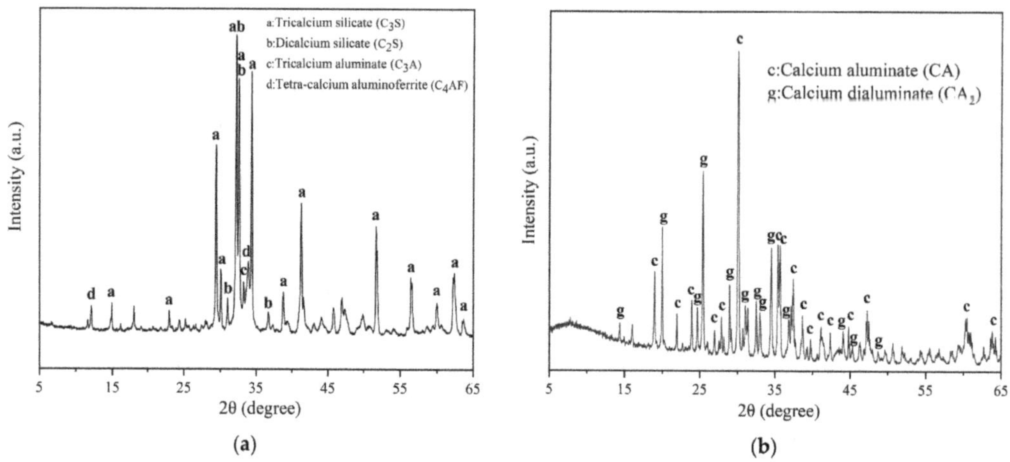

Figure 2. XRD pattern of the cement: (**a**) Portland cement and (**b**) calcium aluminate cement.

Table 2. Performance indices of PVA fiber.

Diameter (μm)	Length (mm)	Tensile Strength (MPa)	Young's Modulus (Gpa)	Elongation (%)	Density (g/cm^3)
31	6	1500	40	6	1.28

2.2. Mixing Ratio of CEAC

To obtain the optimum mixing ratio of CEAC, the control variable method was utilized to first determine the optimum dosing conditions for the water–solid ratio (W/S) and sand–binder ratio (S/B). The effects of the calcium aluminate cement substitution rate (CAC) and emulsified asphalt content (EA) on the CEAC spraying effect were subsequently studied. Finally, the effect of the volume content of PVA fibers (PVA) on the effect of CEAC spraying was studied. Based on the above factors, the mix ratios were optimized step by step to obtain the best mix ratios, and the test mix ratios of CEAC are listed in Table 4, where W/S, S/B, CAC, EA, and PVA stand for the water–solid ratio, sand–binder ratio, substitution rate of calcium aluminate cement, content of emulsified asphalt, and volumetric percentage of PVA fibers, respectively, and the other parameters are fixed when one of them is changed.

Table 3. Test mix ratio of CEAC.

Group	W/S	S/B	CAC (%)	EA (%)	PVA (%)
W/S-0.12	0.12	0.4	20	5	1.0
W/S-0.13	0.13				
W/S-0.14	0.14				
W/S-0.15	0.15				
W/S-0.16	0.16				
S/B-0.2	0.14	0.2	20	5	1.0
S/B-0.3		0.3			
S/B-0.4		0.4			
S/B-0.5		0.5			
S/B-0.6		0.6			
CAC-10	0.14	0.5	10	5	1.0
CAC-15			15		
CAC-20			20		
CAC-25			25		
CAC-30			30		

Table 4. Test mix ratio of CEAC.

Group	W/S	S/B	CAC (%)	EA (%)	PVA (%)
EA-0.0	0.14	0.5	25	0.0	1.0
EA-2.5				2.5	
EA-5.0				5.0	
EA-7.5				7.5	
EA-10.0				10.0	
PVA-0.0	0.14	0.5	25	5	0.0
PVA-0.5					0.5
PVA-1.0					1.0
PVA-1.5					1.5
PVA-2.0					2.0

Note: W/S is the mass ratio of water (sum of water in emulsified asphalt and added water) to solids (sum of cement, sand, and emulsified asphalt solids); S/B is the mass ratio of sand to binder (sum of cement and emulsified asphalt solids); CAC is the mass percentage of calcium aluminate cement to total cement; EA is the mass percentage of emulsified asphalt solids to binder; PVA is the volume of fibers to slurry as a percentage.

2.3. Experiment Methods

2.3.1. Spraying Test

To accurately simulate the negative-temperature environment at high altitudes, the spraying test and curing of specimens in this study were carried out in a plateau environment simulation warehouse independently developed by the Highway Research Institute of the Ministry of Transportation and Communications. The ambient temperature was set at −10 °C, the humidity was set at 50%, and the appearance and operation interface of the simulation warehouse are shown in Figure 3. A 9511 paint spray gun produced by the Taizhou Road and Bridge Construction Pneumatic Tool Factory was used for the CEAC spraying test, and the spraying equipment is shown in Figure 4. The spraying principle is shown in Figure 5.

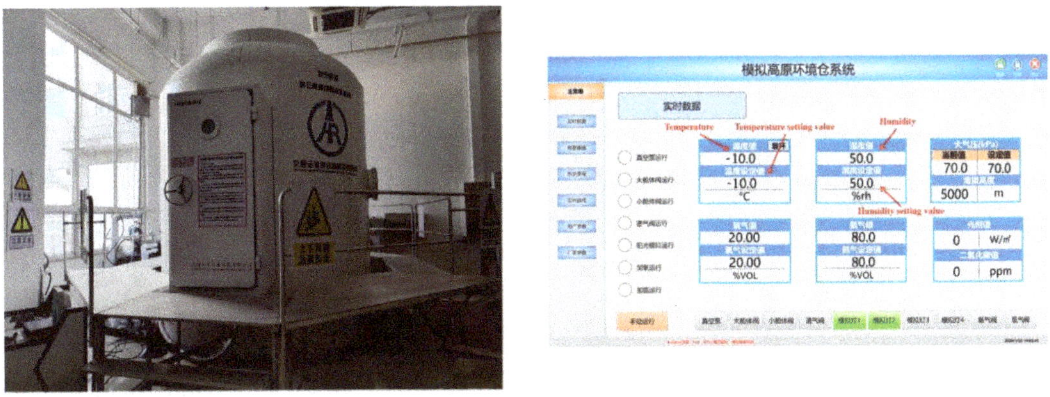

Figure 3. Plateau environment simulation silo.

All the raw materials for the test needed to be placed in the environment simulation silo to keep warm for 24 h. First, cement and sand were mixed with dry mixing in a mixer for 120 s. Then, emulsified asphalt was introduced into the mixer and blended for an additional 60 s. Eventually, PVA fibers were incorporated and mixed for 120 s, resulting in the production of the CEAC. The spraying test of CEAC was sprayed according to Chinese industry standard JGJT 372-2016 [23].

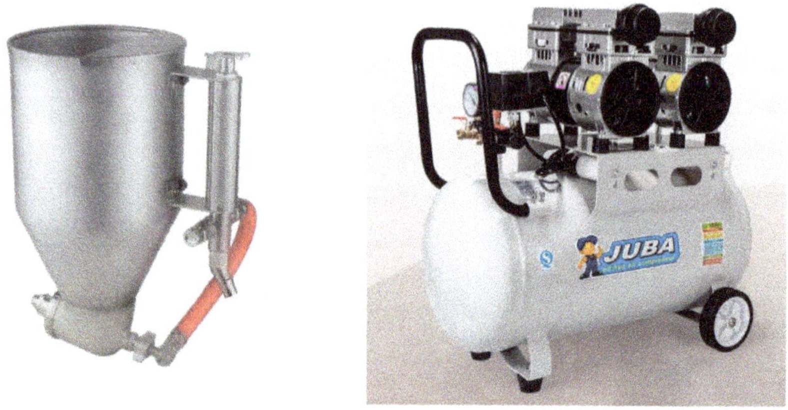

(a) (b)

Figure 4. Spraying equipment. (**a**) Spraying device. (**b**) Air compressor.

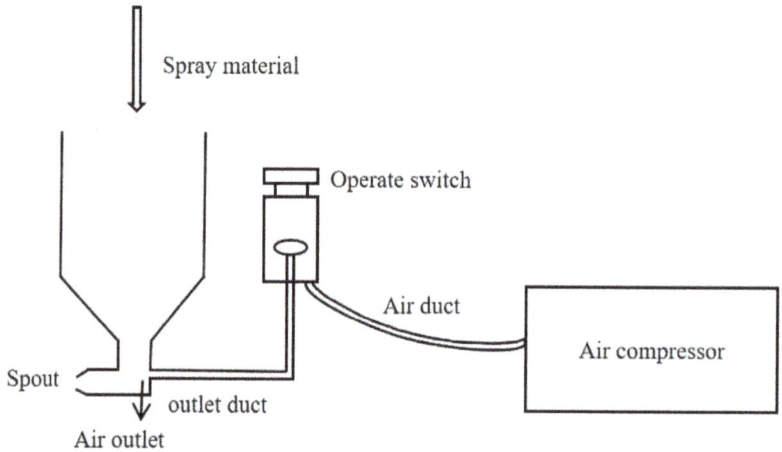

Figure 5. Spraying principle of the spraying equipment.

2.3.2. Spraying Performance Test

The spraying performance was assessed based on two key metrics: the build-up thickness and the rebound rate. The build-up thickness serves as a metric indicative of the cement-based materials' ability to adhere securely to the wall surface. It is measured by spraying continuously at a base point perpendicular to the spraying surface until the material falls off under its own gravity; at this time, the thickness of the material is the build-up thickness, and the method for measuring it is illustrated in Figure 6.

The rebound rate serves as a measure of how well the sprayed material adheres to the wall. A higher rebound rate indicates inferior application of the coating and substandard adhesion properties of the material. Conversely, a lower rebound rate signifies a satisfactory spraying effect and the material's ability to effectively bond with the sprayed structural surface. The rebound rate test method refers to the Japanese standard JSCE-F 563-2005 [24]; the rebound rate test method is shown in Figure 7, and the calculation formula is presented in Equation (1):

$$R = \frac{W_2}{W_1+W_2} \times 100\% \quad (1)$$

where R is the rebound rate (%), W_1 is the mass of CEAC sprayed on the wall (g), and W_2 is the mass of CEAC dropped on the floor (g).

Figure 6. Build-up thickness test.

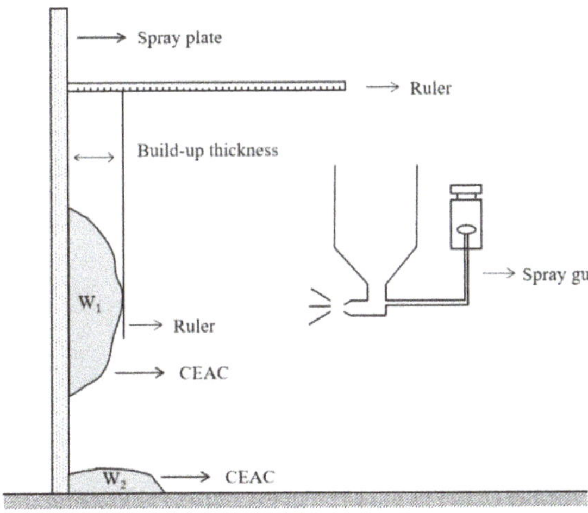

Figure 7. Rebound rate test.

2.3.3. Rheological Test

The Bingham model stands out as the most widely employed model in the rheological examination of cement-based materials, as depicted in Figure 8. According to the Bingham model, when the shear stress is greater than the yield stress, the shear stress of cement-based materials is linearly related to the shear rate. Its intercept is the yield stress, and the slope is the plastic viscosity. In our investigation, the Bingham model was utilized to ascertain the yield stress and plastic viscosity of CEAC [25,26], and the corresponding calculation formula is outlined in Equation (2):

$$\tau = \tau_0 + \eta \cdot \gamma \quad (2)$$

where τ is the shear stress (Pa), τ_0 is the yield stress (Pa), η is the plastic viscosity (Pa·s), and γ is the shear rate (s^{-1}).

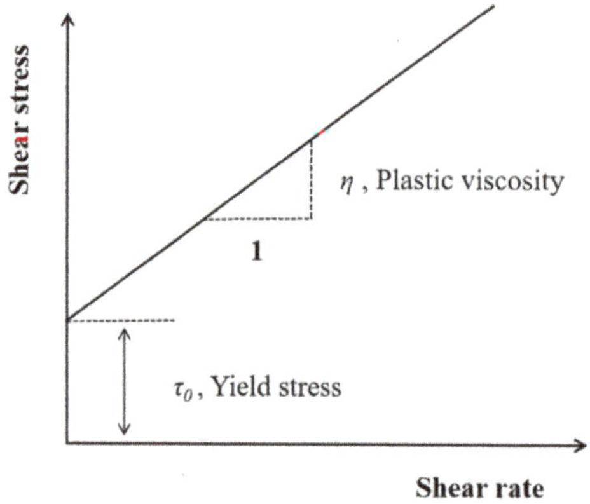

Figure 8. Bingham model.

The rheological parameters of the CEAC were tested using an RSX rotational rheometer manufactured by Brookfield Company, as shown in Figure 9. The holding solution was a mixture of ethanol and water, and the test temperature was set at −10 °C. The test was performed using a measuring rotor with a diameter of 40 mm. First, the CEAC to be tested was placed in an insulation sleeve for 30 min. Then, the CEAC mixture was subjected to a rheological test with a shear rate ranging from 0 to 10 s^{-1}. The first data point was taken at 1 s^{-1}, followed by one data point every 1.5 s^{-1}, and the Bingham model was chosen to fit the rheological curves to obtain τ_0 and η.

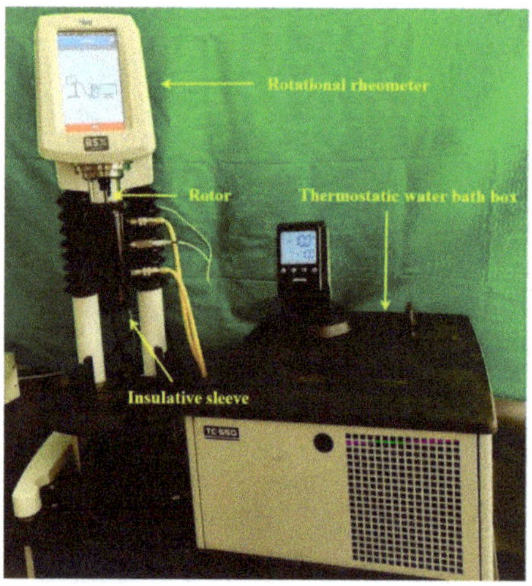

Figure 9. RSX rotational rheometer.

2.3.4. Frost Resistance Test

According to the JGJT 70-2009 [27], CS and CEAC with the best matching ratio were sprayed into a cube mold with a side length of 70.7 mm and then placed in an environment simulation warehouse for negative-temperature maintenance. After undergoing a 28 d curing process with negative-temperature maintenance, the specimens were then subjected to a total of 100 freeze–thaw cycles. Following every 25 cycles of freezing and thawing, the rate of loss in both quality and strength of the specimen was tested.

2.3.5. Mercury Intrusion Porosimetry (MIP) Test

The CS and CEAC specimen blocks, which were cured at a negative temperature until 28 days of age, were cracked and put into an ethanol solution to terminate the hydration, after which the specimen blocks were removed and put into a 60 °C oven to dry for 4 h before the test. The pore structure of the test blocks was tested by a fully automatic mercuric pressure tester (modeled as Autopore IV 9500; manufactured by Mack Corporation, Flagstaff, AZ, USA).

3. Results and Discussion

3.1. Factors Affecting Spraying Performance and Rheological Parameters

3.1.1. Effect of the Water–Solid Ratio (W/S)

Figure 10 illustrates the influence of the W/S on the spraying performance of CEAC. As the W/S increased, the build-up thickness of CEAC sprayed in a single application initially rose and subsequently declined, while the rebound rate initially decreased and later increased. Notably, when the W/S reached 0.13, CEAC exhibited optimal spraying performance.

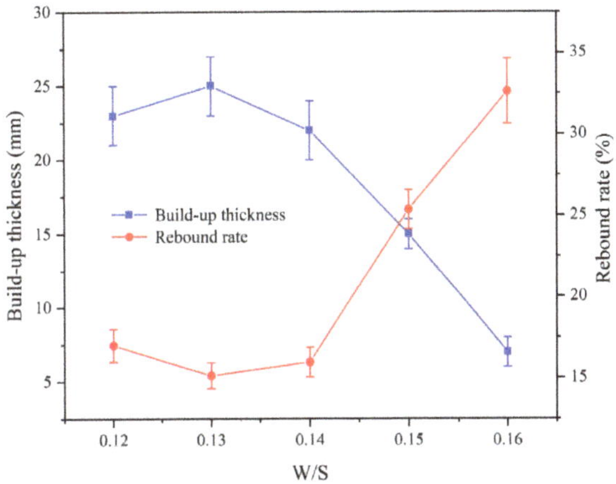

Figure 10. Effect of the W/S on the spraying performance of CEAC.

Figure 11 shows the effect of the water–solid ratio on the rheological properties of CEAC. Figure 11a shows the fitting relationship between shear stress and the shear rate of CEAC at different water–solid ratios. The results showed that there was a good linear relationship between shear rate and shear stress, and R^2 was between 0.93 and 0.99. Figure 11b demonstrates the impact of the W/S on the rheological parameters of CEAC. As the W/S increased, the yield stress and plastic viscosity of CEAC decreased. Generally, a higher W/S leads to improved rheological properties [28]. However, considering the spraying performance of CEAC, when the W/S exceeded 0.14, the slurry's fluidity became excessive, resulting in a significant reduction in the build-up thickness. Conversely, at W/S below 0.14, the slurry exhibited excessive viscosity, compromising its workability. Consequently, an optimal W/S of 0.14 was chosen. Under these conditions, CEAC achieved a build-up thickness of 22 mm and a rebound rate of 15.98% after a single spraying treatment, while its yield stress and plastic viscosity were 88.52 Pa and 2.46 Pa·s, respectively.

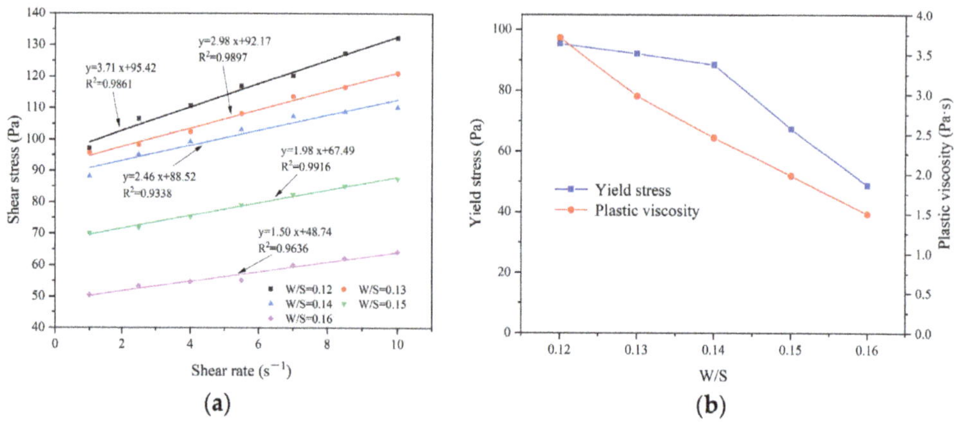

Figure 11. Effect of the W/S on the rheological parameters of CEAC: (**a**) rheological parameter fitting curve and (**b**) yield stress and plastic viscosity.

3.1.2. Effect of the Sand–Binder Ratio (S/B)

Figure 12 illustrates the influence of the S/B on the spraying performance of CEAC. As the S/B rose, the build-up thickness of CEAC after a single spraying session initially increased, reaching a peak, and then gradually decreased. Similarly, the rebound rate initially decreased before gradually increasing. Notably, when the S/B was set at 0.5, CEAC exhibited optimal spraying performance. Under these conditions, the build-up thickness measured 26 mm, while the rebound rate stood at 14.18%.

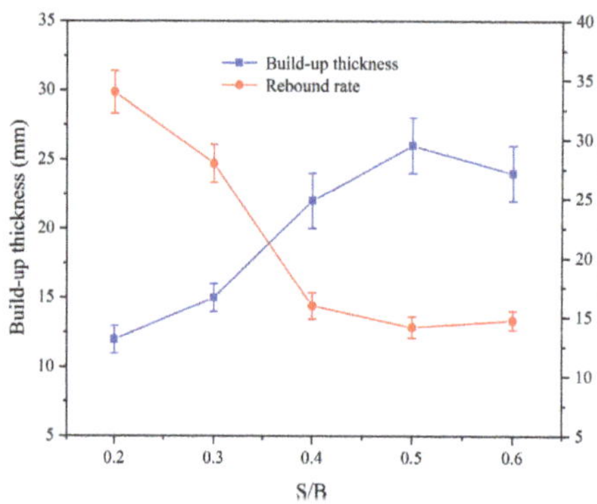

Figure 12. Effect of the S/B on the spraying performance of CEAC.

Figure 13 shows the effect of the sand–binder ratio on the rheological properties of CEAC. Figure 13a shows the fitting relationship between the shear stress and shear rate of CEAC at different sand–binder ratios. The results showed that there was a good linear relationship between the shear rate and shear stress, and R^2 was between 0.92 and 0.99. Figure 13b demonstrates the impact of the S/B on the rheological parameters of CEAC. As the S/B increased, both the yield stress and plastic viscosity of CEAC rose. This trend primarily stems from the enhanced sand content, which elevated the likelihood of particle contact and collision, thereby boosting the yield stress. Additionally, the augmented sand content resulted in a thinner coating of gelling material on the sand particle surfaces, increasing interparticle friction and subsequently leading to higher plastic viscosity [29]. Considering the combined effect of the S/B on CEAC spraying performance, an optimal S/B of 0.5 was established, at which point the yield stress and plastic viscosity were measured as 94.94 Pa and 2.98 Pa·s, respectively.

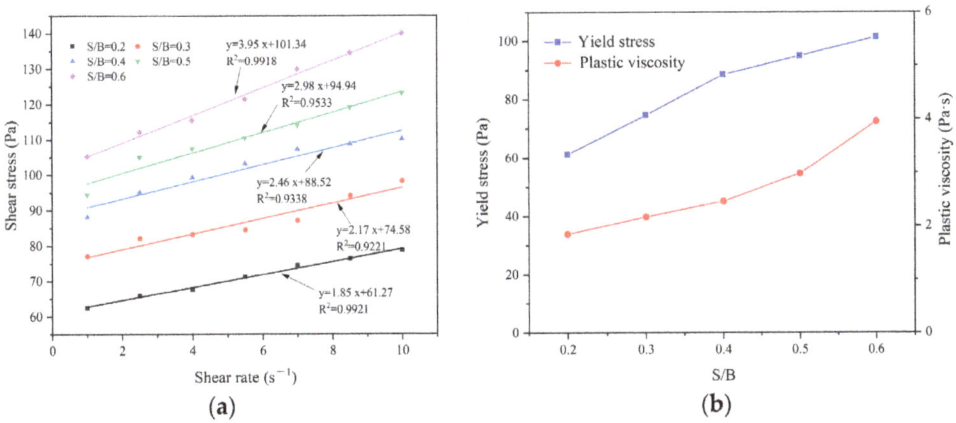

Figure 13. Effect of the S/B on the rheological parameters of CEAC: (**a**) rheological parameter fitting curve and (**b**) yield stress and plastic viscosity.

3.1.3. Effect of the Substitution Rate of Calcium Aluminate Cement(CAC)

Figure 14 shows the effect of the substitution rate of CAC on the spraying performance of CEAC, and it can be seen that with the increase in the substitution rate of CAC, the build-up thickness of CEAC continued to increase, and the rebound rate continued to decrease. This was because, compared with ordinary Portland cement, CAC had a higher chemical reaction activity, which could cause hydration and hardening reactions to occur more quickly, thus forming a compact structure [30]. The best CEAC spraying performance was achieved when the substitution rate of the CAC was 30%.

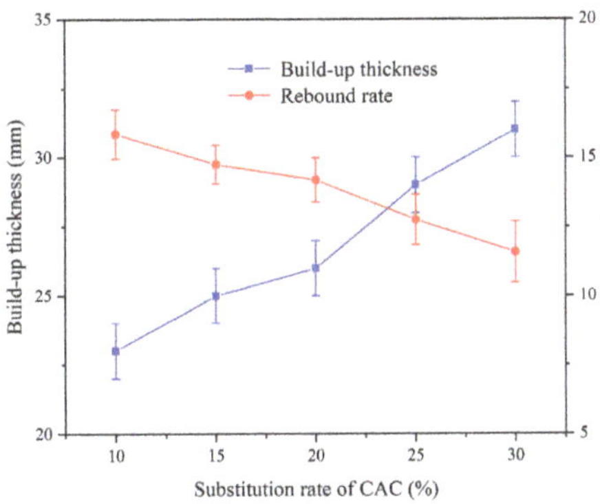

Figure 14. Effect of the CAC substitution rate on the spraying performance of CEAC.

Figure 15 shows the effect of the CAC substitution rate on the rheological properties of CEAC. Figure 15a shows that there was a good linear relationship between shear rate and shear stress at different CAC substitution rates, and R^2 was between 0.95 and 0.99. Figure 15b shows the effect of the substitution rate of CAC on the rheological parameters of the CEAC. From Figure 15b, it can be seen that the yield stress and plastic viscosity of CEAC increased as the substitution rate of CAC increased. This was because the hydration process of CAC released a large amount of heat; it accelerated the hydration process of the whole cement system in a negative-temperature environment, which ultimately led to an increase in yield stress and plastic viscosity [31]. To ensure that the CEAC had good spraying performance and working performance at the same time, the optimal substitution rate of CAC was determined to be 25%. At this time, the build-up thickness and rebound rate of CEAC after one spraying treatment were 29 mm and 12.75%, respectively, and the yield stress and plastic viscosity were 98.12 Pa and 3.34 Pa·s, respectively.

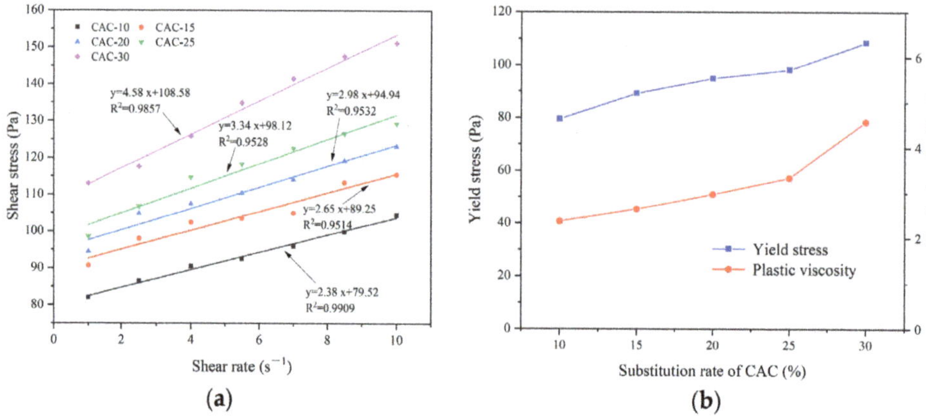

Figure 15. Effect of the CAC substitution rate on the rheological parameters of CEAC: (**a**) rheological parameter fitting curve and (**b**) yield stress and plastic viscosity.

3.1.4. Effect of the Emulsified Asphalt Content

Figure 16 illustrates the influence of emulsified asphalt content on the spraying efficiency of CEAC. It can be observed that as the emulsified asphalt content increased, there was a gradual enhancement in the build-up thickness during the spraying process of CEAC, while the rebound rate experienced a gradual decline. When the content of emulsified asphalt was 10%, the CEAC spraying performance was the best. This was because the hydration process of cement broke the double electron layer structure in emulsified asphalt, thus accelerating its demulsification, which in turn led to the destabilization and dehydration of the emulsified asphalt, forming asphalt particles with a certain viscosity [32]. When CEAC was sprayed onto a wall, the moisture in the material detached due to its impact on the wall, and the asphalt particles aggregated with each other due to extrusion so that the overall cohesion increased, thus improving the build-up thickness and lowering the rebound rate.

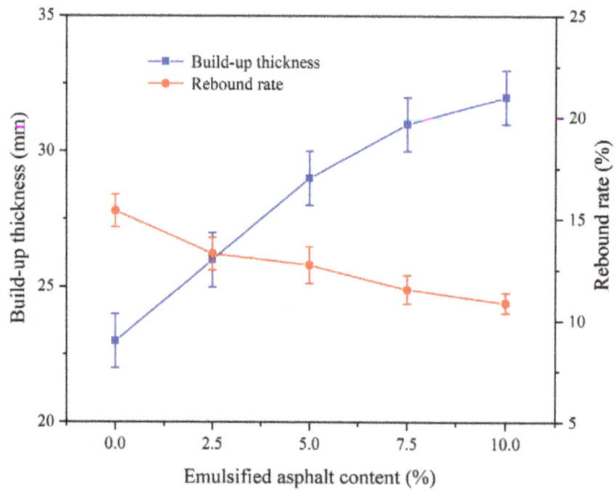

Figure 16. Effect of the emulsified asphalt content on the spraying performance of CEAC.

Figure 17 shows the effect of emulsified asphalt content on the rheological properties of CEAC. Figure 17a shows that there was a good linear relationship between shear rate and shear stress at different emulsified asphalt contents, and R^2 was between 0.95 and 0.99. Figure 17b shows the effect of the emulsified asphalt content on the rheological parameters of the CEAC. With increasing emulsified asphalt content, the yield stress and plastic viscosity of CEAC first increased and then decreased, and the maximum value occurred when the content of emulsified asphalt was 5%. This was because in CEAC, the hydration process of the cement and the breaking process of the asphalt emulsion were carried out simultaneously. When the emulsified asphalt content fell below 5%, the hydration of cement dominated, while the emulsification of asphalt particles resulted in an elevation of the overall system's viscosity. This viscosity increase led to greater friction among particles, subsequently boosting the yield stress and plastic viscosity of the CEAC. When the content of emulsified asphalt was greater than 5%, the emulsification of asphalt dominated, and too many asphalt particles were adsorbed onto the cement particles, thus delaying the hydration and hardening of the cement [33]. In addition, the emulsifier component in emulsified asphalt was a surfactant, which was similar to the water reducer component and had a lubricating effect, thus reducing the yield stress and plastic viscosity of CEAC [34]. Combined with the good rheological properties of CEAC when the emulsified asphalt content was 5%, the optimal content of emulsified asphalt was determined to be 5%, at which time the build-up thickness and rebound rate of CEAC after one spraying treatment were 29 mm and 12.75%, respectively, and the yield stress and plastic viscosity were 98.12 Pa and 3.34 Pa·s, respectively.

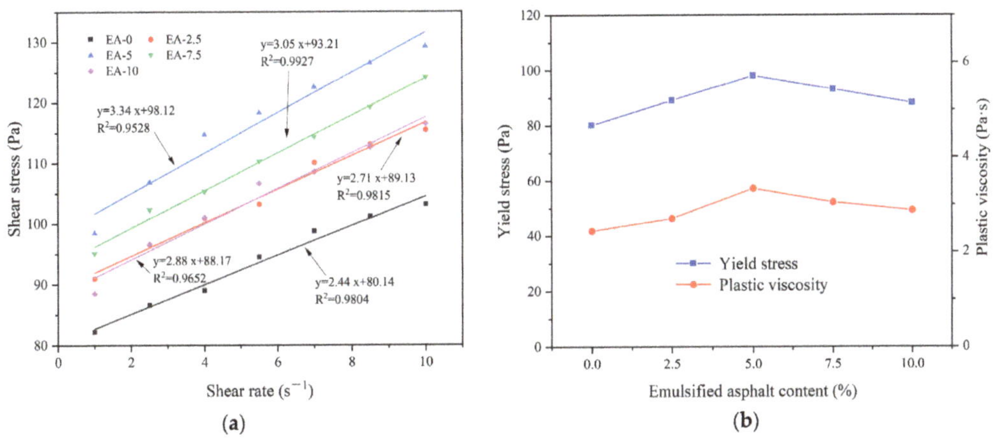

Figure 17. Effect of the emulsified asphalt content on the rheological parameters of CEAC: (**a**) rheological parameter fitting curve and (**b**) yield stress and plastic viscosity.

3.1.5. Effect of PVA Content

Figure 18 illustrates the impact of PVA content on the spraying performance of CEAC. As the PVA content increased, the CEAC's build-up thickness initially rose and then gradually declined. Alternately, the rebound rate initially decreased, but later on, it exhibited a rise. This was due to the friction created by the disordered fiber distribution within the sprayed cement-based material, which interfered with the slurry's flow, decreasing its fluidity while enhancing stability [35]. Furthermore, as PVA fibers are hydrophilic, they absorbed water molecules or attracted them to their surface, which not only accelerated cement hardening but also increased slurry viscosity [36]. However, when the content of PVA exceeded 1.5%, the mixing of CEAC slurry was uneven, resulting in poor fluidity and a decrease in the spraying performance of CEAC [37].

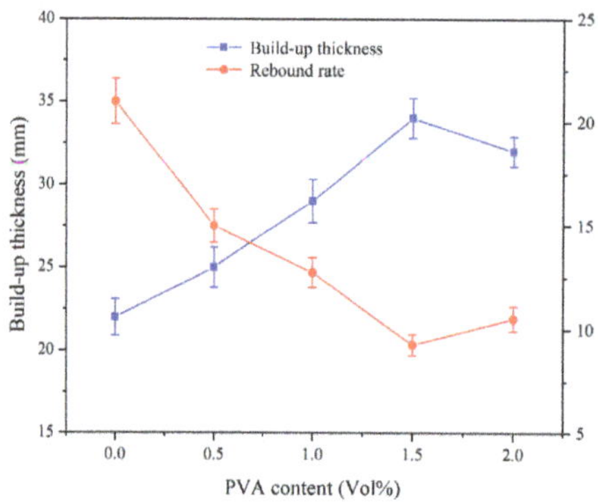

Figure 18. Effect of the PVA content on the spraying performance of CEAC.

Figure 19 shows the effect of PVA content on the rheological properties of CEAC. Figure 19a shows that there was a good linear relationship between shear rate and shear stress at different PVA contents, and R^2 was between 0.92 and 0.99. Figure 19b demonstrates the influence of PVA content on the rheological parameters of CEAC. As the PVA content rose, both the yield stress and plastic viscosity of CEAC increased. This trend was attributed to the water absorption properties of PVA, which reduced the thickness of the water film surrounding solid particles. Consequently, the likelihood of collisions and friction among solid particles intensified. Under these conditions, the yield stress and plastic viscosity of CEAC were enhanced [38]. Given that CEAC exhibited optimal spraying performance when the PVA content was 1.5%, we determined this to be the ideal PVA content. At this point, the build-up thickness and rebound rate of CEAC after one spraying treatment were 34 mm and 9.27%, respectively, and the yield stress and plastic viscosity of CEAC were measured to be 101.30 Pa and 3.61 Pa·s, respectively.

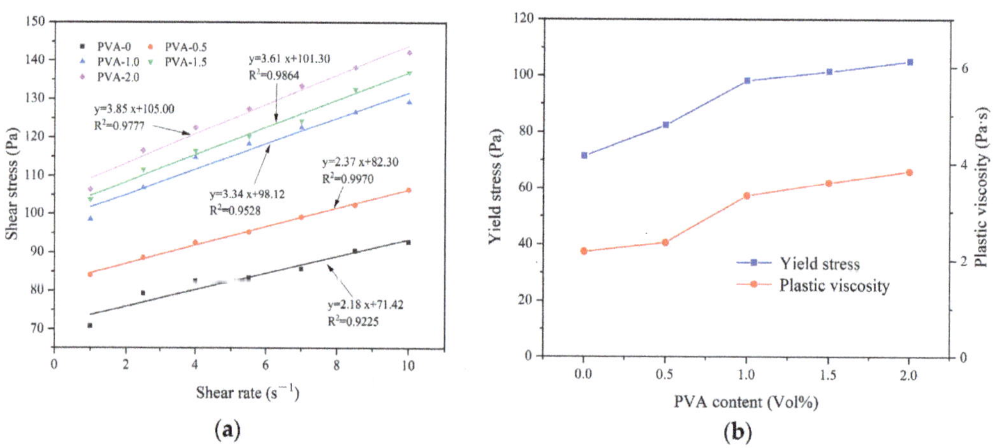

Figure 19. Effect of PVA content on the rheological parameters of CEAC: (**a**) rheological parameter fitting curve and (**b**) yield stress and plastic viscosity.

3.2. Correlation between Spraying Performance and Rheological Parameters

Figure 20 shows the correlation between the spraying performance and rheological parameters of CEAC. As depicted in Figure 20a,c, both the yield stress (τ_0) and plastic viscosity (η) of CEAC exhibited a positive association with the build-up thickness (T) achieved during a single spraying event. The specific equations governing these relationships are presented in Equations (3) and (4):

$$T = -0.003\tau_0^2 + 0.87\tau_0 - 29.03 \quad (3)$$

$$T = 1.51\eta^3 - 18.19\eta^2 + 73.02\eta - 68.75 \quad (4)$$

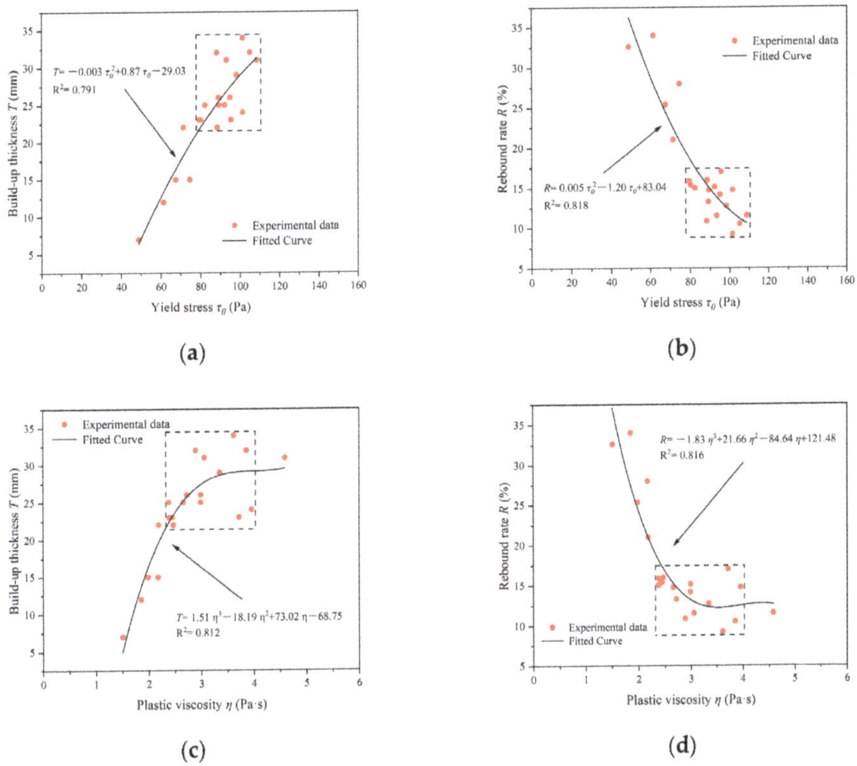

Figure 20. The relationship between spraying performance and rheological parameters of CEAC: (**a**) Yield stress and build-up thickness; (**b**) Yield stress and rebound rate; (**c**) Plastic viscosity and build-up thickness and (**d**) Plastic viscosity and rebound rate.

As depicted in Figure 20b,d, both the τ_0 and η of CEAC exhibited a negative correlation with the rebound rate ©. The specific equations that describe these relationships were provided in Equations (5) and (6):

$$R = 0.005\tau_0^2 - 1.20\tau_0 + 83.04 \quad (5)$$

$$R = -1.83\eta^3 + 21.66\eta^2 - 84.64\eta + 121.48 \quad (6)$$

It is well known that as the τ_0 increases, the material becomes less fluid. When the τ_0 of CEAC increases, its T increases, and the R decreases. As the η rose, the cohesive force of the material was also enhanced, leading to an increase in T and a corresponding decrease in R. From Figures 15d and 20c, it can be seen that there is a clear turning point in plastic viscosity. When the plastic viscosity was near this turning point, CEAC had good spraying performance, and the influence of plastic viscosity on spraying performance was not significant. Figure 20 illustrates that CEAC exhibited superior spraying performance when the τ_0 fell within the range of 79.52–108.58 Pa and the η lay between 2.37 and

3.95 Pa·s. Within this range, the T varied from 22 to 34 mm, while the R remained between 9.27 and 17.01%.

Figure 21 shows the relationship between the R and T of the CEAC. The T of the CEAC was linearly and negatively correlated with the R, and the fitting relationship is shown in Equation (7). Yun et al. [19] also obtained a similar conclusion in their study.

$$R = -0.96T + 39.98 \tag{7}$$

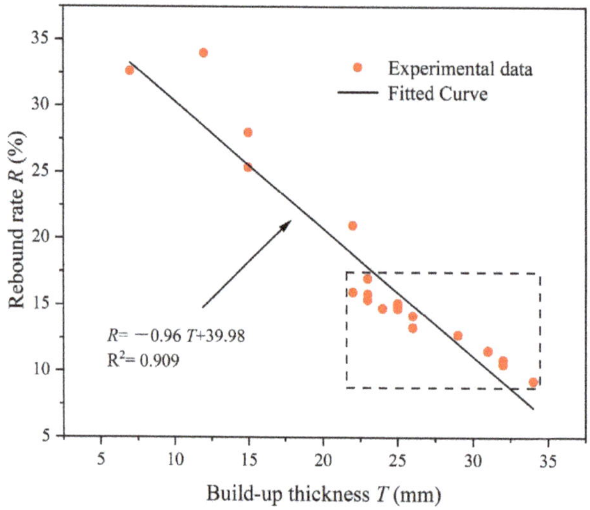

Figure 21. Relationship between the R and T of CEAC.

3.3. Frost Resistance

After analyzing the spraying performance and rheological characteristics of CEAC, we have arrived at an optimal mixing ratio. This includes a water–solid ratio of 0.14, a sand–binder ratio of 0.5, a 25% substitution rate for calcium aluminate cement, a 5% emulsified asphalt content, and a 1.5% volumetric content of PVA fibers. With these settings, the build-up thickness and rebound rate were recorded as 34 mm and 9.27%, respectively, while the yield stress and plastic viscosity measured 101.30 Pa and 3.61 Pa·s, respectively. Using this optimal mix ratio for CEAC, we further investigated the frost resistance of both the cement slurry (CS) without emulsified asphalt and CEAC.

Figure 22a shows the variation curves of the mass loss rates of CS and CEAC under the action of freeze–thaw cycles. The mass loss rates of CS and CEAC increased continuously with an increasing number of freeze–thaw cycles. Under the same number of freeze–thaw cycles, the mass loss rate of the CEAC was lower than that of the CS. The mass loss rate after 100 freeze–thaw cycles was only 2.0%. This was because, in CEAC, the emulsified asphalt could enhance the adhesion between aggregates after emulsification, which improved the integrity of the specimen; on the other hand, the water content of the specimen was reduced due to the water barrier of the asphalt content, which reduced the probability of damage to the microcellular freezing under freeze–thaw cycles. In addition, the mass loss rates of CS and CEAC first increased slowly and then increased rapidly, showing similar patterns of change. This was because during the pre-freeze–thaw cycle period, the freeze–thaw damage only occurred on the specimen surface. After 75 freeze–thaw cycles, the rate of mass loss increased significantly due to the continuous action of freeze–thaw cycles, which led to the intensification of matrix damage [15,39].

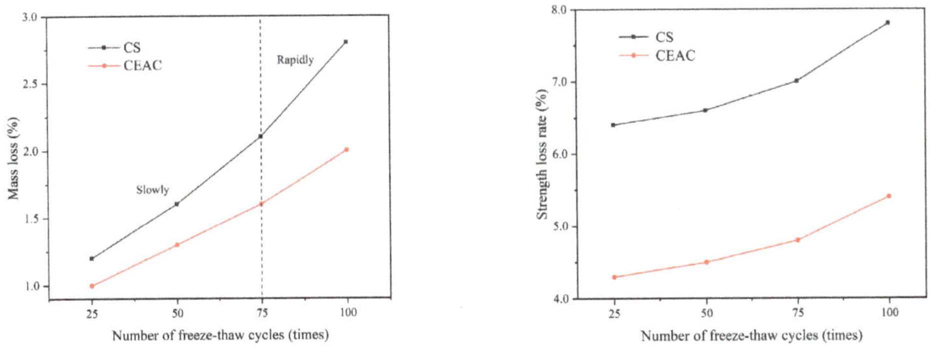

(a) **(b)**

Figure 22. Frost resistance test results of CS and CEAC: (**a**) mass loss and (**b**) strength loss rate.

Figure 22b shows the varying trends of the strength loss rates for both CS and CEAC as they undergo freeze–thaw cycles. It is evident that as the number of freeze–thaw cycles increased, the strength loss rates for both CS and CEAC continued to rise. Under the same number of freeze–thaw cycles, the strength loss rate of the CEAC was lower than that of the CS. On the one hand, the emulsifier component of emulsified asphalt was a surfactant, similar to air-entraining agents, which could introduce stable and small bubbles in concrete to isolate water transport channels, thus relieving water crystallization pressure and inhibiting the formation and growth of ice crystals [40]. On the other hand, due to the simultaneous destruction of the emulsified asphalt emulsion and the cement hydration process, the generated asphalt film and cement hydration products intertwined with the flocculent colloid, and this flocculent colloid filled up the specimen inside some large pores, thereby inhibiting the damage of the specimen during freeze–thaw cycling. In summary, the frost resistance of CEAC was superior to that of CS.

3.4. Hole Structure Analysis

Based on the findings reported by Odler et al. [41], cement concrete exhibits four distinct pore size categories: gel pores (with diameters below 10 nm), transition pores (spanning from 10 to 100 nm), capillary pores (ranging from 100 to 1000 nm), and macropores (exceeding 1000 nm in diameter). Furthermore, the pore structures of CS and CEAC were analyzed through the MIP test, and the outcomes of this test are presented in Figure 23. Figure 23 shows that the volume of capillary pores and macropores in CS reached 44.64% and 39.67%, respectively, of the total pore volume, while the volume of capillary pores and macropores in CEAC reached 40.45% and 37.64%, respectively, of the total pore volume. Hover et al. [42] concluded that capillary pores and macropores mainly affect the frost resistance of concrete. Therefore, the lower capillary pore content and macropore content corroborate that CEAC had better frost resistance.

The above analysis revealed that the emulsified asphalt and cement cemented to each other, generated a flocculent colloid, and spread to the surrounding space to effectively fill the large pores, thus reducing the proportion of large pores. Therefore, mixing emulsified asphalt in cement-based spraying materials can effectively improve the shortcomings of high porosity and poor densification, thus improving frost resistance.

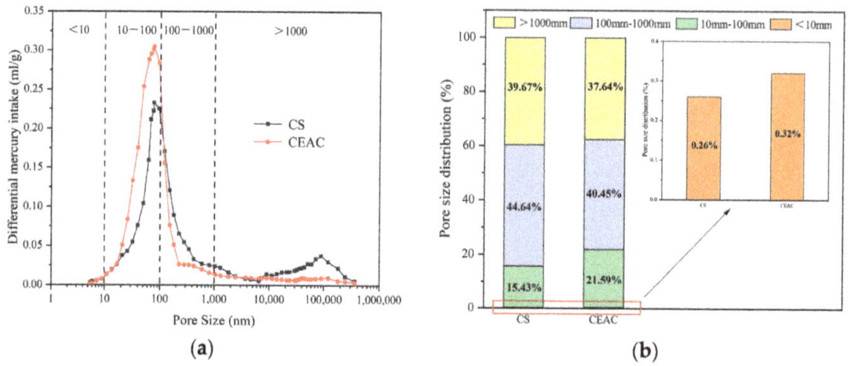

Figure 23. MIP test results of CS and CEAC: (**a**) pore size differential curve and (**b**) pore size distribution.

4. Conclusions

This study comprehensively explores the impact of various factors, including water–solid ratio, sand–binder ratio, calcium aluminate cement substitution rate, emulsified asphalt dosing, and PVA fiber content, on the spraying performance and rheological properties of CEAC. The relationship between spraying performance and rheological parameters was established, and the optimum mixing ratio of CEAC was obtained. In addition, the frost resistance of the cement slurry (CS) without emulsified asphalt and with CEAC was investigated, and the pore structure characteristics of the materials were analyzed via a MIP test. The following conclusions were obtained:

(1) Based on the spray performance and rheological parameters, the optimal mix ratio for CEAC was determined using the control variable method: a water–solid ratio of 0.14, a sand–binder ratio of 0.5, a calcium aluminate cement substitution rate of 25%, an emulsified asphalt dosage of 5%, and a PVA fiber volume content of 1.5%.

(2) Utilizing the correlation between spraying performance and rheological parameters, this study determined the ranges of spraying performance and rheological parameters for CEAC suitable for spraying in cold environments. Specifically, the yield stress ranged from 79.52 to 108.58 Pa, the plastic viscosity varied between 2.37 and 3.95 Pa·s, the build-up thickness was between 22 and 34 mm, and the rebound rate ranged from 9.27 to 17.01%. These ranges provide valuable guidance for the application of CEAC spraying in negative-temperature environments.

(3) The build-up thickness of CEAC was positively correlated with yield stress and plastic viscosity, and the rebound rate was negatively correlated with yield stress and plastic viscosity. The rheological performance parameters of CEAC can be used to characterize its spraying performance within a specific range.

(4) The results of frost resistance testing conducted on CS and CEAC, both prepared using the optimal mix ratio for CEAC spraying in cold environments, revealed that CEAC exhibited lower mass loss and strength loss rates compared to CS after an equal number of freezing and thawing cycles. This indicates that CEAC possesses superior frost resistance properties compared to CS.

(5) The results of the MIP test show that the content of capillary pores and macropores mainly affects the freezing resistance of CEAC, and the smaller their content was, the better the freezing resistance of CEAC was.

Author Contributions: Conceptualization, K.N. and B.T.; methodology, X.L.; software, X.L.; validation, Y.H., K.N. and B.T.; formal analysis, Y.H.; investigation, Y.H.; resources, B.T.; data curation, J.C.; writing—original draft preparation, X.L. and J.C.; writing—review and editing, Y.H.; visualization, X.L.; supervision, X.L.; project administration, J.C.; funding acquisition, K.N. and B.T. All authors have read and agreed to the published version of the manuscript.

Funding: This research was funded by the National Key Research and Development Program (Grant Nos. 2018YFB1600100) and the National Natural Science Foundation of China (Grant Nos. 52178428, 52178427).

Institutional Review Board Statement: Not applicable.

Informed Consent Statement: Not applicable.

Data Availability Statement: Some or all data, models, or code that support the findings of this study are available from the corresponding author upon reasonable request.

Conflicts of Interest: Author Junli Chen was employed by the company Chongqing Communications Construction (Group) Co., Ltd. The remaining authors declare that this research was conducted in the absence of any commercial or financial relationships that could be construed as potential conflicts of interest.

References

1. Choi, P.; Yeon, J.H.; Yun, K.-K. Air-void structure, strength, and permeability of wet-mix shotcrete before and after shotcreting operation: The influences of silica fume and air-entraining agent. *Cem. Concr. Compos.* **2016**, *70*, 69–77. [CrossRef]
2. Galan, I.; Baldermann, A.; Kusterle, W.; Dietzel, M.; Mittermayr, F. Durability of shotcrete for underground support—Review and update. *Constr. Build. Mater.* **2019**, *202*, 465–493. [CrossRef]
3. He, S.; Lai, J.; Wang, L.; Wang, K. A literature review on properties and applications of grouts for shield tunnel. *Constr. Build. Mater.* **2020**, *239*, 468–482. [CrossRef]
4. Steindl, F.R.; Galan, I.; Baldermann, A.; Sakoparnig, M.; Briendl, L.; Juhart, J.; Thumann, M.; Dietzel, M.; Röck, R.; Kusterle, W.; et al. Sulfate durability and leaching behaviour of dry- and wet-mix shotcrete mixes. *Cem. Concr. Res.* **2020**, *137*, 106180. [CrossRef]
5. Chen, L.; Ma, G.; Liu, G.; Liu, Z. Effect of pumping and spraying processes on the rheological properties and air content of wet-mix shotcrete with various admixtures. *Constr. Build. Mater.* **2019**, *225*, 311–323. [CrossRef]
6. Li, P.; Zhou, Z.; Chen, L.; Liu, G.; Xiao, W. Research on Dust Suppression Technology of Shotcrete Based on New Spray Equipment and Process Optimization. *Adv. Civ. Eng.* **2019**, *2019*, 4831215. [CrossRef]
7. Liu, G.; Cheng, W.; Chen, L. Investigating and optimizing the mix proportion of pumping wet-mix shotcrete with polypropylene fiber. *Constr. Build. Mater.* **2017**, *150*, 14–23. [CrossRef]
8. Cui, S.; Liu, P.; Li, Z.; Xu, X.; Ju, J.W. Shotcrete performance-loss due to seepage and temperature coupling in cold-region tunnels. *Constr. Build. Mater.* **2020**, *246*, 118488. [CrossRef]
9. Ning, F.W.; Cai, Y.B.; Bai, Y.; Ding, J.T.; Zhang, F. Research on rebound rate of nanoscale admixture wet-mix shotcrete. *IOP Conf. Ser. Mater. Sci. Eng.* **2019**, *678*, 012109. [CrossRef]
10. Fu, Z.; Li, T.; Jiang, Y. Shotcrete and rock bolt support technique used in construction of Beijing Badaling Tunnel under cold weather Condition. *Railw. Constr. Technol.* **2001**, *2*, 15–17. (In Chinese)
11. Zhang, D.; Wang, M. New technique for construction of Fenghuoshan Tunnel the highest tunnel in the world of Qinghai-Tibet Railway. *J. Eng. Geol.* **2003**, *11*, 213–216. (In Chinese)
12. Yang, A.; Zhou, D. Tunnel wet concrete-shot technology in permafrost region. *China Railw. Sci.* **2004**, *25*, 141–144. (In Chinese)
13. Kim, D.; Lee, G.; Bae, G. Compressive and adhesive strengths of shotcrete deteriorated by hazardous components in the groundwater. *Tunn. Uudergr. Space Technol.* **2006**, *21*, 323–332. [CrossRef]
14. Haldane, J. Heythrop, Copleston, and the Jesuit Contribution to Philosophy. *Philosophy* **2016**, *91*, 559–589. [CrossRef]
15. Wan, Z.; He, T.; Chang, N.; He, S.; Shao, Z.; Ma, X.; Yang, R. Study on frost resistance of shotcrete by micro-nano bubble water and admixture. *Ceram. Int.* **2023**, *49*, 11123–11139. [CrossRef]
16. Hu, Z.; Ding, H.; Lai, J.; Wang, H.; Wang, X.; He, S. The durability of shotcrete in cold region tunnel: A review. *Constr. Build. Mater.* **2018**, *185*, 670–683. [CrossRef]
17. Holter, K.G.; Smeplass, S.; Jacobsen, S. Freeze-thaw resistance of sprayed concrete in tunnel linings. *Mater. Struct.* **2015**, *49*, 3075–3093. [CrossRef]
18. Kalhori, H.; Bagherzadeh, B.; Bagherpour, R.; Akhlaghi, M.A. Experimental study on the influence of the different percentage of nanoparticles on strength and freeze-thaw durability of shotcrete. *Constr. Build. Mater.* **2020**, *256*, 119470. [CrossRef]
19. Yun, K.K.; Choi, P.; Yeon, J.H. Correlating rheological properties to the pumpability and shootability of wet-mix shotcrete mixtures. *Constr. Build. Mater.* **2015**, *98*, 884–891. [CrossRef]
20. Pan, G.; Li, P.; Chen, L.; Liu, G. A study of the effect of rheological properties of fresh concrete on shotcrete-rebound based on different additive components. *Constr. Build. Mater.* **2019**, *224*, 1069–1080. [CrossRef]
21. Feng, H.; Zheng, X.; Yu, Z.; Chen, B.; Zhu, P.; Yu, J.; Luo, Z. Development of sprayable ultra-high ductility magnesium phosphate cement-based composites based on the rheological properties. *Constr. Build. Mater.* **2023**, *377*, 131113. [CrossRef]
22. Li, Z.; Tian, B.; Li, L.; Quan, L.; Li, D. Effect of phosphates on early-age hydration process and products of calcium aluminate cement at −10 °C. *Constr. Build. Mater.* **2022**, *346*, 128405. [CrossRef]

23. *JGJT 372-2016;* Technical Specification for Application of Sprayed Concrete. Ministry of Housing and Urban-Rural Development of the People's Republic of China: Beijing, China, 2016.
24. *JSCE-F 563-2005;* Test Method for Rebound Percentage of Sprayed Concrete (Mortar). The Concrete Committee of Japan Society of Civil Engineers: Tokyo, Japan, 2005.
25. Yun, K.K.; Kim, J.B.; Song, C.S.; Hossain, M.S.; Han, S. Rheological behavior of high-performance shotcrete mixtures containing colloidal silica and silica fume using the Bingham model. *Materials* **2022**, *15*, 428. [CrossRef] [PubMed]
26. Papo, A.; Piani, L. Effect of various superplasticizers on the rheological properties of Portland cement pastes. *Cem. Concr. Res.* **2004**, *34*, 2097–2101. [CrossRef]
27. *JGJT 70-2009;* Standard for Test Methods of Basic Properties of Construction Mortar. Ministry of Housing and Urban-Rural Development of the People's Republic of China: Beijing, China, 2009.
28. Zhao, Z.; Chen, J.; Zhang, Y.; Jiang, T.; Wang, W. Study on preparation and performance of foamed lightweight soil grouting material for goaf treatment. *Materials* **2023**, *16*, 4325. [CrossRef] [PubMed]
29. Ren, Q.; Tao, Y.; Jiao, D.; Jiang, Z.; Ye, G.; Schutter, D.D. Plastic viscosity of cement mortar with manufactured sand as influenced by geometric features and particle size. *Cem. Concr. Compos.* **2021**, *122*, 104163. [CrossRef]
30. Gu, P.; Fu, Y.; Xie, P.; Beaudoin, J.J. A study of the hydration and setting behaviour of OPC-HAC pastes. *Cem. Concr. Res.* **1994**, *24*, 682–694. [CrossRef]
31. Morris, E.F. Evaluation of cement systems for permafrost. In *Annual Meeting of the American Institute of Mining, Metallurgical, and Petroleum Engineers;* SPE: Denver, CO, USA, 1970.
32. Zhang, P.; Hou, Y.; Niu, K.; Tian, B.; Wang, H. Effects of Anionic Emulsifiers and Emulsified Asphalt on Hydration and Microstructure of Cement. *Materials* **2024**, *17*, 36. [CrossRef] [PubMed]
33. Li, W.; Mao, Z.; Xu, G.; Chang, H.; Hong, J.; Zhao, H.; Xu, J.; Liu, Z. Study on the early cement hydration process in the presence of cationic asphalt emulsion. *Constr. Build. Mater.* **2020**, *261*, 120025. [CrossRef]
34. Li, W.; Zhu, X.; Hong, J.; She, W.; Wang, P.; Zuo, W. Effect of anionic emulsifier on cement hydration and its interaction mechanism. *Constr. Build. Mater.* **2015**, *93*, 1003–1011. [CrossRef]
35. Xin, Q.; Gao, R.; Shao, P.; Lyu, W.; Wei, H. Effect of Different Water-Binder Ratios and Fiber Contents on the Fluidity and Mechanical Properties of PVA-ECC Materials. *Teh. Vjesn.* **2023**, *30*, 1366–1372.
36. Zhang, W.; Yin, C.; Ma, F.; Huang, Z. Mechanical properties and carbonation durability of engineered cementitious composites reinforced by polypropylene and hydrophilic polyvinyl alcohol fibers. *Materials* **2024**, *11*, 1147. [CrossRef] [PubMed]
37. Kaufmann, J.; Frech, K.; Schuetz, P.; Münch, B. Rebound and orientation of fibers in wet sprayed concrete applications. *Constr. Build. Mater.* **2013**, *49*, 15–22. [CrossRef]
38. Martinie, L.; Rossi, P.; Roussel, N. Rheology of fiber reinforced cementitious materials: Classification and prediction. *Cem. Concr. Res.* **2010**, *40*, 226–234. [CrossRef]
39. Zhang, S.; Zhao, B. Research on the performance of concrete materials under the condition of freeze-thaw cycles. *Eur. J. Environ. Civ. Eng.* **2013**, *17*, 860–871. [CrossRef]
40. Łaźniewska-Piekarczyk, B. The type of air-entraining and viscosity modifying admixtures and porosity and frost durability of high performance self-compacting concrete. *Constr. Build. Mater.* **2013**, *40*, 659–671. [CrossRef]
41. Odler, I.; Rößler, M. Investigations on the relationship between porosity, structure and strength of hydrated Portland cement pastes. II. Effect of pore structure and of degree of hydration. *Cem. Concr. Res.* **1985**, *15*, 401–410. [CrossRef]
42. Hover, K. Why is there air in concrete? *Concr. Constr.* **1993**, *38*, 11–15.

MDPI AG
Grosspeteranlage 5
4052 Basel
Switzerland
Tel.: +41 61 683 77 34

Materials Editorial Office
E-mail: materials@mdpi.com
www.mdpi.com/journal/materials

www.ingramcontent.com/pod-product-compliance
Lightning Source LLC
Chambersburg PA
CBHW040747250926
51850CB00007BB/1433